**Second
Edition**

*A Transition
to Advanced
Mathematics*

Douglas Smith

*University of North Carolina
at Wilmington*

Maurice Eggen

Trinity University

Richard St. Andre

Central Michigan University

*Brooks/Cole Publishing Company
Monterey, California*

Brooks/Cole Publishing Company
A Division of Wadsworth, Inc.

Printed in the United States of America

10 9 8 7 6 5 4 3 2 1

Library of Congress Cataloging in Publication Data

Smith, Douglas, 1943–
   A transition to advanced mathematics.

   Includes index.
   1. Mathematics—1961–   . I. Eggen, Maurice,
1942–   . II. St. Andre, Richard, 1945–
III. Title.
QA37.2.S575   1985   510   85-11392

ISBN 0-534-05796-9

Sponsoring Editor: *Craig Barth*
Editorial Assistant: *Amy Mayfield*
Production Editor: *Joan Marsh*
Manuscript Editor: *Charles Cox*
Permissions Editor: *Carline Haga*
Interior & Cover Design: *John Edeen*
Art Coordinator: *Sue C. Shepherd*
Interior Illustration: *Reese Thornton*
Typesetting: *Omegatype Typography, Inc.*
Printing and Binding: *R. R. Donnelley & Sons Co., Crawfordsville, Indiana*

To Karen, Karen, and Karen

Preface

to the First Edition

"I understand mathematics but I just can't do proofs."

Our experience has led us to believe that the remark above, though contradictory, expresses the frustration many students feel as they pass from beginning calculus to a more rigorous level of mathematics. This book developed from a series of lecture notes for a course at Central Michigan University that was designed to address this lament. The text is intended to bridge the gap between calculus and advanced courses in at least three ways. First, it provides a firm foundation in the major ideas needed for continued work. Second, it guides students to think and to express themselves mathematically— to analyze a situation, extract pertinent facts, and draw appropriate conclusions. Finally, we present introductions to modern algebra and analysis of sufficient depth to capture some of their spirit and characteristics.

We begin in Chapter 1 with a study of the logic required by mathematical arguments, discussing not formal logic but rather the standard methods of mathematical proof and their validity. Methods of proof are examined in detail, and examples of each method are analyzed carefully. Denials are given special attention, particularly those involving quantifiers. Techniques of proof given in this chapter are used and referred to later in the text. Although the chapter was written with the idea that it may be assigned as out-of-class reading, we find that most students benefit from a thorough study of logic.

Much of the material in Chapters 2, 3, and 4 on sets, relations, and functions, will be familiar to the student. Thus, the emphasis is on enhancing the student's ability to write and understand proofs. The pace is deliberate. The rigorous approach requires the student to deal precisely with these concepts.

Chapters 5, 6, and 7 make use of the skills and techniques the student has acquired in Chapters 1 through 4. These last three chapters are a cut above the earlier chapters in terms of level and rigor. *Chapters 1 through 4 and **any one of** Chapters 5, 6, or 7 provide sufficient material for a one-semester course.* An alternative is to choose among topics by selecting, for example, the first two sections of Chapter 5, the first three sections of Chapter 6, and the first two sections of Chapter 7.

Chapter 5 begins the study of cardinality by examining the properties of finite and infinite sets and establishing countability or uncountability for the familiar number systems. The emphasis is on a working knowledge of cardinality—particularly countable sets, the ordering of cardinal numbers, and applications of the Cantor-Schröeder-Bernstein Theorem. We include a brief discussion of the Axiom of Choice and relate it to the comparability of cardinals.

Chapter 6, which introduces modern algebra, concentrates on the concept of a group and culminates in the Fundamental Theorem of Group Homomorphisms. The idea of an operation preserving map is introduced early and developed throughout the section. Permutation groups, cyclic groups, and modular arithmetic are among the examples of groups presented.

Chapter 7 begins with a description of the real numbers as a complete ordered field. We continue with the Heine-Borel Theorem, the Bolzano-Weierstrass Theorem, and the Bounded Monotone Sequence Theorem (each for the real number system), and then return to the concept of completeness.

Exercises marked with a solid star ★ have complete answers at the back of the text. Open stars ☆ indicate that a hint or a partial answer is provided. "Proofs to Grade" are a special feature of most of the exercise sets. We present a list of claims with alleged proofs, and the student is asked to assign a letter grade to each "proof" and to justify the grade assigned. Spurious proofs are usually built around a single type of error, which may involve a mistake in logic, a common misunderstanding of the concepts being studied, or an incorrect symbolic argument. Correct proofs may be straightforward, or they may present novel or alternate approaches. We have found these exercises valuable because they reemphasize the theorems and counterexamples in the text and also provide the student with an experience similar to grading papers. Thus, the student becomes aware of the variety of possible errors and develops the ability to read proofs critically.

In summary, our main goals in this text are to improve the student's ability to think and write in a mature mathematical fashion and to provide a solid understanding of the material most useful for advanced courses. Student readers, take comfort from the fact that we do not aim to turn you into theorem-proving wizards. Few of you will become research mathematicians. Nevertheless, in almost any mathematically related work you may do, the kind of reasoning you need to be able to do is the same reasoning you use in proving theorems. You must first understand exactly what you want to prove (verify, show, or explain), and you must be familiar with the logical steps that allow you to get from the hypothesis to the conclusion. Moreover, a proof is the ultimate test of your understanding of the subject matter and of mathematical reasoning.

We are grateful to the many students who endured earlier versions of the manuscript and gleefully pointed out misprints. We acknowledge also the helpful comments of Edwin H. Kaufman, Melvin Nyman, Mary R. Wardrop, and especially Douglas W. Nance, who saw the need for a course of this kind at CMU and did a superb job of reviewing the manuscript.

We thank our reviewers: William Ballard of the University of Montana, Sherralyn Craven of Central Missouri State University, Robert Dean of Stephen F. Austin State University, Harvey Elder of Murray State University, Hoseph H. Oppenheim of San Francisco State University, Joseph Teeters of the University of Wisconsin, Dale Schoenefeld of the University of Tulsa, Kenneth Slonnegar of State University of New York at Fredonia, and Douglas Smith of University of the Pacific. And we wish to thank Karen St. Andre for her superb and expeditious typing of the manuscript.

Second Edition

The response to our first edition was gratifyingly enthusiastic. We are pleased that so many faculty who adopted *A Transition to Advanced Mathematics* as a text reported that students found the discussion, examples, and exercises helpful and understandable. We are grateful for the suggestions and corrections we have received. This response has encouraged us to proceed with this second edition, incorporating many of these suggestions and correcting, we hope, the errors.

Beyond the addition of further explanation, examples, and exercises, the major changes in the second edition are reworkings of Chapters 1 and 5, where our intent is still to introduce logic informally as a framework for understanding mathematical arguments, an added section on principles of counting, and an added section on graph theory. The counting section includes applications of the sum and product rules, permutations, and binomial coefficients. Directed graphs are incorporated into the study of relations as another way of representing a relation. The section on simple undirected graphs is optional.

We find that most instructors follow the development of the text, covering either Chapters 1–4, or 1–5, and use portions of the last independent chapters to introduce topics from algebra on analysis as time permits. Many instructors prefer to treat selectively the cardinality topics in Chapter 5. One common approach is to treat the definitions and results in the first two sections on finite and countable sets, the definition of cardinal number and Cantor's Theorem from the next section, and the facts about countable sets in the last section of Chapter 5.

We would like to thank our reviewers for the second edition: David Barnette, University of California at Davis; Michael J. Evans, North Carolina State University; Robert Gamble, Winthrop College; Robert P. Hunter, Pennsylvania State University; Jack Johnson, Brigham Young University-Hawaii; and Daniel Kocan, State University of New York, Potsdam. We also thank Craig Barth, Joan Marsh, and all the staff at Brooks/Cole for their professional work and friendly encouragement.

Richard St. Andre
Douglas Smith

Contents

1

Logic and Proofs   1

1.1   Propositions and
      Connectives   *2*
1.2   Conditionals and
      Biconditionals   *7*
1.3   Quantifiers   *14*
1.4   Mathematical Proofs   *21*
1.5   Proofs Involving
      Quantifiers   *32*

2

Set Theory   41

2.1   Basic Notions of Set
      Theory   *41*
2.2   Set Operations   *50*
2.3   Extended Set Operations and
      Indexed Families of Sets   *56*
2.4   Induction   *64*
2.5   Principles of Counting   *77*

3

Relations   89

3.1   Cartesian Products and
       Relations   *89*
3.2   Equivalence Relations   *101*
3.3   Partitions   *108*
3.4   Graphs   *113*

4

Functions   125

4.1   Functions as Relations   *126*
4.2   Constructions of
       Functions   *133*
4.3   Onto Functions; One-to-One
       Functions   *142*
4.4   Induced Set Functions   *150*

5

Cardinality   157

5.1   Equivalent Sets; Finite
       Sets   *158*
5.2   Infinite Sets   *164*
5.3   The Ordering of Cardinal
       Numbers   *173*
5.4   Comparability of Cardinals and
       the Axiom of Choice   *179*
5.5   Countable Sets   *183*

6

Concepts of Algebra   187

6.1   Algebraic Structures   *187*
6.2   Groups   *194*
6.3   Examples of Groups   *199*
6.4   Subgroups   *204*
6.5   Cosets and Lagrange's
       Theorem   *211*

6.6   Quotient Groups   *215*
6.7   Isomorphism; The Fundamental
       Theorem of Group
       Homomorphisms   *219*

7

Real Analysis   *225*

7.1   Field Properties of the Real
       Numbers   *225*
7.2   The Hein-Borel
       Theorem   *231*
7.3   The Bolzano-Weierstrass   *239*
7.4   The Bounded Monotone
       Sequence Theorem   *242*
7.5   Equivalents of
       Completeness   *249*

Answers to Selected Exercises   *253*
Index   *272*
List of Symbols   *276*

1

Logic and Proofs

Although mathematics is both a science and an art, special characteristics distinguish mathematics from the humanities and from other sciences. Particularly important is the kind of reasoning that typifies mathematics. The natural or social scientist generally makes observations of particular cases or phenomena and seeks a general theory that describes or explains the observations. This approach is called **inductive reasoning,** and it is tested by making further observations. If the results are incompatible with theoretical expectations, the scientist usually must reject or modify his theory.

The mathematician, too, frequently uses inductive reasoning in attempting to describe patterns and relationships among quantities and structures. The characteristic thinking of the mathematician, however, is **deductive reasoning,** in which one uses logic to draw conclusions based on statements accepted as true. The conclusions of a mathematician are proved to be true, *provided that the assumptions are true*. If a mathematical theory predicts results incompatible with reality, the fault will lie not in the theory but with the inapplicability of the theory to that portion of reality. Indeed, the mathematician is not restricted to the study of observable phenomena, even though mathematics can trace its development back to the need to describe spatial relations (geometry) and motion (calculus) or to solve numerical problems (algebra). Using logic, the mathematician can draw conclusions about any mathematical structure imaginable.

The goal of this chapter is to provide a working knowledge of the basics of logic and the idea of proof, which are fundamental to deductive reasoning.

This knowledge is important in many areas other than mathematics. For example, the thought processes used to construct an algorithm for a computer program are much like those used to develop the proof of a theorem.

1.1

Propositions and Connectives

A **proposition** is a sentence that is either true or false. Some examples of propositions are

(a) $\sqrt{2}$ is irrational.
(b) $1 + 1 = 5$.
(c) In the year 2020, more peach ice cream than fudge ripple ice cream will be sold.
(d) Julius Caesar had two eggs for breakfast on his tenth birthday.

We are not concerned here with the difficulty of establishing the actual truth value of a proposition. It will take several years to determine whether proposition (c) is true or false. There may be no way ever to determine whether proposition (d) is true. Nevertheless, each is either true or false, hence is a proposition.

Here are some sentences that are not propositions:

(e) What did you say?
(f) $x = 6$.
(g) This sentence is false.†

Propositions (a)–(d) are **simple** or **atomic** in the sense that they do not have any other propositions as components. **Compound** propositions can be formed by using logical connectives with simple propositions.

Definitions.   Given propositions P and Q,

- The **conjunction** of P and Q, denoted $P \wedge Q$, is the proposition "P and Q." $P \wedge Q$ is true exactly when *both* P and Q are true.
- The **disjunction** of P and Q, denoted $P \vee Q$, is the proposition "P or Q." $P \vee Q$ is true exactly when *at least one* of P or Q is true.
- The **negation** of P, denoted $\sim P$, is the proposition "not P." $\sim P$ is true exactly when P is false.

---

†This is an example of a sentence that is neither true nor false. The study of paradoxes such as this has played a key role in the development of modern mathematical logic.

If P is "$1 \neq 3$" and Q is "7 is odd," then P $\wedge$ Q is "$1 \neq 3$ and 7 is odd." $P \vee Q$ is "$1 \neq 3$ or 7 is odd" and $\sim Q$ is "It is not the case that 7 is odd," which is to say "7 is even." Since in this example both P and Q are true, $P \wedge Q$ and $P \vee Q$ are true, while $\sim Q$ is false.

All of the following are true propositions:

"It is not the case that $\sqrt{10} > 4$."
"$\sqrt{2} < \sqrt{3}$ or chickens have lips."
"Venus is smaller than Earth or $1 + 4 = 5$."
"$6 < 7$ and $7 < 8$."

All of the following are false:

"1955 was a bad year for wine and π is rational."
"It is not the case that 10 is divisible by 2."
"$2^4 = 16$ and a quart is larger than a liter."

Other connectives commonly used in English are "but," "while," and "although," each of which would normally be translated symbolically with the conjunction connective. A variant of the connective "or" is discussed in the exercises.

A technical distinction must be made between a proposition and the form of the proposition. When we study propositions, it is important to know about the **form** of a proposition, meaning how it is put together. Thus $P \wedge Q$ and $P \vee Q$ are two different propositional forms. They may also be viewed as propositions, provided that both P and Q represent propositions. *As a propositional form $P \wedge Q$ is neither true nor false*, because its truth value depends on the truth values of P and of Q, according to the definition given.

The truth values of a compound proposition are readily obtained by exhibiting all possible combinations of the truth values for its components in a truth table. Since each connective $\wedge$ and $\vee$ involves two components, their truth tables must list the four possible combinations of the truth values of those components. When we use T for true and F for false, the truth tables for $P \wedge Q$ and $P \vee Q$ are

P	Q	$P \wedge Q$
T	T	T
F	T	F
T	F	F
F	F	F

P	Q	$P \vee Q$
T	T	T
F	T	T
T	F	T
F	F	F

Since the value of $\sim P$ depends only on the two possible values for P, its truth table is

P	$\sim P$
T	F
F	T

Frequently you will encounter compound propositions with more than two simple components. The proposition $(\sim Q \vee P) \wedge (R \vee S)$ has four simple components; it follows that there are $2^4 = 16$ possible combinations of values for P, Q, R, S. Two main components are $\sim Q \vee P$ and $R \vee S$. One way to make the table for $(\sim Q \vee P) \wedge (R \vee S)$ is to make tables first for both of these components and then combine those values by using the truth table for $\wedge$.

P	Q	R	S	$\sim Q$	$\sim Q \vee P$	$R \vee S$	$(\sim Q \vee P) \wedge (R \vee S)$
T	T	T	T	F	T	T	T
F	T	T	T	F	F	T	F
T	F	T	T	T	T	T	T
F	F	T	T	T	T	T	T
T	T	F	T	F	T	T	T
F	T	F	T	F	F	T	F
T	F	F	T	T	T	T	T
F	F	F	T	T	T	T	T
T	T	T	F	F	T	T	T
F	T	T	F	F	F	T	F
T	F	T	F	T	T	T	T
F	F	T	F	T	T	T	T
T	T	F	F	F	T	F	F
F	T	F	F	F	F	F	F
T	F	F	F	T	T	F	F
F	F	F	F	T	T	F	F

Two propositions P and Q are **equivalent** if and only if they have the same truth value. The propositions "$1 + 1 = 2$" and "$6 < 10$" are equivalent (even though they have nothing to do with each other) because both are true. The ability to write equivalent statements from a given statement is an important skill in writing proofs. Of course, in a proof we expect some logical connection between such statements. This connection may be based on the form of the propositions.

Definition.   Two propositional forms are **equivalent** if and only if they have the same truth tables.

For example, the propositional forms $P \vee (Q \wedge P)$ and P are equivalent. To show this, we examine their truth tables.

P	Q	$Q \wedge P$	$P \vee (Q \wedge P)$
T	T	T	T
F	T	F	F
T	F	F	T
F	F	F	F

Since the P column and the $P \vee (Q \wedge P)$ column are identical, the propositional forms are equivalent. This means that, whatever propositions we choose to use for P and for Q, the results will be equivalent. If we let P be "91 is prime" and Q be "1 + 1 = 2," then "91 is prime" is equivalent to the proposition "91 is prime, or 1 + 1 = 2 and 91 is prime." With these propositions for P and Q, Q is true and both P and $P \vee (Q \wedge P)$ are false. Thus we have an instance of the second line of the truth table.

Any proposition P is equivalent to itself. Also the propositional forms P and $\sim(\sim P)$ are equivalent. Their tables are

P	$\sim P$	$\sim(\sim P)$
T	F	T
F	T	F

> **Definition.**   A **denial** of a proposition S is any proposition equivalent to $\sim S$.

By definition the negation $\sim P$ is a denial of the proposition P, but a denial need not be the negation. The ability to rewrite the negation of a proposition into a useful denial will be very important for writing indirect proofs (see section 1.4).

The proposition "The water is cold and the soap is not here" has the negation "It is not the case that the water is cold and the soap is not here." Some other denials are

"Either the water is not cold or the soap is here"

and

"It is not the case that the water is cold and the soap is not here and the water is cold."

The propositional forms $\sim C \vee H$ and $\sim[(C \wedge \sim H) \wedge C]$ are equivalent to $\sim(C \wedge \sim H)$.

To avoid writing large numbers of parentheses, we use the rule that, first, $\sim$ applies to the smallest proposition following it, then $\wedge$ connects the smallest propositions surrounding it, and finally, $\vee$ connects the smallest propositions surrounding it. Thus $\sim P \vee Q$ is an abbreviation for $(\sim P) \vee Q$. The negation of the disjunction $P \vee Q$ must be written with parentheses $\sim(P \vee Q)$. The propositional form $P \wedge \sim Q \vee R$ abbreviates $[P \wedge (\sim Q)] \vee R$.

When the same connective is used several times in succession, parentheses may also be omitted. We reinsert parentheses from the left, so that $P \vee Q \vee R$ is really $(P \vee Q) \vee R$. Leaving out parentheses is not required; some propositions are easier to read with a few well-chosen "unnecessary" parentheses.

Conjunction, disjunction, and negation are very important in mathematics. Two other important connectives, the conditional and biconditional, will be studied in the next section. Other connectives having two components are not as useful in mathematics, but some are extremely important in digital computer circuit design.

Exercises 1.1

1. Make truth tables for each of the following propositions.
    - ★ (a) $P \wedge \sim P$
    - (b) $P \vee \sim P$
    - ★ (c) $P \wedge (Q \vee R)$
    - (d) $(P \wedge Q) \vee (P \wedge R)$
    - ★ (e) $P \wedge \sim Q$
    - (f) $P \wedge (Q \vee \sim Q)$
    - ★ (g) $(P \wedge Q) \vee \sim Q$
    - (h) $\sim (P \wedge Q)$
    - −(i) $(P \vee \sim Q) \wedge R$
    - (j) $\sim P \wedge \sim Q$
    - (k) $P \wedge P$
    - (l) $(P \wedge Q) \vee (R \wedge \sim S)$

2. Which of the following pairs of propositions are equivalent?
    - ★ (a) $P \wedge P,$ P
    - (b) $P \vee P,$ P
    - ★ (c) $P \wedge Q,$ $Q \wedge P$
    - (d) $P \vee Q,$ $Q \vee \sim P$
    - ★ (e) $(P \wedge Q) \wedge R, P \wedge (Q \wedge R)$
    - (f) $\sim (P \wedge Q),$ $\sim P \wedge \sim Q$
    - ★ (g) $\sim P \wedge \sim Q,$ $\sim (P \wedge \sim Q)$
    - (h) $(P \vee Q) \vee R, P \vee (Q \vee R)$
    - ★ (i) $(P \wedge Q) \vee R, P \wedge (Q \vee R)$
    - −(j) $\sim (P \vee Q),$ $\sim P \wedge \sim Q$
    - ★ (k) $\sim (P \wedge Q),$ $\sim P \vee \sim Q$
    - (l) $(P \wedge Q) \vee R, P \vee (Q \wedge R)$
    - ★ (m) $P \wedge (Q \vee R), (P \wedge Q) \vee (P \wedge R)$
    - (n) $\sim P \vee \sim Q,$ $\sim (P \vee \sim Q)$

3. If P, Q, and R are true while S and T are false, which of the following are true?
    - ★ (a) $Q \wedge (R \wedge S)$
    - (b) $Q \vee (R \wedge S)$
    - ★ (c) $(P \vee Q) \wedge (R \vee S)$
    - −(d) $(\sim P \vee \sim Q) \vee (\sim R \vee \sim S)$
    - ★ (e) $\sim P \vee (Q \wedge \sim Q)$
    - (f) $\sim P \vee \sim Q$
    - ★ (g) $(\sim Q \vee S) \wedge (Q \vee S)$
    - (h) $(S \wedge R) \vee (S \wedge T)$
    - ★ (i) $(P \vee S) \wedge (P \vee T)$
    - (j) $(\sim T \wedge P) \vee (T \wedge P)$
    - ★ (k) $\sim P \wedge (Q \vee \sim Q)$
    - (l) $\sim R \wedge \sim S$

4. Give a useful denial of each statement.
    - ★ (a) x is a positive integer. (Assume that x is some fixed integer.)
    - (b) We will win the first game or the second one.
    - ★ (c) $5 \geqslant 3$.
    - (d) 641371 is a composite integer.
    - ★ (e) Roses are red and violets are blue.
    - −(f) $x < y$ or $m^2 < 1$. (Assume that x, y, and m are fixed real numbers.)
    - ★ (g) T is not green or T is yellow.
    - (h) She will choose yogurt but will not choose ice cream.

5. If P, Q, and R are propositions, and P is equivalent to Q, and Q is equivalent to R, prove that
    - ★ (a) Q is equivalent to P.
    - (b) P is equivalent to R.
    - (c) $\sim Q$ is equivalent to $\sim P$.
    - (d) $P \wedge Q$ is equivalent to $Q \wedge R$.
    - −(e) $P \vee Q$ is equivalent to $Q \vee R$.

6. Let P be the sentence "Q is true" and Q be the sentence "P is false." Is P a proposition? Explain.

7. The word *or* is used in two different ways in English. We have presented the truth table for $\vee$, the **inclusive or** whose meaning is "one or the other or both." The **exclusive or,** meaning "one or the other but not both" and denoted $\underline{\vee}$, has its uses in English, as in "She will marry Heckle or she will marry Jeckle." The inclusive or is much more useful in mathematics and is the accepted meaning unless there is a statement to the contrary.

★   (a)   Make a truth table for the "exclusive or" connective, $\underline{\vee}$.

   (b)   Show that A $\underline{\vee}$ B is equivalent to (A $\vee$ B) $\wedge$ ~(A $\wedge$ B).

1.2

Conditionals and Biconditionals

It is fair to say that the most important kind of proposition in mathematics is a sentence of the form "If P, then Q." Examples include "If a natural number is written in two ways as a product of primes, then the two factorizations are identical except for the order in which the prime factors are written"; "If two lines in a plane have the same slope, then they are parallel"; and "If f is differentiable at x_0 and $f(x_0)$ is a relative minimum for f, then $f'(x_0) = 0$."

Definition.   Given propositions P and Q, the **conditional sentence** $P \Rightarrow Q$ (read "P implies Q") is the proposition "If P, then Q." The proposition P is the **antecedent** and Q is the **consequent.** The conditional sentence $P \Rightarrow Q$ is true whenever the antecedent is false or the consequent is true.

The truth table for $P \Rightarrow Q$ is

P	Q	$P \Rightarrow Q$
T	T	T
F	T	T
T	F	F
F	F	T

This table gives $P \Rightarrow Q$ the value F only when P is true and Q is false, and thus it agrees with the meaning of "if . . . , then . . ." in promises. For example, the person who promises, "If Lincoln was the second U.S. President, I'll give you a dollar" would not be a liar for failing to give you a dollar. In fact, he could give you a dollar and still not be a liar. In both cases we say the statement is true because the antecedent is false.

The situation in mathematics is similar. We all agree that "If x is odd, then $x + 1$ is even" is a true statement about any integer x. It would be hopeless to protest that in the case where x is 6, then $x + 1$ is 7, which is not even, for after all, we claim only that *if x is odd, then $x + 1$ is even.*

One curious consequence of the truth table for $P \Rightarrow Q$ is that conditional sentences may be true even when there is no connection between the antecedent and the consequent. For example, all of the following are true:

$\sin 30° = \frac{1}{2} \Rightarrow 1 + 1 = 2$.
Mars has ten moons $\Rightarrow 1 + 1 = 2$.
Mars has ten moons $\Rightarrow$ Paul Revere made plastic spoons.

And the following are false:

$1 + 2 = 3 \Rightarrow 1 < 0$.
Ducks have webbed feet $\Rightarrow$ The U.S. national debt is \$915.45.

Note that in the truth table of $P \Rightarrow Q$ the only line in which both P and $P \Rightarrow Q$ are true is the first line, in which case Q is also true. In other words if we know that both P and $P \Rightarrow Q$ are true, then we know that Q must be true. This deduction, called **modus ponens,** is one of several we will discuss in section 1.4.

Two propositions closely related to the conditional sentence $P \Rightarrow Q$ are its converse and its contrapositive.

Definition.   For propositions P and Q, the **converse** of $P \Rightarrow Q$ is $Q \Rightarrow P$, and the **contrapositive** of $P \Rightarrow Q$ is $\sim Q \Rightarrow \sim P$.

For the conditional sentence "If a function f is differentiable at x_0, then f is continuous at x_0," its converse is "If f is continuous at x_0, then f is differentiable at x_0, while the contrapositive is "If f is not continuous at x_0, then f is not differentiable at x_0." Calculus students know that the converse is a false statement.

If P is the proposition "It is raining here" and Q is "It is cloudy overhead," then $P \Rightarrow Q$ is true. Its contrapositive is "If it is not cloudy overhead, then it is not raining here," which is also true. However, the converse "If it is cloudy overhead, then it is raining here" is not a true proposition. We describe the relationships between a conditional sentence and its contrapositive and converse in the following theorem.

Theorem 1.1   The propositional form $P \Rightarrow Q$ is equivalent to its contrapositive $\sim Q \Rightarrow \sim P$ and not equivalent to its converse $Q \Rightarrow P$.

Proof.   A proof requires examining the truth tables:

P	Q	$P \Rightarrow Q$	$\sim Q$	$\sim P$	$\sim Q \Rightarrow \sim P$	$Q \Rightarrow P$
T	T	T	F	F	T	T
F	T	T	F	T	T	F
T	F	F	T	F	F	T
F	F	T	T	T	T	T

Comparing the third and sixth columns we conclude $P \Rightarrow Q$ is equivalent to $\sim Q \Rightarrow \sim P$. Comparing the third and seventh columns, we see they differ in the second and third lines. Thus $P \Rightarrow Q$ and $Q \Rightarrow P$ are not equivalent.   ■

The equivalence of a conditional sentence and its contrapositive will be the basis for an important proof technique developed in section 1.4 (proof by contrapositive). The truth or falsity of a conditional sentence has no influence on the truth value of its converse. The converse cannot be used to prove a conditional sentence.

Closely related to the conditional sentence is the **biconditional sentence** $P \Leftrightarrow Q$. The double arrow $\Leftrightarrow$ reminds one of both $\Leftarrow$ and $\Rightarrow$, and this is no accident, for $P \Leftrightarrow Q$ is equivalent to $(P \Rightarrow Q) \wedge (Q \Rightarrow P)$.

Definition.   For propositions P and Q, the **biconditional sentence** $P \Leftrightarrow Q$ is the proposition "P if and only if Q." The sentence $P \Leftrightarrow Q$ is true exactly when P and Q have the *same truth values.*

The truth table for $P \Leftrightarrow Q$ is

P	Q	$P \Leftrightarrow Q$
T	T	T
F	T	F
T	F	F
F	F	T

As a form of shorthand the words "if and only if" are frequently abbreviated to "iff" in mathematics. The statements

"A rectangle is a square iff the rectangle's diagonals are perpendicular"

and

"$1 + 7 = 6$ iff $\sqrt{2} + \sqrt{3} = \sqrt{5}$"

are both true biconditional sentences, while

"Lake Erie is in Peru iff π is an irrational number"

is a false biconditional sentence.

Any properly stated definition is an example of a biconditional sentence. Although a definition might not include the iff wording, biconditionality does provide a good test of whether a statement could serve as a definition or just a description. The sentence "A diameter of a circle is a chord of maximum length" is a correct definition of diameter because "A chord is a diameter iff the chord has maximum length" is a true proposition. However, the sentences "A sundial is an instrument for measuring time" and "A square is a quadrilateral whose interior angles are right angles" can be recognized as descriptions rather than definitions.

Because the biconditional sentence $P \Leftrightarrow Q$ has the value T exactly when the values of P and Q are the same, we can use the biconditional connective to restate the meaning of equivalent propositional forms. That is,

> The propositional forms P and Q are equivalent precisely when the truth table for $P \Leftrightarrow Q$ has all T's.

One key to success in mathematics is the ability to replace a statement by a more useful or enlightening one. This is precisely what you do to "solve" $x^2 - 7x = -12$.

$$
\begin{aligned}
x^2 - 7x = -12 \ &\Leftrightarrow\ x^2 - 7x + 12 = 0 \\
&\Leftrightarrow\ (x - 3)(x - 4) = 0 \\
&\Leftrightarrow\ x - 3 = 0 \text{ or } x - 4 = 0 \\
&\Leftrightarrow\ x = 3 \text{ or } x = 4.
\end{aligned}
$$

Each statement is simply an equivalent of its predecessor but is more illuminating as to the solution. The ability to write equivalents is crucial in writing proofs. The next theorem contains seven important equivalences.

Theorem 1.2   For propositions P and Q,

(a) $P \Leftrightarrow Q$ is equivalent to $(P \Rightarrow Q) \wedge (Q \Rightarrow P)$.
(b) $\sim(P \wedge Q)$ is equivalent to $\sim P \vee \sim Q$.
(c) $\sim(P \vee Q)$ is equivalent to $\sim P \wedge \sim Q$.
(d) $\sim(P \Rightarrow Q)$ is equivalent to $P \wedge \sim Q$.
(e) $\sim(P \wedge Q)$ is equivalent to $P \Rightarrow \sim Q$.
(f) $P \wedge (Q \vee R)$ is equivalent to $(P \wedge Q) \vee (P \wedge R)$.
(g) $P \vee (Q \wedge R)$ is equivalent to $(P \vee Q) \wedge (P \vee R)$.

You will be asked to give a proof of this theorem in exercise 6. Before giving the proof you should think about the meaning behind each equivalence. For example, in (d), if $\sim(P \Rightarrow Q)$ is true, then $P \Rightarrow Q$ is false, which forces P to be true and Q to be false. But this means both P and $\sim Q$ are true, and so $P \wedge \sim Q$ is true. This reasoning can be reversed to show that if $P \wedge \sim Q$ is true, then $\sim(P \Rightarrow Q)$ is true. We conclude $\sim(P \Rightarrow Q)$ is true precisely when $P \wedge \sim Q$ is true, and thus they are equivalent. For example, given any fixed triangle ABC, the statement "It is not the case that if triangle ABC has a right angle, then it is equilateral" is equivalent to "Triangle ABC has a right angle and is not equilateral."

Recognizing the structure of a sentence and translating the sentence into symbolic form using logical connectives is an aid in determining its truth or falsity. The translation of sentences into propositional symbols is sometimes very complicated because English is such a rich and powerful language, with many nuances, and because the ambiguities we tolerate in English would destroy structure and usefulness if we allowed them in mathematics. Consider the sentence "The Dolphins won't make the playoffs unless the Rams win Sunday or the Bears win all the rest of their games." In conversation an explanation can clarify the meaning. As it stands, there are at least three nonequivalent ways in which people translate the sentence. The word "unless" is variously used to mean a conditional or its converse or a biconditional. Ordinary English is particularly flexible in using conditional connectives. For example, "I will pay you only if you fix my car" has the meaning of a biconditional.

Here are some phrases in English that are ordinarily translated by using the connectives $\Rightarrow$ and $\Leftrightarrow$.

Use $P \Rightarrow Q$ to translate

If P, then Q.
P implies Q.
P is sufficient for Q.
P only if Q.
Q, if P.
Q whenever P.
Q is necessary for P.

Use $P \Leftrightarrow Q$ to translate

P if and only if Q.
P is equivalent to Q.
P is necessary and sufficient for Q.

In the following examples of sentence translations it is not essential to know the meaning of all the words because the logical connectives are what concern us.

Examples.   Assume that S and G have been specified. The sentence

"S is compact is sufficient for S to be bounded"

is translated

S is compact $\Rightarrow$ S is bounded

"A necessary condition for a group G to be cyclic is that G is abelian"

is translated

G is cyclic $\Rightarrow$ G is abelian

"A set S is infinite if S has an uncountable subset"

is translated

S has an uncountable subset $\Rightarrow$ S is infinite

If we let P denote the proposition "Roses are red" and Q denote the proposition "Violets are blue," we can translate the sentence "It is not the case that roses are red, nor that violets are blue" in at least two ways: $\sim(P \vee Q)$ or $\sim P \wedge \sim Q$. Fortunately these are equivalent by Theorem 1.2(c). Note that the proposition "Violets are purple" requires a new symbol, say R, since it expresses a new idea which cannot be formed from the components P and Q.

The sentence "17 and 35 have no common divisors" shows that the meaning, and not just the form of the sentence, must be considered in translating; it cannot be broken up into the two propositions: "17 has no common divisors" and "35 has no common divisors." Compare this with the proposition "17 and 35 have digits totaling 8," which can be written as a conjunction.

Example.   Suppose b is a fixed real number. "If b is an integer, then b is either even or odd" may be translated into $P \Rightarrow (Q \vee R)$, where P is "b is an integer," Q is "b is even," and R is "b is odd."

Example.   Suppose a, b, and p are fixed integers. "If p is a prime number that divides ab, then p divides a or b" becomes $(P \wedge Q) \Rightarrow (R \vee S)$, when P is "p is prime," Q is "p divides ab," R is "p divides a," and S is "p divides b."

The convention governing use of parentheses, adopted at the end of section 1.1, can be extended to the connectives $\Rightarrow$ and $\Leftrightarrow$. The connectives $\sim, \wedge, \vee, \Rightarrow$, and $\Leftrightarrow$ are applied in the order listed. That is, $\sim$ applies to the smallest possible proposition, and so forth. For example, $P \Rightarrow \sim Q \vee R \Leftrightarrow S$ is an abbreviation for $(P \Rightarrow [(\sim Q) \vee R]) \Leftrightarrow S$.

Exercises 1.2

1. Identify the antecedent and the consequent for each of the following conditional sentences. Assume that a, b, and f represent some fixed sequence, integer, or function, respectively.
    ★ (a) If squares have three sides, then triangles have four sides.
        (b) If the moon is made of cheese, then 8 is an irrational number.
        (c) b divides 3 only if b divides 9.
    ★ (d) The differentiability of f is sufficient for f to be continuous.
    —(e) A sequence a is bounded whenever a is convergent.
    ★ (f) A function f is bounded if f is integrable.
    —(g) $1 + 2 = 3$ is necessary for $1 + 1 = 2$.

☆ 2. Write the converse and contrapositive of each conditional sentence in exercise 1.

3. Which of the following conditional sentences are true?
    ★ (a) If triangles have three sides, then squares have four sides.
    — (b) If a hexagon has six sides, then the moon is made of cheese.
    ★ (c) If $7 + 6 = 14$, then $5 + 5 = 10$.
        (d) If $5 < 2$ then $10 < 7$.
    ★ (e) If one interior angle is a right triangle is 92°, then the other interior angle is 88°.
        (f) If Euclid's birthday was April 2, then rectangles have four sides.

4. Which of the following are true?
    ★ (a) Triangles have three sides iff squares have four sides.
        (b) $7 + 5 = 12$ iff $1 + 1 = 2$.
    ★ (c) b is even iff $b + 1$ is odd. (Assume that b is some fixed integer.)
    — (d) $5 + 6 = 6 + 5$ iff $7 + 1 = 10$.
        (e) A parallelogram has three sides iff 27 is prime.

5. Make truth tables for these propositions.
        (a) $P \Rightarrow (Q \wedge P)$.
    ★ (b) $(\sim P \Rightarrow Q) \vee (Q \Leftrightarrow P)$.
    ★ (c) $(\sim Q) \Rightarrow (Q \Leftrightarrow P)$.
        (d) $[(Q \Rightarrow S) \wedge (Q \Rightarrow R)] \Rightarrow [(P \vee Q) \Rightarrow (S \vee R)]$.

☆ 6. Prove Theorem 1.2 by constructing truth tables for each equivalence.

7. Rewrite each of the following sentences using logical connectives. Assume that each symbol f, n, x, S, B represents some fixed object.
    ★ (a) If f has a relative minimum at x_0 and if f is differentiable at x_0, then $f'(x_0) = 0$.
        (b) If n is prime, then $n = 2$ or n is odd.

- (c)   A number x is real and not rational whenever x is irrational.
★ (d)   If $x = 1$ or $x = -1$, then $|x| = 1$.
★ (e) f has a critical point at x_0 iff $f'(x_0) = 0$ or $f'(x_0)$ does not exist.
- (f) S is compact iff S is closed and bounded.
- (g) B is invertible is a necessary and sufficient condition for det $B \neq 0$.

8. Show that the following pairs of statements are equivalent.
- (a) $P \lor Q \Rightarrow R$ and $\sim R \Rightarrow \sim P \land \sim Q$.
★ (b) $P \land Q \Rightarrow R$ and $P \land \sim R \Rightarrow \sim Q$.
- (c) $P \Rightarrow Q \land R$ and $\sim Q \lor \sim R \Rightarrow \sim P$.
- (d) $P \Rightarrow Q \lor R$ and $P \land \sim R \Rightarrow Q$.
- (e) $(P \Rightarrow Q) \Rightarrow R$ and $(P \land \sim Q) \lor R$.
- (f) $P \Leftrightarrow Q$ and $(\sim P \lor Q) \land (\sim Q \lor P)$.

9. Give, if possible, an example of a true conditional sentence for which
★ (a)   the converse is true.   (b)   the converse is false.
★ (c)   the contrapositive is false.   (d)   the contrapositive is true.

10. Give, if possible, an example of a false conditional sentence for which
- (a)   the converse is true.   (b)   the converse is false.
- (c)   the contrapositive is true.   (d)   the contrapositive is false.

11. Give the converse and contrapositive of each sentence of exercise 7 (a), (b), (c), and (d). Tell whether each converse and contrapositive is true or false.

12. The **inverse** of the conditional sentence $P \Rightarrow Q$ is $\sim P \Rightarrow \sim Q$.
- (a)   Show that $P \Rightarrow Q$ and its inverse are not equivalent forms.
- (b)   For what values of the propositions P and Q are $P \Rightarrow Q$ and its inverse both true?
- (c)   Which is equivalent to the converse of a conditional sentence, the contrapositive of its inverse, or the inverse of its contrapositive?

1.3

Quantifiers

Unless some particular value of x has been assigned, the sentence "$x \geq 3$" is not a proposition because it is neither true nor false. When the variable x is replaced by certain values (for example, 7), the resulting proposition is true while for other values of x (for example, 2) it is false. This is an example of an **open sentence**—that is, a sentence containing one or more variables—which becomes a proposition only when the variables are replaced by the names of particular objects. For notation, if an open sentence is called P and the variables are $x_1, x_2, \ldots, x_k$, we write $P(x_1, x_2, \ldots, x_k)$, and in the case of a single variable x, we write $P(x)$.

The sentence "x_1 is equal to $x_2 + x_3$" is an open sentence with three variables. If we denote this sentence by $P(x_1, x_2, x_3)$, then $P(7, 3, 4)$ is true since $7 = 3 + 4$, while $P(1, 2, 3)$ is false.

The collection of objects that may be substituted to make an open sentence a true proposition is called the **truth set** of the sentence.† Before the truth set can be determined we must know what objects are available for consideration. That is, we must have specified a **universe** of discourse. In many cases the universe will be understood from the context. However, there are times when it must be specified. For the sentence "$x < 5$," with the universe all the natural numbers, the truth set is $\{1, 2, 3, 4\}$, while with the universe all integers, the truth set is $\{\ldots, -2, -1, 0, 1, 2, 3, 4\}$. For a sentence like "Some intelligent people admire musicians," the universe is presumably the entire human population of the earth.

Let $Q(x)$ be the sentence "$x^2 = 4$." With the universe specified as all real numbers, the truth set for $Q(x)$ is $\{2, -2\}$. With the universe the set of natural numbers, the truth set is $\{2\}$.

With a particular universe in mind, we say two open sentences $P(x)$ and $Q(x)$ are **equivalent** iff they have the *same truth set*.

> **Example.** The sentences "$x^2 = 4$" and "$|x| = 2$" are equivalent regardless of what number system we use as the universe. On the other hand, "$x^2 = 4$" and "$x = 2$" are equivalent when the universe of discourse is the set of all *positive* integers but are not equivalent for the universe consisting of *all* integers.

An open sentence $P(x)$ is not a proposition, but $P(a)$ is a proposition for any a in the universe. Another way to construct a proposition from $P(x)$ is to modify it with a quantifier.

Definitions. For an open sentence $P(x)$ with variable x, the sentence $(\forall x)P(x)$ is read "for all x, $P(x)$" and is true precisely when the truth set for $P(x)$ is the *entire universe*. The symbol $\forall$ is called the **universal quantifier.**

The sentence $(\exists x)P(x)$ is read "there exists x such that $P(x)$" and is true precisely when the truth set for $P(x)$ is *nonempty*. The symbol $\exists$ is called the **existential quantifier.**

If the universe is the set of all real numbers, then

$(\exists x)(x \geq 3)$ is true and $(\forall x)(x \geq 3)$ is false.
$(\exists x)(|x| > 0)$ is true and $(\forall x)(|x| > 0)$ is false.
$(\exists x)(x^2 = -1)$ is false and $(\forall x)(x + 2 > x)$ is true.

†The reader unfamiliar with the basic concepts of set theory is encouraged to read the first few pages of section 1 of chapter 2 before proceeding.

We agree that "All apples have spots" is quantified with ∀, but what form does it have? If we use $A(x)$ to represent "x is an apple" and $S(x)$ to represent "x has spots," and consider the universe to be the set of all fruits, should we write $(\forall x)[A(x) \wedge S(x)]$ or $(\forall x)[A(x) \Rightarrow S(x)]$? First, $(\forall x)[A(x) \wedge S(x)]$ says "For all objects x in the universe, x both is an apple and has spots." Since our universe is not spotted apples, this is not the meaning we want. Our other choice, $(\forall x)[A(x) \Rightarrow S(x)]$, says "For all objects x, if x is an apple, then x has spots," which is the meaning we seek. In general a sentence of the form "All $P(x)$ are $Q(x)$" should be symbolized $(\forall x)[P(x) \Rightarrow Q(x)]$.

Now consider "Some apples have spots." Should this be symbolized $(\exists x)[A(x) \wedge S(x)]$ or $(\exists x)[A(x) \Rightarrow S(x)]$? The first translates as "There is an object x such that it is an apple and has spots," and thus $(\exists x)[A(x) \wedge S(x)]$ is correct. On the other hand, $(\exists x)[A(x) \Rightarrow S(x)]$ reads "There is an object x such that, if it is an apple, then it has spots." Thus $(\exists x)[A(x) \Rightarrow S(x)]$ is not a correct symbolic translation, as it does not ensure the existence of apples with spots; it ensures only the existence of spots if there are any apples. In general a sentence of the form "Some $P(x)$ are $Q(x)$" should be symbolized $(\exists x)[P(x) \wedge Q(x)]$.

Some statements have the meaning of a quantified sentence even when the words "for all" or "there exists" are not present. You should be on the alert for hidden quantifiers.

> **Examples.**   When we are talking about integers, the sentence "If x is prime, then $x > 1$" has the meaning "For all integers x, if x is a prime, then $x > 1$" and is symbolically written $(\forall x)(x$ is prime $\Rightarrow x > 1)$.
> 　　　"Some real numbers have real multiplicative inverses" becomes $(\exists x)[x$ is a real number $\wedge (\exists y)(y$ is a real number $\wedge xy = 1)]$. "Some people dislike taxes" is symbolized by $(\exists x)[x$ is a person $\wedge (\forall y)(y$ is a tax $\Rightarrow x$ dislikes $y)]$.

Let us consider the translation of "Some integers are even and some are odd." One correct translation is $(\exists x)(x$ is even$) \wedge (\exists x)(x$ is odd$)$, because the first quantifier $(\exists x)$ extends only as far as the word "even." After that, any variable (even x again) may be used to express "some are odd." It is equally correct and sometimes preferable to write $(\exists x)(x$ is even$) \wedge (\exists y)(y$ is odd$)$, but it would be wrong to write $(\exists x)(x$ is even $\wedge x$ is odd$)$.

You will be reminded in chapter 2 that the symbolism "$x \in A$" stands for "The object x is an element of the set A." Since this occurs so frequently in combinations with quantifiers, we adopt two abbreviations. The sentence "Every $x \in A$ has the property P" could be restated as "If $x \in A$, then x has property P" and symbolized by $(\forall x)[x \in A \Rightarrow P(x)]$. This is abbreviated to $(\forall x \in A)P(x)$. Likewise, "Some $x \in A$ has the property P" can be restated as "There is an x that is in A and that has property P" and symbolized by $(\exists x)[x \in A \wedge P(x)]$. This is abbreviated by $(\exists x \in A)P(x)$.

Example.   When **N** is the set of natural numbers and **R** is the set of real numbers, "For every natural number there is a real number greater than the natural number" may be symbolized by

$$(\forall n \in \mathbf{N})(\exists r \in \mathbf{R})(r > n).$$

Definition.   Two quantified sentences are **equivalent** (for a particular universe) iff they have the *same truth value.*

Example. $(\forall x)(x > 3)$ is equivalent to $(\forall x)(x \geqslant 4)$ in the universe of integers. Again, it is necessary to make a distinction between a sentence and its logical form. With universe the set of all integers, the sentence "All integers are odd" is an instance of the logical form $(\forall x)P(x)$, where $P(x)$ is "x is odd." The form $(\forall x)P(x)$ is neither true nor false but becomes false in this case when "x is odd" is used for $P(x)$. Two logical forms of quantified sentences are equivalent if the truth of one implies the truth of the other and conversely, for every possible meaning of the open sentences in every universe. For example, the forms $(\forall x)[P(x) \wedge Q(x)]$ and $(\forall x)[Q(x) \wedge P(x)]$ are equivalent.

There is a natural interplay between negation and the two quantifiers $\forall$ and $\exists$.

Theorem 1.3   If $A(x)$ is an open sentence with variable x, then

(a) $\sim(\forall x)A(x)$ is equivalent to $(\exists x)\sim A(x)$.
(b) $\sim(\exists x)A(x)$ is equivalent to $(\forall x)\sim A(x)$.

Proof.

(a)   The sentence $\sim(\forall x)A(x)$ is true
      iff $(\forall x)A(x)$ is false
      iff the truth set of $A(x)$ is not the universe
      iff the truth set of $\sim A(x)$ is nonempty
      iff $(\exists x)\sim A(x)$ is true.
    Thus $\sim(\forall x)A(x)$ is true if and only if $(\exists x)\sim A(x)$ is true, so the propositions are equivalent.
(b)   The proof of this part is exercise 3. There is a proof similar to part (a) and another proof that uses part (a).   ∎

Example.   Find a denial of "All primes are odd." Use the natural numbers as the universe. The sentence may be symbolized

$$(\forall x)(x \text{ is prime} \Rightarrow x \text{ is odd}).$$

The negation is

$$\sim(\forall x)(x \text{ is prime} \Rightarrow x \text{ is odd})$$

which by Theorem 1.3(a) is equivalent to

$$(\exists x)[\sim(x \text{ is prime} \Rightarrow x \text{ is odd})].$$

By Theorem 1.2(d) this is equivalent to

$$(\exists x)[x \text{ is prime} \wedge \sim(x \text{ is odd})].$$

Thus a denial is "There exists a prime number and it is not odd" or "Some prime number is even."

Example.   Find a denial of "Every positive real number has a multiplicative inverse." Let the universe be the set of all real numbers. The sentence is symbolized by

$$(\forall x)[x > 0 \Rightarrow (\exists y)(xy = 1)].$$

The negation is

$$\sim(\forall x)[x > 0 \Rightarrow (\exists y)(xy = 1)]$$

which may be rewritten successively as

$$(\exists x)\sim[x > 0 \Rightarrow (\exists y)(xy = 1)]$$
$$(\exists x)[x > 0 \wedge \sim(\exists y)(xy = 1)]$$
$$(\exists x)[x > 0 \wedge (\forall y)(xy \neq 1)].$$

In English the last of these statements would read "There is a positive real number for which there is no multiplicative inverse."

Example.   Let f be a function whose domain is the real numbers. Find a denial of the definition of "f is continuous at a." Use the reals as the universe. A translation of the definition is

$$(\forall \epsilon)(\epsilon > 0 \Rightarrow (\exists \delta)[\delta > 0 \wedge (\forall x)(|x - a| < \delta \Rightarrow |f(x) - f(a)| < \epsilon)])$$

Make several applications of Theorems 1.2 and 1.3. (You are expected to carry out the steps.) A denial is

$$(\exists \epsilon)(\epsilon > 0 \land (\forall \delta)[\delta > 0 \Rightarrow (\exists x)(|x - a| < \delta \land |f(x) - f(a)| \ge \epsilon)])$$

In words this denial is "There is a positive ϵ such that, for all positive δ, there exists x such that $|x - a| < \delta$, but $|f(x) - f(a)| \ge \epsilon$."

Example.   Find a denial of "Some intelligent people admire musicians." The sentence is symbolized by

$(\exists x)[x$ is an intelligent person $\land (\forall y)(y$ is a musician $\Rightarrow x$ admires $y)]$.

By Theorems 1.2 and 1.3, its negation is equivalent to

$(\forall x)[x$ is an intelligent person $\Rightarrow (\exists y)(y$ is a musician $\land x$ does not admire $y)]$

which is translated into "Every intelligent person has some musician whom he or she does not admire."

We often hear statements like the complaint one fan had after a great ball game. The game was fine, he said, but "Everybody didn't get to play." We easily understand the complaint to be that "Not everyone got to play" or "Some team members didn't get into the game." This meaning is obviously intended because if it were true that everybody didn't get to play, there would have been no game.

Such imprecision is always to be avoided in mathematics. For example, in a vector space three vectors **x, y,** and **z** are dependent if there exist scalars a, b, and c, not all zero, such that $a\mathbf{x} + b\mathbf{y} + c\mathbf{z} = \mathbf{0}$. It would be incorrect to say that the scalars a, b, and c must be all nonzero, because $\sim(\forall x)A(x)$ is not equivalent to $(\forall x)\sim A(x)$.

The last quantifier we consider is unique existence.

Definition.   For an open sentence $P(x)$, the proposition $(\exists !x)P(x)$ is read "There exists a unique x such that $P(x)$." The sentence $(\exists !x)P(x)$ is true when the truth set for $P(x)$ contains *exactly one element* from the universe, hence the designation of $\exists !$ as the **unique existence quantifier.**

In the universe of natural numbers, $(\exists !x)(x$ is even and prime) is true, since the truth set of "x is even and prime" contains only the number 2.

The sentence "$(\exists!x)(x^2 = 4)$" is true when the universe is the natural numbers, and false when the universe is the integers.

A useful equivalent form for $(\exists!x)P(x)$ is

$$(\exists x)(P(x) \wedge (\forall y)[P(y) \Rightarrow x = y])$$

for if there is exactly one x such that $P(x)$, then $P(x)$ is true; and for every y, if $P(y)$ is true, then y must be identical to x, and conversely.

Using this form we see that a denial of $(\exists!x)P(x)$ is

$$(\forall x)(\sim P(x) \vee (\exists y)[P(y) \wedge x \neq y])$$

That is, $(\exists!x)P(x)$ is false when, for every x, either $P(x)$ is false or $P(y)$ is true for some y different from x.

Exercises 1.3

1. Translate the following English sentences into symbolic sentences with quantifiers. The universe for each is given in parentheses.
    ★ (a) Not all precious stones are beautiful. (All stones)
    ☆ (b) All precious stones are not beautiful. (All stones)
    ★ (c) There is a smallest positive integer. (Real numbers)
        (d) There is a rational number between any two real numbers. (Real numbers)
    — (e) Not all drunkards are truculent. (All people)
    ★ (f) No one loves everybody. (All people)
    —(g) For every positive real number x, there is a unique real number y such that $2^y = x$. (Real numbers)
    ★ (h) At least somebody cares about me. (All people)
        (i) All people are honest or no one is honest. (All people)
    —(j) Some people are honest and some people aren't honest. (All people)

☆ 2. For each of the propositions in exercise 1, write a useful denial, and give an idiomatic English version.

☆ 3. Give two proofs of Theorem 1.3(b).

4. Which of the following are true for the universe of all real numbers?
    ★ (a) $(\forall x)(\exists y)(x + y = 0)$.
        (b) $(\exists x)(\forall y)(x + y = 0)$.
        (c) $(\exists x)(\exists y)(x^2 + y^2 = -1)$.
    ★ (d) $(\forall x)[x > 0 \Rightarrow (\exists y)(y < 0 \wedge xy > 0)]$.
    — (e) $(\forall y)(\exists x)(\forall z)(xy = xz)$.
    ★ (f) $(\exists x)(\forall y)(x \leq y)$.
        (g) $(\forall y)(\exists x)(x \leq y)$.
    —(h) $(\exists!y)(y < 0 \wedge y + 3 > 0)$.
    ★ (i) $(\exists!x)(\forall y)(x = y^2)$.
    — (j) $(\forall y)(\exists!x)(x = y^2)$.
        (k) $(\exists!x)(\exists!y)(\forall w)(w^2 > x - y)$.

5.   Which of the following are denials of $(\exists!x)P(x)$?
    (a) $(\forall x)(P(x)) \lor (\forall x)(\sim P(x))$.
    (b) $(\forall x)(\sim P(x)) \lor (\exists y)(\exists z)(y \neq z \land P(y) \land P(z))$.
  ― (c) $(\forall x)[P(x) \Rightarrow (\exists y)(P(y) \land x \neq y)]$.
  ★ (d) $\sim(\forall x)(\forall y)[(P(x) \land P(y)) \Rightarrow x = y]$.

6.   Give a denial of "You can fool some of the people all of the time and all of the people some of the time, but you cannot fool all of the people all of the time."

★   7.   Riddle: What is the English translation of the symbolic statement $\forall\exists\exists\forall$?

1.4

Mathematical Proofs

You have undoubtedly written some proofs but the proofs you will be working with in this book are not like the high school geometry proofs that may have conformed to a special format that included reasons beside each step. *A proof is a complete justification of the truth of a statement* called a **theorem.** It generally begins with some hypotheses stated in the theorem and proceeds by correct reasoning to the claimed statement. Along the way it may draw upon other hypotheses, previously defined concepts, or some basic axioms setting forth properties of the concepts being considered. A **proof,** then, is a logically valid deduction of a theorem, from axioms or the theorem's premises, and may use previously proved theorems.

   As you write a proof, be sure it is not just a string of symbols. *Every step of your proof should express a complete sentence.* It is perfectly acceptable and at times advantageous to write the sentence symbolically, but be sure to include important connective words to complete the meaning of the symbols.

   The truth of any statement in a proof can eventually be traced back to some initial set of concepts and assumptions. We cannot define all terms, or else we would have a circular set of definitions. Neither can we prove all statements from previous ones. There must be an initial set of statements called **axioms** or **postulates** that are assumed true, and an initial set of concepts called **undefined terms,** from which new concepts can be introduced by means of definitions and from which new statements (theorems) can be deduced.

   The validity of a proof is based on the notion of a **tautology**—that is, a propositional form that is true for every assignment of truth values to its components. A few of the basic tautologies we shall refer to are

$$P \lor \sim P \qquad\qquad\qquad\qquad \text{(Excluded Middle)}$$
$$P \Rightarrow Q \Leftrightarrow \sim Q \Rightarrow \sim P \qquad\qquad \text{(Contrapositive)}$$
$$\left. \begin{array}{l} P \lor (Q \lor R) \Leftrightarrow (P \lor Q) \lor R \\ P \land (Q \land R) \Leftrightarrow (P \land Q) \land R \end{array} \right\} \quad \text{(Associativity)}$$

$$P \wedge (Q \vee R) \Leftrightarrow (P \wedge Q) \vee (P \wedge R)$$
$$P \vee (Q \wedge R) \Leftrightarrow (P \vee Q) \wedge (P \vee R)$$
$\left.\vphantom{\begin{matrix}a\\a\end{matrix}}\right\}$ (Distributivity)

$$(P \Leftrightarrow Q) \Leftrightarrow [(P \Rightarrow Q) \wedge (Q \Rightarrow P)]$$
$$\sim(P \Rightarrow Q) \Leftrightarrow P \wedge \sim Q$$
$$\sim(P \wedge Q) \Leftrightarrow \sim P \vee \sim Q$$
$$\sim(P \vee Q) \Leftrightarrow \sim P \wedge \sim Q$$
$\left.\vphantom{\begin{matrix}a\\a\end{matrix}}\right\}$ (De Morgan's Laws)

$$P \Leftrightarrow (\sim P \Rightarrow Q \wedge \sim Q)$$ (Contradiction)
$$[(P \Rightarrow Q) \wedge (Q \Rightarrow R)] \Rightarrow (P \Rightarrow R)$$ (Transitivity)
$$P \wedge (P \Rightarrow Q) \Rightarrow Q$$ (Modus Ponens)

Each may be verified as a tautology by its truth table, but you should also examine each to see that it expresses a logical relationship that is always true. See the discussion following Theorem 1.2 (p. 10). De Morgan's laws are named in honor of the English logician Augustus De Morgan (1806–1873).

In writing proofs, a working knowledge of tautologies is helpful for several reasons. One reason is that *a sentence whose symbolic translation is a tautology may be used at any time in a proof.* For example, if a proof involves a number *x,* one could at any time correctly assert "Either $x = 0$ or $x \neq 0$" since this is an instance of the tautology $P \vee \sim P.$

In a proof *you may at any time state any assumption or axiom.* The introduction of an assumption *P* generally takes the form "Assume *P.*"

Most steps in a proof follow from earlier lines or other known results. We may use the modus ponens tautology to make deductions. That is,

A statement *Q* may be deduced as true in a proof if both the statements *P* and $P \Rightarrow Q$ have been deduced earlier in the proof.

We refer to this rule of deduction as the **modus ponens rule.** As a useful extension of the modus ponens rule we allow *P* to be a compound statement whose components are either hypotheses, axioms, earlier statements in the proof, or statements of previously proved theorems. Likewise, the conditional sentence $P \Rightarrow Q$ may have been deduced earlier in the proof or may be a previous theorem, an axiom, or an instance of a tautology.

Example.   Suppose it has been proved that "If *F* is a finite integral domain, then *F* is a field." Suppose that one step in a proof is the statement "*F* is finite" and another step is the statement "*F* is an integral domain." Then a later step may be the statement "*F* is a field." We have applied the modus ponens rule to deduce *R* from $(S \wedge D)$ and $[(S \wedge D) \Rightarrow R)]$, where *S* is "*F* is finite," *D* is "*F* is an integral domain," and *R* is "*F* is a field."

Example.   Suppose a proof contains the statements (1) "If the crime did not take place in the billiard room, then Colonel Mustard is guilty"; (2) "The lead pipe was not the weapon"; and (3) "Either Colonel Mustard is not guilty or the weapon used was the lead pipe." Then we may deduce that "The crime took place in the billiard room." In this application of the modus ponens rule to deduce Q from P and $P \Rightarrow Q$, P is the conjunction of the three statements from the proof (which we symbolize below), and Q is the sentence deduced. You should verify that $P \Rightarrow Q$ (symbolized below) is a tautology, because that fact is what makes this deduction a legitimate use of the modus ponens rule. It is not necessary that the statements P and Q be included when this proof is written out. What matters is that Q follows from the other statements by the modus ponens rule.

$$\begin{array}{ll}
\text{Statement 1 is} & \sim B \Rightarrow M \\
\text{Statement 2 is} & \sim L \\
\text{Statement 3 is} & \sim M \vee L \\
P \text{ is} & (\sim B \Rightarrow M) \wedge \sim L \wedge (\sim M \vee L) \\
Q \text{ is} & B \\
P \Rightarrow Q \text{ is} & [(\sim B \Rightarrow M) \wedge \sim L \wedge (\sim M \vee L)] \Rightarrow B
\end{array}$$

If a sentence R appears in a proof and another sentence S is equivalent to R, then S may be deduced in the proof. This **replacement** of R by its equivalent S is justified because $(R \wedge (R \Leftrightarrow S)) \Rightarrow S$ is a tautology. Thus by modus ponens, S may be asserted.

For instance, a proof containing the line "The product of real numbers a and b is zero" could later have the statement "a is zero or b is zero." In this example the equivalence of the two statements comes from our knowledge of real numbers, not from the form of the statements.

As another use of replacement, we may deduce the contrapositive of a conditional sentence in a proof. For instance, since we know "If $f'(x_0) = 0$, then f has a critical point at x_0," we could assert in a proof that "If f does not have a critical point at x_0, then $f'(x_0) \neq 0$."

Replacement can also be applied to components of statements in a proof. For instance, the statement "If x is an integer greater than 2 and x is prime, then x does not divide 32" may be replaced by "If x is an odd prime, then x does not divide 32."

Example.   We may not deduce the statement "n is a prime number" from the statements "If n is a prime number, then n is an integer" and "n is an integer," since this could employ only the proposition $(P \Rightarrow Q) \wedge Q \Rightarrow P$, which is not a tautology.

The strategy to construct a proof of a given theorem depends greatly on the logical form of the theorem's statement and on the particular concepts involved. As a general rule, when you write a step in a proof, ask yourself if *making that assertion is valid in the sense that some tautology permits you to deduce it*. It is not necessary to cite the tautology in your proof. In fact, with practice you should eventually come to write proofs without purposefully thinking about tautologies. What is necessary is that *every step must be justifiable.*

Great latitude is allowed for differences in taste and style among proof writers. Generally the further you go in mathematics the less justification you will find given, because with more advanced topics more is expected of the reader. In this text our proofs will be complete and concise, but we shall on occasion insert parenthetical comments (offset by ⟨ ⟩ and in italics) to explain how and why a proof is proceeding as it is. Such comments should not be taken as part of the proof but are inserted to help clarify the workings of the proof.

The first method of proof we will examine is the **direct proof** of a conditional sentence. How do you prove a sentence of the form $P \Rightarrow Q$? This implication is false only when P is true and Q is false, so it suffices to show this situation is an impossibility. The simplest way to proceed, then, is to assume that P is true, and show (deduce) that Q is also true. A direct proof of $P \Rightarrow Q$ will have the following form:

Direct Proof of $P \Rightarrow Q$

Proof.
Assume P.

    •

    •

    •

Therefore, Q.
Thus $P \Rightarrow Q$. ∎

Some examples we consider in this section actually involve quantified sentences. Since we will consider proofs with quantifiers in the next section, in this section we will imagine that a variable represents some fixed object. That is, when you read "If x is odd, then $x + 1$ is even," you should think of x as being some particular integer.

Example.   Suppose x is an integer. Prove that if x is odd, then $x + 1$ is even.

Proof. ⟨*The proof consists of assuming the antecedent and then writing four equivalent statements, the last being the consequent.*⟩

Suppose x is an odd integer.
Then $x = 2r + 1$ for some integer r.
Thus $x + 1 = (2r + 1) + 1 = 2(r + 1)$ for some integer r.
Since $x + 1$ is twice the integer $r + 1$, $x + 1$ is even. ∎

Your strategy for developing a direct proof of a conditional sentence should involve these steps:

1. Determine precisely the antecedent and consequent.
2. Replace (if necessary) the antecedent with a more usable equivalent.
3. Replace (if necessary) the consequent by something equivalent and more readily shown.
4. Develop a chain of statements, each deducible from its predecessors or other known results, that leads from the antecedent to the consquent.

Our second example has to do with a point (x, y) in the Cartesian plane (figure 1.1).

Example.   Prove that if $x < -4$ and $y > 2$, then the distance from (x, y) to $(1, -2)$ is at least 6.

Proof.   If we assume that $x < -4$ and $y > 2$, then $x - 1 < -5$, so $(x - 1)^2 > 25$. Also $y + 2 > 4$, so $(y + 2)^2 > 16$. Therefore

$$\sqrt{(x - 1)^2 + (y + 2)^2} > \sqrt{25 + 16} > \sqrt{36}$$

so the distance from (x, y) to $(1, -2)$ is at least 6.   ∎

Our next example of a direct proof is drawn from number theory. It makes use of the definition that the integer x **divides** the integer y iff $y = xn$ for some integer n.

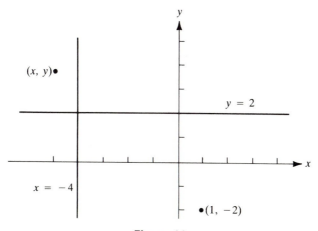

Figure 1.1

Example.   Let a, b, and c be integers. Prove that if a divides b and a divides c, then a divides $b - c$.

Proof.   Suppose a divides b and a divides c. ⟨*Now we use a more useful equivalent of "divides."*⟩ Then $b = an$ for some integer n and $c = am$ for some integer m. Thus $b - c = an - am = a(n - m)$. Since $n - m$ is an integer ⟨*we use the fact that the difference of two integers is an integer*⟩ a divides $b - c$. ∎

Direct proofs of statements of the form $P \Rightarrow Q$ are not quite so straightforward when either P or Q is itself a compound proposition.

Example.   To prove $P \Rightarrow (Q \vee R)$, one often proves either the equivalent $(P \wedge \sim Q) \Rightarrow R$ or the equivalent $(P \wedge \sim R) \Rightarrow Q$. For instance, to prove "If the polynomial f has degree 4, then f has a real zero or f can be written as the product of two irreducible quadratics," we would prove "If f has degree 4 and no real zeros, then f can be written as the product of two irreducible quadratics."

Example.   To prove $(P \vee Q) \Rightarrow R$, one could proceed by cases, first proving $P \Rightarrow R$, then proving $Q \Rightarrow R$. This is valid because of the tautology $[(P \vee Q) \Rightarrow R] \Leftrightarrow [(P \Rightarrow R) \wedge (Q \Rightarrow R)]$. The statement "If a quadrilateral has opposite sides equal or opposite angles equal, then it is a parallelogram" is proved by showing both "A quadrilateral with opposite sides equal is a parallelogram" and "A quadrilateral with opposite angles equal is a parallelogram."

Example.   A proof of a statement symbolized by $P \Rightarrow (Q \wedge R)$ would probably also have two parts. We first show $P \Rightarrow Q$ and then show $P \Rightarrow R$. We would use this method to prove the statement "If two parallel lines are cut by a transversal, then corresponding angles are equal and alternate interior angles are equal."

A second form of proof for a conditional sentence is **proof by contrapositive.** The idea here is that since $P \Rightarrow Q$ is equivalent to its contrapositive, $\sim Q \Rightarrow \sim P$, we first give a direct proof of $\sim Q \Rightarrow \sim P$ and then conclude $P \Rightarrow Q$. This method works well when the connection between the denials of P and Q is easier to understand than the connection between P and Q themselves. The format of a proof by contrapositive is as follows:

Contrapositive Proof of P ⇒ Q

Proof.
Suppose $\sim Q$.
 •
 •
 •

Therefore $\sim P$. (Via a direct proof)
Thus $\sim Q \Rightarrow \sim P$.
Therefore $P \Rightarrow Q$. ∎

Example. Let m be an integer. Prove that if m^2 is odd, then m is odd.

Proof. Suppose m is not odd. ⟨*Suppose $\sim Q$.*⟩ Then m is even. Thus $m = 2k$ for some integer k. ⟨*Equivalent statement.*⟩ Then $m^2 = (2k)^2 = 4k^2 = 2(2k^2)$. Since m^2 is twice the integer $2k^2$, m^2 is even. ⟨*Deduce $\sim P$.*⟩ Thus if m is even, then m^2 is even; so, by the contrapositive, if m^2 is odd, then m is odd. ∎

A **proof by contradiction** makes use of the tautology $P \Leftrightarrow [\sim P \Rightarrow (Q \wedge \sim Q)]$. To prove a proposition P, it is sufficient to prove $\sim P \Rightarrow (Q \wedge \sim Q)$. Two aspects about this form of proof are especially noteworthy. First, this method of proof can be applied to any proposition P, whereas direct proofs and proofs by contrapositive can be used only for conditional sentences. Second, the proposition Q does not even appear on the left side of the tautology. The idea of proving $\sim P \Rightarrow (Q \wedge \sim Q)$ then has an advantage and a disadvantage. We don't know what proposition to use for Q, but any proposition that will do the job is a good one. This means a proof by contradiction will require a "spark of insight" to determine a useful Q. A proof by contradiction has the following form:

Proof of P by Contradiction

Proof.
Suppose $\sim P$.
 •
 •
 •

Therefore Q.
 •
 •
 •

Therefore $\sim Q$.
Hence $Q \wedge \sim Q$, a contradiction.
Thus P. ∎

Example. Prove that $\sqrt{2}$ is an irrational number.

Proof. Suppose that $\sqrt{2}$ is a rational number. ⟨*Assume* $\sim P.$⟩ Then $\sqrt{2} = s/t$ where s and t are integers. Thus $2 = s^2/t^2$, and $2t^2 = s^2$. Since s^2 and t^2 are squares, s^2 contains an even number of 2's as factors ⟨*This is our Q statement.*⟩, and t^2 contains an even number of 2's. But then $2t^2$ contains an odd number of 2's as factors. Since $s^2 = 2t^2$, s^2 has an odd number of 2's. ⟨*This is the statement* $\sim Q.$⟩ This is a contradiction. We conclude that $\sqrt{2}$ is irrational. ∎

As a second example of a proof by contradiction we give a proof (attributed to Euclid) that there are an infinite number of primes.

Example. Prove that the set of primes is infinite.

Proof. Suppose the set of primes is finite. ⟨*Suppose* $\sim P.$⟩ Let the primes be $p_1, p_2, p_3, \ldots, p_k$ and consider the number $n = (p_1 p_2 \cdots p_k) + 1$. Since n is a natural number, n has a prime divisor q where $q > 1$. ⟨*The Q statement is* $q > 1.$⟩ Since q is a prime and $p_1, p_2, \ldots, p_k$ is the list of all primes, q is one of the p_i and thus q divides the product $p_1 p_2 \cdots p_k$. Since q also divides n, q divides $n - p_1 p_2 \cdots p_k$. ⟨*If a divides b and c, then a divides* $b - c.$⟩ But $n - p_1 p_2 \cdots p_k = 1$ and so $q = 1$. ⟨*This is* $\sim Q.$⟩ From this contradiction we conclude that the set of primes is infinite. ∎

Proofs of biconditional sentences are often based on the equivalence of $P \Leftrightarrow Q$ with $(P \Rightarrow Q) \wedge (Q \Rightarrow P)$. Many proofs of $P \Leftrightarrow Q$ will have the following form:

Two-Part Proof of $P \Leftrightarrow Q$

Proof.

(i) Show $P \Rightarrow Q$ by any method.
(ii) Show $Q \Rightarrow P$ by any method.

Therefore $P \Leftrightarrow Q$. ∎

Of course, the two proofs in (i) and (ii) may use different methods. Frequently the proof of one part is more difficult than the other. This is true, for example, of the proof that "The natural number x is prime iff no positive integer greater than 1 and less than or equal to $\sqrt{x}$ divides x." It is immediate from the definition that "If x is prime, then no positive integer greater than 1 and less than or equal to $\sqrt{x}$ divides x," but the converse requires a little reasoning.

In some cases it is possible to prove a biconditional sentence $P \Leftrightarrow Q$ that uses the "iff" connective throughout. This amounts to starting with P and then replacing it with a sequence of equivalent statements, the last one being Q.

Example.   Consider a triangle with sides of length a, b, c. Use the Law of Cosines to prove that the triangle is a right triangle with hypotenuse c if and only if $a^2 + b^2 = c^2$ (figure 1.2).

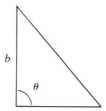

Figure 1.2

Proof.   By the Law of Cosines, $a^2 + b^2 = c^2 - 2ab \cos \theta$, where θ is the angle between the sides of length a and b. Thus

$$a^2 + b^2 = c^2 \quad \text{iff} \quad 2ab \cos \theta = 0$$
$$\text{iff} \quad \cos \theta = 0$$
$$\text{iff} \quad \theta = 90°.$$

Thus $a^2 + b^2 = c^2$ iff the triangle is a right triangle with hypotenuse c.   ∎

One form of proof is known as ***proof by exhaustion.*** Such a proof consists of examining every possible case. This was our method in Theorem 1.1 where we examined all four combinations of truth values for two propositions. Naturally, the idea of proof by exhaustion is appealing only when there is a small number of cases, or when large numbers of cases can be systematically handled, but there have been instances of truly exhausting proofs involving great numbers of cases. In 1976 Kenneth Appel and Wolfgang Haken announced a proof of the long-standing Four-Color Conjecture. The original version of their proof contains 1879 cases, took 3½ years to develop, required 1200 hours of computer time, and is the result of over 10,000,000,000 calculations.

By now you may have the impression that given a set of axioms and definitions of a mathematical system, any correctly stated proposition in that system can be proved true or proved false. This is not the case. There are important examples in mathematics of **consistent axiom systems** (so that there exist structures satisfying all the axioms) that include statements such that neither the statement nor its negation can be proved. Such statements are called **undecidable** in the system because their truth is independent of the truth of the axioms. The classic case involves the fifth of five postulates that Euclid (circa 300 B.C.) set forth as his basis for plane geometry: "Given a line and a point not on the line, exactly one line can be drawn through the point parallel to the line." For centuries some thought Euclid's axioms were not independent, believing that the fifth postulate could be proved from the other four. It was not until the nineteenth century that it became clear that the fifth postulate was undecidable. There are now theories of Euclidean geometry where the fifth postulate is assumed true and non-Euclidean geometries where it is assumed false. Both are perfectly reasonable subjects for mathematical study.

Exercises 1.4

1. Using truth tables, verify that each of the basic tautologies of this section is a tautology.

2. A *contradiction* is a propositional form that is false for every assignment of truth values to its components. For example, $Q \wedge \sim Q$ is a contradiction. Which of the following are tautologies, contradictions, or neither?
   ★ (a) $[(P \Rightarrow Q) \Rightarrow P] \Rightarrow P.$
   (b) $P \Leftrightarrow P \wedge (P \vee Q).$
   (c) $P \Rightarrow Q \Leftrightarrow P \wedge \sim Q.$
   ★ (d) $P \Rightarrow [P \Rightarrow (P \Rightarrow Q)].$
   (e) $P \wedge (Q \vee \sim Q) \Leftrightarrow P.$
   (f) $[Q \wedge (P \Rightarrow Q)] \Rightarrow P.$
   (g) $(P \wedge Q) \vee (\sim P \wedge Q) \vee (P \wedge \sim Q) \vee (\sim P \wedge \sim Q).$
   (h) $(P \Leftrightarrow Q) \Leftrightarrow \sim (\sim P \vee Q) \vee (\sim P \wedge Q).$
   (i) $[P \Rightarrow (Q \vee R)] \Rightarrow [(Q \Rightarrow R) \vee (R \Rightarrow P)].$
   (j) $P \wedge (P \Leftrightarrow Q) \wedge \sim Q.$

3. **Proofs to Grade.** Problems with this title throughout this book ask you to analyze an alleged proof of a claim and to give one of three grades. Assign a grade of A (excellent) if the claim and proof are correct, even if the proof is not the simplest or the proof you would have given. Assign an E (failure) if the claim is incorrect, if the main idea of the proof is incorrect, or if most of the statements in it are incorrect. Assign a grade of C (partial credit) for a proof that is largely correct, but contains one or two incorrect statements or justifications. Whenever the proof is incorrect, explain your grade. Tell what is incorrect and why.
   ★ (a) Suppose m is an integer.
      Claim. If m^2 is odd, then m is odd.
      "Proof." Assume m is odd. Then $m = 2k + 1$ for some integer k. Therefore $m^2 = (2k + 1)^2 = 4k^2 + 4k + 1 = 2(2k^2 + 2k) + 1$, which is odd. Therefore if m^2 is odd, then m is odd. ∎
   (b) Suppose m is an integer.
      Claim. If m^2 is odd, then m is odd.

"Proof."   Assume that m^2 is not odd. Then m^2 is even and $m^2 = 2k$ for some integer k. Thus $2k$ is a perfect square; that is, $\sqrt{2k}$ is an integer. If $\sqrt{2k}$ is odd, then $\sqrt{2k} = 2n + 1$ for some integer n, which means $m^2 = 2k = (2n + 1)^2 = 4n^2 + 4n + 1 = 2(2k^2 + 2k) + 1$. Thus m^2 is odd, contrary to our assumption. Therefore $\sqrt{2k} = m$ must be even. Thus if m^2 is not odd, then m is not odd. Hence if m^2 is even, then m is even.   ■

(c) Suppose a, b, and c are integers.
  Claim. If a divides both b and c, then a divides $b + c$.
  "Proof."   Assume that a does not divide $b + c$. Then there is no integer k such that $ak = b + c$. However, a divides b, so $am = b$ for some integer m; and a divides c, so $an = c$ for some integer n. Thus $am + an = a(m + n) = b + c$. Therefore $k = m + n$ is an integer satisfying $ak = b + c$. Thus the assumption that a does not divide $b + c$ is false, and a does divide $b + c$.   ■

★ (d) Suppose t is a real number.
  Claim. If t is irrational, then $5t$ is irrational.
  "Proof."   Suppose $5t$ is rational. Then $5t = p/q$ where p and q are integers and $q \neq 0$. Therefore $t = p/5q$ where p and $5q$ are integers and $5q \neq 0$, so t is rational. Therefore if t is irrational, then $5t$ is irrational.   ■

4. Construct an outline of the proof for the given statement by the indicated method. Do not include details of the proof. You are not expected to give a complete proof nor to understand all the terms involved.
  ★ (a) Outline a direct proof that if $(G, *)$ is a cyclic group, then $(G, *)$ is abelian.
  (b) Outline a direct proof that if B is a nonsingular matrix, then the determinant of B is not zero.
  ★ (c) Outline a contrapositive proof of the statement in (a).
  (d) Outline a contrapositive proof of the statement in (b).
  (e) Suppose A, B, and C are sets. Outline a direct proof that if A is a subset of B and B is a subset of C, then A is a subset of C.
  (f) Outline a direct proof that if the maximum value of the differentiable function $f(x)$ on the closed interval $[a, b]$ occurs at x_0, then either $x_0 = a$ or $x_0 = b$ or $f'(x_0) = 0$.
  (g) Outline a proof by contradiction that the set of natural numbers is not finite.

5. Suppose d and n are natural numbers. Prove that if $d \neq 1$ and d is the smallest divisor of n other than 1, then d is prime.

6. Suppose a, b, and c are positive integers. Prove that
  (a) if a divides b and b divides c, then a divides c.
  (b) if $ab = 1$, then $a = 1$ and $b = 1$.
  (c) if a divides b and b divides a, then $a = b$.
  (d) if a divides b, then $a \leq b$.
  (e) if ac divides bc, then a divides b.

7. In this exercise the proofs involve verifying an algebraic expression. In such proofs (and many others) it is a good idea to "work backwards." That is, to show that an equation is true, decide what other equation it could be proved from, and what that equation could be proved from, and so forth. After doing such preliminary work, remember to write your proof forwards, so that your conclusion is the statement to be proved.

(a) Let x and y be positive real numbers. Prove that $\dfrac{x + y}{2} \geqslant \sqrt{xy}$. Where in the proof is it essential that x and y are positive?

(b) Suppose a right triangle has hypotenuse c and legs a and b. Prove that the triangle is isosceles if and only if its area is $\frac{1}{4}c^2$.

8. Recall that except for degenerate cases, the graph of
$$Ax^2 + Bxy + Cy^2 + Dx + Ey + F = 0 \text{ is}$$
        an ellipse iff $B^2 - 4AC < 0$
        a parabola iff $B^2 - 4AC = 0$
        a hyperbola iff $B^2 - 4AC > 0$

(a) Prove that the equation is a hyperbola if $AC < 0$ or $B < C < 4A < 0$.

(b) Prove that if the graph is a parabola, then $BC = 0$ or $A = B^2/4C$.

1.5

Proofs Involving Quantifiers

Most theorems in mathematics are quantified sentences even though the quantifier may not actually appear in the statement. For example, proving "If x is an odd integer, then $x + 1$ is even," as we did in section 1.4, actually involves a quantifier, since the sentence has the symbolic translation $(\forall x)(x$ is odd $\Rightarrow x + 1$ is even). In this section we present methods of proof for theorems of the form $(\exists x)P(x)$, $(\forall x)P(x)$, and $(\exists ! x)P(x)$. The validity of the proof techniques still depends upon employing statements that are always true.

There are several ways to prove **existence theorems**—that is, propositions of the form $(\exists x)P(x)$. The most direct, called a **constructive proof,** is to name or describe some object in the universe that actually makes $P(x)$ true. For example, stating "2 is a prime" is a proof of the theorem "There is an even prime integer." Here is another example.

Example. Prove the number $x = 4294967297$ is not a prime.

Proof. ⟨*It suffices to show the existence of natural numbers a and b both greater than 1, such that x = ab.*⟩ We note that $x = (641)(6700417)$. Thus x is not prime. ∎

This factorization of x was first observed by Leonard Euler (1707–1783), one of the greatest mathematicians of all time. Euler's work pervades almost every area of mathematics. The symbol e for the constant used as the base for natural logarithms is used in tribute to Euler.

It is also possible to give a proof of $(\exists x)P(x)$ by **contradiction.** The proof has the following form:

Proof of (∃x)P(x) by Contradiction

> **Proof.**
> Suppose $\sim(\exists x)P(x)$.
> Then $(\forall x)\sim P(x)$.
>
> •
> •
> •
>
> Therefore $Q \wedge \sim Q$, a contradiction.
> Hence $\sim(\exists x)P(x)$ is false; so $(\exists x)P(x)$ is true. ∎

The heart of such a proof generally involves making deductions from the universally quantified statement $(\forall x)\sim P(x)$.

Other proofs of existence theorems $(\exists x)P(x)$ show that there must be some object for which $P(x)$ is true, without ever actually producing a particular object. Both Rolles' Theorem and the Mean Value Theorem from calculus are good examples of this. Here is another.

Example.   Prove that the polynomial $r(x) = x^{71} - 2x^{39} + 5x - 0.3$ has a real zero.

Proof.   ⟨*What we must show has the form $(\exists t)(t$ is real and $r(t) = 0).$*⟩ By the Fundamental Theorem of Algebra, $r(x)$ has 71 zeros that are either real or complex. Since the polynomial has real coefficients, its complex zeros come in pairs ⟨*by the Complex Root Theorem*⟩. Hence there are an even number of nonreal roots, and that leaves an odd number of real roots. Therefore $r(x)$ has at least one real root. ∎

It is often necessary to prove that a statement of the form $(\forall x)P(x)$ is false. Since $\sim(\forall x)P(x)$ is equivalent to $(\exists x)\sim P(x)$, this amounts to a proof that $(\exists x)\sim P(x)$ is true. Any object t in the universe for which $\sim P(t)$ is true is called a **counterexample** to $(\forall x)P(x)$. For example, $f(x) = |x|$ is a counterexample to "Every function continuous at 0 is differentiable at 0." The number 2 is a counterexample to "All primes are odd." Some counterexamples have eluded mathematicians for centuries before being observed.

Finally, if an existentially quantified statement appears as a premise of a theorem, it should be assumed and then applied. For example, a proof of a statement of the form $(\exists x)P(x) \Rightarrow R$ could proceed as follows:

Proof of (∃x)P(x) ⇒ R

> **Proof.**
> By hypothesis, $(\exists x)P(x)$.
> Let t be an object such that $P(t)$ is true.
>
> •
> •
> •
>
> Therefore R.
> Hence $(\exists x)P(x) \Rightarrow R$.   ■

To prove a proposition of the form $(\forall x)P(x)$, we must show that $P(x)$ is true for every object x in the universe. A direct proof of this is done by letting x represent an arbitrary object in the universe, and showing that $P(x)$ is true for that object without using any special properties of the object x. Then, since x is arbitrary, we can conclude that $(\forall x)P(x)$ is true.

Thus a **direct proof** of $(\forall x)P(x)$ has the following form:

Direct Proof of (∀x)P(x)

> **Proof.**
> Let x be an arbitrary object in the universe. (Actually name the universe.)
>
> •
> •
> •
>
> Hence $P(x)$ is true.
> Since x is arbitrary, $(\forall x)P(x)$ is true.   ■

The open sentence $P(x)$ will often be a combination of other open sentences, and thus a deduction of $P(x)$ will require the selection of an appropriate proof technique.

Example.   If x is an even integer, then x^2 is an even integer.

Proof.   ⟨*The universe is evidently the integers. Here $P(x)$ denotes "x is even implies that x^2 is even."*⟩ Let x be an integer. Suppose x is even. ⟨*We are giving a direct proof of a conditional sentence.*⟩ Then $x = 2k$ for some integer k. Therefore $x^2 = (2k)^2 = 2(2k^2)$. Since $2k^2$ is an integer, x^2 is even. Hence if x is even, then x^2 is even.   ■

Many of the statements we considered in section 1.4 could easily have been treated as universally quantified statements. See, for example, the proof that if x is an odd integer, then $x + 1$ is even. The only difference is that in that section we were viewing the symbol x as a fixed but unknown number. Usually we would view x as a variable and write the proof in the style shown above.

Another method of proof of a statement of the form $(\forall x)P(x)$ is by **contradiction.** The form is as follows:

Proof of (∀x)P(x) by Contradiction

Proof.
Suppose $\sim(\forall x)P(x)$.
Then $(\exists x)\sim P(x)$.
Let t be an object such that $\sim P(t)$.

•

•

•

Therefore $Q \wedge \sim Q$.
Thus $(\exists x)\sim P(x)$ is false; so its denial $(\forall x)P(x)$ is true. ∎

The last proof method we give is for the $\exists !$ quantifier. The standard technique for proving a proposition of the form $(\exists ! x)P(x)$ is based on its expression in the form $(\exists x)P(x) \wedge (\forall y)(\forall z)[(P(y) \wedge P(z)) \Rightarrow y = z]$.

Proof of (∃!x)P(x)

Proof.

(i)   Prove $(\exists x)P(x)$ is true by any method.
(ii) Assume that t_1 and t_2 are objects in the universe such that $P(t_1)$ and $P(t_2)$ are true.

•

•

•

Therefore $t_1 = t_2$.
We conclude $(\exists ! x)P(x)$. ∎

Example.   Prove that the polynomial $r(x) = x - 3$ has a unique zero.

Proof.

(i)   Since $r(3) = 3 - 3 = 0$, 3 is a zero of $r(x)$.
(ii) Suppose that t_1 and t_2 are two zeros of $r(x)$. Then $r(t_1) = 0 = r(t_2)$. Therefore $t_1 - 3 = t_2 - 3$. Thus $t_1 = t_2$. Therefore $r(x) = x - 3$ has a unique zero. ∎

Example.   Every nonzero real number has a unique multiplicative inverse.

Proof.   ⟨*The theorem is symbolized* $(\forall x)(x \neq 0) \Rightarrow (\exists! y)(y$ *is real and* $xy = 1).$⟩ Let x be a real number ⟨*We start this way because the statement to be proved is quantified with* $\forall$⟩. Suppose $x \neq 0$. ⟨*Assume the antecedent.*⟩ We must show that $xy = 1$ for exactly one real number y.

(i)   Let $y = 1/x$. ⟨*This is a constructive proof that there exists an inverse.*⟩ Then y is a real number since $x \neq 0$, and $xy = x(1/x) = x/x = 1$. Thus x has a multiplicative inverse.

(ii)   Now suppose y and z are two real multiplicative inverses for x. ⟨*This y is not necessarily the $y = 1/x$ in part (i).*⟩ Then $xy = 1$ and $xz = 1$. Thus $xz = zx$, and $xy - xz = x(y - z) = 0$. Since $x \neq 0$, $y - z = 0$. Therefore $y = z$.

Thus every nonzero real number has a unique multiplicative inverse.   ∎

Many statements have more than one quantifier, so we must deal with each quantifier in succession.

Example: For every natural number n, there is a natural number M such that for all natural numbers $m > M$, $\dfrac{1}{m} < \dfrac{1}{3n}$.

Proof.   ⟨*With the natural numbers as the universe the statement may be symbolized*

$$(\forall n)(\exists M)(\forall m)\left(m > M \Rightarrow \frac{1}{m} < \frac{1}{3n} \right).$$

We consider quantifiers in order from the left.⟩ Let n be a natural number. Choose M to be $3n$. Let m be a natural number, and suppose $m > M$. Then $m > 3n$, and $3mn > 0$, so dividing by $3mn$ we have $\dfrac{1}{m} < \dfrac{1}{3n}$. ⟨*The choice of $3n$ for M is the result of some scratchwork, working backwards from the intended conclusion* $\dfrac{1}{m} < \dfrac{1}{3n}$.⟩   ∎

Great care must be taken in proofs that contain expressions involving more than one quantifier. Here are some manipulations of quantifiers that permit valid deductions.

1. $(\forall x)(\forall y)P(x, y) \Leftrightarrow (\forall y)(\forall x)P(x, y)$.
2. $(\exists x)(\exists y)P(x, y) \Leftrightarrow (\exists y)(\exists x)P(x, y)$.
3. $[(\forall x)P(x) \vee (\forall x)Q(x)] \Rightarrow (\forall x)[P(x) \vee Q(x)]$.
4. $(\forall x)[P(x) \Rightarrow Q(x)] \Rightarrow [(\forall x)P(x) \Rightarrow (\forall x)Q(x)]$.
5. $(\forall x)[P(x) \wedge Q(x)] \Leftrightarrow [(\forall x)P(x) \wedge (\forall x)Q(x)]$.
6. $(\exists x)(\forall y)P(x, y) \Rightarrow (\forall y)(\exists x)P(x, y)$.

You should convince yourself that each of these is a logically valid conditional or biconditional. For example, the last on the list is always true because if $(\exists x)(\forall y)P(x, y)$ is true, then there is (at least) one x that makes $P(x, y)$ true no matter what y is. Therefore for any y, $(\exists x)P(x, y)$ is true because the one x exists.

It is important to be aware of the most common *incorrect deductions* making use of quantifiers. We list four here and give counterexamples.

1. $(\exists x)P(x) \Rightarrow (\forall x)P(x)$ is not valid.
   If the universe is all integers and $P(x)$ is the sentence "x is odd," then $P(5)$ is true and $P(8)$ is false. Thus $(\exists x)P(x)$ is true and $(\forall x)P(x)$ is false, so the implication fails.
2. $(\forall x)[P(x) \vee Q(x)] \Rightarrow [(\forall x)P(x) \vee (\forall x)Q(x)]$ is not valid.
   We use the example in 1 and also let $Q(x)$ be "x is even." Then it is true that "All integers are either odd or even" but false that "Either all integers are odd or all integers are even."
3. $[(\forall x)P(x) \Rightarrow (\forall x)Q(x)] \Rightarrow (\forall x)[P(x) \Rightarrow Q(x)]$ is not valid.
   The example in 2 can be used here. Because $(\forall x)P(x)$ is false, $(\forall x)P(x) \Rightarrow (\forall x)Q(x)$ is true. However, $(\forall x)[P(x) \Rightarrow Q(x)]$ is false.
4. $(\forall y)(\exists x)P(x, y) \Rightarrow (\exists x)(\forall y)P(x, y)$ is not valid.
   This is probably the most troublesome of all the possibilities for dealing with quantifiers. Let the universe be the set of all married people and $P(x, y)$ be the sentence "x is married to y." Then $(\forall y)(\exists x)P(x, y)$ is true, since everyone is married to someone. But $(\exists x)(\forall y)P(x, y)$ would be translated as "There is some married person who is married to every married person," which is clearly false.

Exercises 1.5

1. **Proofs to Grade.**
★ (a) **Claim.**   Every polynomial of degree 3 with real coefficients has a real zero.
   "Proof."   We note that the polynomial $p(x) = x^3 - 8$ has degree 3, real coefficients, and a real zero ($x = 2$). Thus the statement "Every polynomial

of degree 3 with real coefficients does not have a real zero" is false, and hence its denial, "Every polynomial of degree 3 with real coefficients has a real zero," is true.   ■

★ (b) **Claim.**   There is a unique polynomial whose first derivative is $2x + 3$ and which has a zero at $x = 1$.

   "Proof."   The antiderivative of $2x + 3$ is $x^2 + 3x + C$. If we let $p(x) = x^2 + 3x - 4$, then $p'(x) = 2x + 3$ and $p(1) = 0$. So $p(x)$ is the desired polynomial.   ■

(c) **Claim.**   There exists an integer x such that $x + 13$ is a perfect square.

   "Proof."   Let the universe be the set of all integers x such that $x + 13$ is a perfect square. Then the following sentence, $(\forall x)(x + 13$ is a perfect square), is true. Thus we may deduce the sentence $(\exists x)(x + 13$ is a perfect square) as true.   ■

★ (d) **Claim.**   There exists an irrational number r such that $r^{\sqrt{2}}$ is rational.

   "Proof."   If $\sqrt{3}^{\sqrt{2}}$ is rational, then $r = \sqrt{3}$ is the desired example. Otherwise, $\sqrt{3}^{\sqrt{2}}$ is irrational and $(\sqrt{3}^{\sqrt{2}})^{\sqrt{2}} = (\sqrt{3})^2 = 3$, which is rational. Therefore either $\sqrt{3}$ or $\sqrt{3}^{\sqrt{2}}$ is an irrational number r such that $r^{\sqrt{2}}$ is rational.   ■

(e) **Claim.**   Every real function is continuous at $x = 0$.

   "Proof."   Since either a sentence or its negation is true, we know that for every real function either it is continuous at $x = 0$ or it is not continuous at $x = 0$. Thus for every real function it is continuous at $x = 0$ or for every real function it is not continuous at $x = 0$. The latter half of this sentence is false, since $f(x) = x^2$ is a real function that is continuous at $x = 0$. Since the sentence is a disjunction, the first half is true. Thus for every real function, it is continuous at $x = 0$.   ■

(f) **Claim.**   If x is a prime, then $x + 7$ is composite.

   "Proof."   Let x be a prime number. If $x = 2$, then $x + 7 = 9$, which is composite. If $x \neq 2$, then x is odd, so $x + 7$ is even and greater than 2. In this case, too, $x + 7$ is composite. Therefore if x is prime, then $x + 7$ is composite.   ■

(g) **Claim.**   For all irrational numbers t, $t - 8$ is irrational.

   "Proof."   Suppose there exists an irrational number t such that $t - 8$ is rational. Then $t - 8 = p/q$, where p and q are integers and $q \neq 0$. Then $t = p/q + 8 = (p + 8q)/q$, with $p + 8q$ and q integers and $q \neq 0$. This is a contradiction because t is irrational. Therefore for all irrational numbers t, $t - 8$ is irrational.   ■

2. Prove that
    (a)   there exist integers m and n such that $2m + 7n = 1$.
    (b)   there do not exist integers m and n such that $2m + 4n = 7$.

3. Prove that if every even natural number greater than 2 is the sum of two primes, then every odd natural number greater than 5 is the sum of three primes.

4. Prove that if p is a prime number and $p \neq 3$, then 3 divides $p^2 + 2$. (*Hint:* When p is divided by 3, the remainder is either 0, 1, or 2. That is, for some integer k, $p = 3k$ or $p = 3k + 1$ or $p = 3k + 2$.)

★ 5. The numbers 3, 5, 7 are a **prime triple**—that is, a set of three consecutive odd numbers that are all prime. Prove that there are no other prime triples. (*Note:* 3 and 5, 11 and 13, and 29 and 31 are among the many pairs of primes, known as **twin primes.** It is not known whether there are only finitely many twin primes.)

6. The **conjugate** of the complex number $a + bi$ is $\overline{a + bi} = a - bi$. Prove that the conjugate of the sum of any two complex numbers is the sum of their conjugates.

7. Prove that

(a) for every natural number n, $\dfrac{1}{n} \leq 1$. (*Hint:* Use the fact that $n \geq 1$ and divide by the positive number n.)

(b) there is a natural number M such that for all natural numbers $n > M$, $\dfrac{1}{n} < 0.13$.

★ (c) for every natural number n, there is a natural number M such that $2n < M$.

(d) there is a natural number M such that for every natural number n, $\dfrac{1}{n} < M$.

(e) there is no largest natural number.

(f) for every natural number n, there is an integer M such that for every natural number $m > n$, $\dfrac{3}{m} + \dfrac{2}{n} > M$.

★ (g) for every real number $\epsilon > 0$, there is a natural number M such that for all natural numbers $n > M$, $\dfrac{1}{n} < \epsilon$.

(h) for every real $\epsilon > 0$, there is a natural number M such that if $m > n > M$, then $\dfrac{1}{n} - \dfrac{1}{m} < \epsilon$.

2

Set Theory

Much of mathematics is written in terms of sets. We assume that you have had some contact with the basic notions of sets, unions, and intersections. Sections 1 and 2 consist of a brief review of sets and operations and present the notation we shall use. Section 2 provides the opportunity for you to prove some set-theoretical results. In section 3 we extend the set operations of union and intersection and encounter indexed collections of sets. Section 4 deals with inductive sets and proof by induction. Basic techniques of counting and some applications of these techniques appear in Section 5.

2.1

Basic Notions of Set Theory

We shall understand a **set** to be any specified collection of objects. The objects in a given set are called the **elements** (or members) of the set. By saying that a set is a specified collection we mean that, for any object, there must be a definite yes or no answer to the question whether the object is a member of the set.

In general, capital letters are used to denote sets and lowercase letters to denote objects. If the object x is an element of set A, we write $x \in A$; if not—that is, if $\sim(x \in A)$—we write $x \notin A$. For example, if B is the set of all signers of the Declaration of Independence, we write John Hancock $\in B$ and Joe Slobotnik $\notin B$.

Sets can be described in words, such as "the set of odd integers between 0 and 12," or the elements may be listed, as in {1, 3, 5, 7, 9, 11}, or even partially listed, as in {1, 3, 5, . . . , 11}. Such explicit descriptions of sets are often impractical and sometimes impossible. To designate most sets we will use the following notation:

$$\{x: P(x)\}$$

where $P(x)$ is an open sentence description of the property that defines the set. The variable x in $\{x: P(x)\}$ is a dummy variable in the sense that any letter or symbol serves equally well. If $P(x)$ is the sentence "x is an odd integer between 0 and 12," then $\{x: P(x)\} = \{y: P(y)\} = \{1, 3, 5, . . . , 11\}$.

A bit of caution should be observed in the type of property used in this notation. It is not true that for every open sentence $P(x)$, there corresponds a set $\{x: P(x)\}$. For more on this, see exercise 16.

Special notation will be used for the following sets of numbers:

> **N** $= \{1, 2, 3, . . .\}$, the natural numbers
> **Z** $= \{. . . , -3, -2, -1, 0, 1, 2, 3, . . . \}$, the integers
> **Q** $=$ the set of rational numbers
> **R** $=$ the set of real numbers

In addition, for $a, b \in \mathbf{R}$ with $a < b$, we will use

$$[a, b] = \{x: x \in \mathbf{R} \text{ and } a \leqslant x \leqslant b\}$$

and

$$(a, b) = \{x: x \in \mathbf{R} \text{ and } a < x < b\}$$

to represent the **closed** (and **open,** respectively) **interval from a to b.** The **half-open** (or **half-closed**) intervals $[a, b)$ and $(a, b]$ are defined similarly. Also,

$$(a, \infty) = \{x: x \in \mathbf{R} \text{ and } a < x\}$$

and

$$(-\infty, a) = \{x: x \in \mathbf{R} \text{ and } x < a\}$$

will be called **open rays,** while

$$[a, \infty) = \{x: x \in \mathbf{R} \text{ and } a \leqslant x\}$$

and

$$(-\infty, a] = \{x: x \in \mathbf{R} \text{ and } x \leqslant a\}$$

are called **closed rays.**

One should be careful not to confuse (1, 6) with {2, 3, 4, 5} since (1, 6) is defined as *all real numbers between 1 and 6* and contains, for example, 2, π, 3, log(15), and $\frac{11}{5}$.

Definition.   Let $\varnothing = \{x: x \neq x\}$. Then $\varnothing$ is *a set with no elements* and is called an **empty set.**

There should be no confusion whether we have defined a set. The open sentence $x \neq x$ is false for any object x; that is, the sentence $x \in \varnothing$ is false for every object x. We can define other empty sets, such as

$$\{x: x \in \mathbf{R} \text{ and } x = x + 1\}$$

or

$$\{x: x \in \mathbf{N} \text{ and } x < 0\}.$$

However, we shall soon see that these are all the same so that there is in fact only one empty set.

Definition.   Let A and B be sets. We say A is a **subset** of B iff *every* element of A is also an element of B. In symbols this is

$$A \subseteq B \Leftrightarrow (\forall x)(x \in A \Rightarrow x \in B).$$

Since the statement $A \subseteq B$ is symbolized with the quantifier $(\forall x)$, a direct proof of $A \subseteq B$ has the following form.

Direct Proof of $A \subseteq B$

Proof.
Let x be any object. ⟨*Show that $x \in A \Rightarrow x \in B$.*⟩
Suppose $x \in A$.

    •

    •

    •

Thus $x \in B$.
Therefore $A \subseteq B$.   ∎

In practice, the first sentences of this outline are often combined into one: "Suppose x is any object in A," or "Suppose $x \in A$."

Theorem 2.1   For any set A, $\varnothing$ is a subset of A.

Proof.   Let A be any set. ⟨*We now give a direct proof that* $\varnothing \subseteq A$.⟩ Let x be any object. Because any conditional sentence is true when the antecedent is false, $(x \in \varnothing \Rightarrow x \in A)$ is true. Therefore $\varnothing \subseteq A$.   ■

Theorem 2.2   For any set B, $B \subseteq B$.

Proof.   Let B be any set. To prove $B \subseteq B$, we must show that, for all objects x, if $x \in B$, then $x \in B$. Let x be any object. Then $x \in B \Rightarrow x \in B$ is true. ⟨*Here we use the tautology* $P \Rightarrow P$.⟩ Therefore $(\forall x)(x \in B \Rightarrow x \in B)$, and so $B \subseteq B$.   ■

Theorem 2.3   If $A \subseteq B$ and $B \subseteq C$, then $A \subseteq C$.

Proof.   Exercise 11.   ■

If A is not a subset of B, we write $A \not\subseteq B$. A denial of $A \subseteq B$ is the proposition $(\exists x)(x \in A$ and $x \notin B)$. To prove $A \not\subseteq B$, all that is required is to show that some element of A is not an element of B. For example, $\{1, 5, 9, 6\} \not\subseteq \{2, 7, 6\}$ because $5 \in \{1, 5, 9, 6\}$ and $5 \notin \{2, 7, 6\}$.

For a given set A, the subsets $\varnothing$ and A are called **improper subsets** of A, while any subset of A other then $\varnothing$ or A is called a **proper subset.** Next we define a set whose elements are themselves sets.

Definition.   Let A be a set. The **power set** of A is the set whose elements are the subsets of A and is denoted $\mathscr{P}(A)$. Thus

$$\mathscr{P}(A) = \{B : B \subseteq A\}.$$

Example.   Let $A = \{a, b, c\}$. Then

$$\mathscr{P}(A) = \{\varnothing, \{a\}, \{b\}, \{c\}, \{a, b\}, \{a, c\}, \{b, c\}, A\}.$$

You should always remember that the elements of the set $\mathscr{P}(A)$ are themselves sets, specifically the subsets of A. Also, in working with sets whose elements are sets, it is important to recognize the distinction between "is an element of" and "is a subset of." To use $A \in B$ correctly, we must consider whether the object A (which happens to be a set) is an element of the set B, whereas $A \subseteq B$ requires determining whether *all* objects in the set A are also in B.

Let $A = \{a, b, c\}$. Then $a \in A$, $b \in A$, $\{a, b\} \subseteq A$, and $\{a, b\} \in \mathcal{P}(A)$. Notice that $\{a, b\}$ is an element of $\mathcal{P}(A)$ because $\{a, b\}$ is a subset of A. Also, $\{c\} \not\subseteq A$ but $\{c\} \in \mathcal{P}(A)$.

Example.   Let $X = \{\{1, 2, 3\}, \{4, 5\}, 6\}$. Then X is a set with three elements, namely, the set $\{1, 2, 3\}$, the set $\{4, 5\}$, and the number 6. The set $\{\{4, 5\}\}$ has one element; it is $\{4, 5\}$. All of the following are true:

$$\{4\} \subseteq \{4, 5\}$$
$$\{4, 5\} \in X$$
$$\{4, 5\} \not\subseteq X$$
$$6 \in X$$
$$\{6\} \subseteq X$$

and

$$\mathcal{P}(X) = \{\varnothing, \{\{1, 2, 3\}\}, \{\{4, 5\}\}, \{6\}, \{\{1, 2, 3\}, \{4, 5\}\},$$
$$\{\{1, 2, 3\}, 6\}, \{\{4, 5\}, 6\}, X\}.$$

Notice that for the set $A = \{a, b, c\}$ in this example, A has three elements and $\mathcal{P}(A)$ has $2^3 = 8$ elements. As we see in the next theorem, this observation can be generalized to any set of n elements. It is for this reason that 2^A is sometimes used to denote the power set of A.

Theorem 2.4   If A is a set with n elements, then $\mathcal{P}(A)$ has 2^n elements.

Proof.   ⟨*The number of elements in $\mathcal{P}(A)$ is the number of subsets of A. Thus to prove this result we must count all of the subsets of A.*⟩ Suppose A has n elements. We may write A as $A = \{x_1, x_2, \ldots, x_n\}$. To describe a subset B of A, we need to know for each $x_i \in A$ whether the element is in B. For each x_i, there are two possibilities ($x_i \in B$ or $x_i \not\in B$), so there are $2 \cdot 2 \cdot 2 \cdots \cdot 2$ (n factors) different ways of making a subset of A. Therefore $\mathcal{P}(A)$ has 2^n elements. ∎

Theorem 2.5   Let A and B be sets. Then $A \subseteq B$ iff $\mathcal{P}(A) \subseteq \mathcal{P}(B)$.

Proof.   ⟨*This is a good example of a two-part proof of a biconditional, which is easier than an iff proof.*⟩

(i)   We must show that $A \subseteq B$ implies $\mathcal{P}(A) \subseteq \mathcal{P}(B)$. Assume that $A \subseteq B$, and suppose $X \in \mathcal{P}(A)$. We must show that

$X \in \mathscr{P}(B)$. But $X \in \mathscr{P}(A)$ implies $X \subseteq A$. Since $X \subseteq A$ and $A \subseteq B$, then $X \subseteq B$ by Theorem 2.3. But $X \subseteq B$ implies $X \in \mathscr{P}(B)$. Therefore $X \in \mathscr{P}(A)$ implies $X \in \mathscr{P}(B)$. Thus $\mathscr{P}(A) \subseteq \mathscr{P}(B)$.

(ii) We must show that $\mathscr{P}(A) \subseteq \mathscr{P}(B)$ implies $A \subseteq B$. Assume that $\mathscr{P}(A) \subseteq \mathscr{P}(B)$. By Theorem 2.2, $A \subseteq A$; so $A \in \mathscr{P}(A)$. Since $\mathscr{P}(A) \subseteq \mathscr{P}(B)$, $A \in \mathscr{P}(B)$. Therefore $A \subseteq B$. ∎

The second half of the proof could have been done by showing directly that if $x \in A$, then $x \in B$. Such proofs that concentrate on the whereabouts of individual members of sets are often called "element chasing" proofs. The given proof is preferable because it makes use of a theorem we already know. Both proofs are correct. When you write proofs, you may choose one method of proof over another because it is shorter, or easier to understand, or for any other reason.

In proving that $A \subseteq B$, we do not assume that A has elements. We begin by supposing that $x \in A$ and use that supposition to show that $x \in B$. The phrase "Let $x \in A$" is usually reserved for a situation in which it is known that A has at least one element. For example, in the proof that $\mathscr{P}(A) \subseteq \mathscr{P}(B)$ in part (i) of Theorem 2.5, we can be sure that $\mathscr{P}(A)$ is not empty because $\varnothing \in \mathscr{P}(A)$ for every set A.

We have seen that a set may be described in different ways. Often it becomes necessary to know whether two descriptions of sets do in fact yield the same set. Intuitively, two sets A and B are "equal" if they contain exactly the same elements. We might say, then, that

$$A = B \text{ iff } (\forall x)(x \in A \Leftrightarrow x \in B).$$

However, this is equivalent to

$$(\forall x)(x \in A \Rightarrow x \in B \text{ and } x \in B \Rightarrow x \in A)$$

which is longer but more natural to use in proving that $A = B$.

Definition.   Let A and B be sets. Then $A = B$ iff $A \subseteq B$ and $B \subseteq A$.

According to this definition, the task of proving an equality between two sets A and B is accomplished by showing that each set is a subset of the other, that is, by showing

(i) $A \subseteq B$,
(ii) $B \subseteq A$,

and then concluding $A = B$. The proofs of (i) and (ii) need not be by the same methods. Occasionally, $A = B$ can be proved by a chain of equivalent statements showing that $x \in A$ iff $x \in B$.

Example.   Prove that $X = Y$ where $X = \{x : x \text{ is a solution to } x^2 - 1 = 0\}$ and $Y = \{-1, 1\}$.

Proof.   We must show (i) $Y \subseteq X$ and (ii) $X \subseteq Y$.

(i)   We show $Y \subseteq X$ by individually checking each element of Y. By substitution we see that both 1 and -1 are solutions to $x^2 - 1 = 0$. Thus $Y \subseteq X$.

(ii)   Next, we must show $X \subseteq Y$. Let $t \in X$. Then, by definition of X, t is a solution to $x^2 - 1 = 0$. Thus $t^2 - 1 = 0$. Factoring, we have $(t - 1)(t + 1) = 0$. This product is 0 exactly when $t - 1 = 0$ or $t + 1 = 0$. Therefore $t = 1$ or $t = -1$. Thus if t is a solution, then $t = 1$ or $t = -1$; so $t \in Y$. This proves $X \subseteq Y$.

By (i) and (ii), $X \subseteq Y$ and $Y \subseteq X$; so $X = Y$.   ∎

We are now in a position to prove that there is only one empty set.

Theorem 2.6   If A and B are sets with no elements, then $A = B$.

Proof.   Since A is empty, the sentence $(\forall x)(x \in A \Rightarrow x \in B)$ is true. Therefore $A \subseteq B$. Similarly $(\forall x)(x \in B \Rightarrow x \in A)$ is true, so $B \subseteq A$. Therefore by definition of set equality, $A = B$.   ∎

It should now be clear that if we want to prove $A \subseteq B$, the simplest case would be that A is empty. This is because if A is empty, then $A \subseteq B$ by Theorem 2.1. In the chapters that follow are some more interesting cases where special care must be taken to be sure that a statement about a set is true, even when the set happens to be empty.

Exercises 2.1

1.   Write the following sets by using the set notation $\{x : P(x)\}$.
★   (a)   The set of natural numbers strictly less than 6.
     (b)   The set of integers whose square is less than 17.
★   (c) $[2, 6]$
     (d) $(-1, 9]$
     (e) $[-5, -1)$
     (f)   The set of rational numbers less than -1.

☆   2.   Write each of the sets in exercise 1 by listing (if possible) all its elements.

3.   True or false?
★ (a) $\mathbf{N} \subseteq \mathbf{Q}$.   (b) $\mathbf{Q} \subseteq \mathbf{Z}$.
★ (c) $\mathbf{N} \subseteq \mathbf{R}$.   (d) $[\frac{1}{2}, \frac{5}{2}] \subseteq \mathbf{Q}$.
★ (e) $[\frac{1}{2}, \frac{5}{2}] \subseteq (\frac{1}{2}, \frac{5}{2})$.   (f) $\mathbf{R} \subseteq \mathbf{Q}$.
★ (g) $[7, 10] \subseteq \mathbf{R}$.   (h) $[2, 5] = \{2, 3, 4, 5\}$.
★ (i) $[7, 10) \subseteq \{7, 8, 9, 10\}$.   (j) $(6, 9] \subseteq [6, 10)$.

4.   Write the power set, $\mathcal{P}(X)$, for each of the following sets.
★ (a) $X = \{0, \triangle, \square\}$ (b) $X = \{S, \{S\}\}$
★ (c) $X = \{\varnothing, \{a\}, \{b\}, \{a, b\}\}$ (d) $X = \{1, \{2, \{3\}\}\}$
   (e) $X = \{1, 2, 3, 4\}$

5.   List all of the proper subsets for each of the following sets.
★ (a) $\varnothing$ (b) $\{1\}$ ★ (c) $\{1, 2\}$
   (d) $\{\{\varnothing\}\}$ (e) $\{\varnothing, \{\varnothing\}\}$ (f) $\{0, \triangle, \square\}$

6.   True or false?
★ (a) $\varnothing \in \{\varnothing, \{\varnothing\}\}$.   (b) $\varnothing \subseteq \{\varnothing, \{\varnothing\}\}$.
★ (c) $\{\varnothing\} \in \{\varnothing, \{\varnothing\}\}$.   (d) $\{\varnothing\} \subseteq \{\varnothing, \{\varnothing\}\}$.
★ (e) $\{\{\varnothing\}\} \in \{\varnothing, \{\varnothing\}\}$.   (f) $\{\{\varnothing\}\} \subseteq \{\varnothing, \{\varnothing\}\}$.
★ (g) For every set A, $\varnothing \in A$.   (h) For every set A, $\{\varnothing\} \subseteq A$.
★ (i) $\{\varnothing, \{\varnothing\}\} \subseteq \{\{\varnothing, \{\varnothing\}\}\}$.   (j) $\{1, 2\} \in \{\{1, 2, 3\}, \{1, 3\}, 1, 2\}$.
★ (k) $\{1, 2, 3\} \subseteq \{1, 2, 3, \{4\}\}$.   (l) $\{\{4\}\} \subseteq \{1, 2, 3, \{4\}\}$.

7.   If possible, give an example of each of the following:
★ (a) sets A, B, and C such that $A \subseteq B$, $B \not\subseteq C$, and $A \subseteq C$.
   (b) sets A, B, and C such that $A \subseteq B$, $B \subseteq C$, and $C \subseteq A$.
★ (c) sets A, B, and C such that $A \not\subseteq B$, $B \not\subseteq C$, and $A \subseteq C$.
   (d) sets A and B such that $A \subseteq B$, and $\mathcal{P}(B) \subseteq \mathcal{P}(A)$.
★ (e) a set A such that $\mathcal{P}(A) = \varnothing$.
   (f) sets A, B, and C such that $A \not\subseteq B$, $B \subseteq C$, and $\mathcal{P}(A) \subseteq \mathcal{P}(C)$.

8.   True or false?
★ (a) $\varnothing \in \mathcal{P}(\{\varnothing, \{\varnothing\}\})$.   (b) $\{\varnothing\} \in \mathcal{P}(\{\varnothing, \{\varnothing\}\})$.
★ (c) $\{\{\varnothing\}\} \in \mathcal{P}(\{\varnothing, \{\varnothing\}\})$.   (d) $\varnothing \subseteq \mathcal{P}(\{\varnothing, \{\varnothing\}\})$.
★ (e) $\{\varnothing\} \subseteq \mathcal{P}(\{\varnothing, \{\varnothing\}\})$.   (f) $\{\{\varnothing\}\} \subseteq \mathcal{P}(\{\varnothing, \{\varnothing\}\})$.
★ (g) $3 \in \mathbf{Q}$.   (h) $\{3\} \subseteq \mathcal{P}(\mathbf{Q})$.
★ (i) $\{3\} \in \mathcal{P}(\mathbf{Q})$.   (j) $\{\{3\}\} \subseteq \mathcal{P}(\mathbf{Q})$.
★ (k) $\{3\} \subseteq \mathbf{Q}$.   (l) $\{\{3\}\} \in \mathcal{P}(\mathbf{Q})$.

9.   Let $A = \{x : P(x)\}$ and $B = \{x : Q(x)\}$.
★ (a) Prove that if $(\forall x)(P(x) \Rightarrow Q(x))$, then $A \subseteq B$.
   (b) Prove that if $(\forall x)(P(x) \Leftrightarrow Q(x))$, then $A = B$.

10.   Prove that if $x \notin B$ and $A \subseteq B$, then $x \notin A$.

╱ 11.   Prove Theorem 2.3.

☆ 12.   Prove that if $A \subseteq B$, $B \subseteq C$, and $C \subseteq A$, then $A = B$ and $B = C$.

╱ 13.   Prove that $X = Y$, where $X = \{x : x \in \mathbf{R}$ and x is a solution to $x^2 - 7x + 12 = 0\}$ and $Y = \{3, 4\}$.

14. Prove that $X = Y$, where $X = \{x \in \mathbf{Z}: |x| \leqslant 3\}$ and $Y = \{-3, -2, -1, 0, 1, 2, 3\}$.

15. Prove that $X = Y$, where $X = \{x \in \mathbf{N}: x^2 < 30\}$ and $Y = \{1, 2, 3, 4, 5\}$.

★ 16. (Russell paradox.) A logical difficulty arises from the idea, which at first appears natural, of calling any collection of objects a set. A set B is **ordinary** if $B \notin B$. For example, if B is the set of all chairs, then $B \notin B$, for B is not a chair. It is only in the case of very unusual collections that we are tempted to say that a set is a member of itself. (The collection of all abstract ideas certainly is an abstract idea.) Let $X = \{x: x \text{ is an ordinary set}\}$. Is $X \in X$? Is $X \notin X$? What should we say about the collection of all ordinary sets?

17. **Proofs to Grade.**

★ (a) **Claim.** If A and B are sets and $\mathcal{P}(A) \subseteq \mathcal{P}(B)$, then $A \subseteq B$.
   "Proof." $x \in A \Rightarrow \{x\} \subseteq A$
   $\qquad\qquad\quad \Rightarrow \{x\} \in \mathcal{P}(A)$
   $\qquad\qquad\quad \Rightarrow \{x\} \in \mathcal{P}(B)$
   $\qquad\qquad\quad \Rightarrow \{x\} \subseteq B$
   $\qquad\qquad\quad \Rightarrow x \in B.$
   Therefore $x \in A \Rightarrow x \in B$. Thus $A \subseteq B$. ∎

(b) **Claim.** If A, B, and C are sets, and $A \subseteq B$, and $B \subseteq C$, then $A \subseteq C$.
   "Proof."   If $x \in C$, then, since $B \subseteq C$, $x \in B$. Since $A \subseteq B$ and $x \in B$, it follows that $x \in A$. Thus $x \in C$ implies $x \in A$. Therefore $A \subseteq C$. ∎

★ (c) **Claim.** If A, B, and C are sets, and $A \subseteq B$, and $B \subseteq C$, then $A \subseteq C$.
   "Proof."   Suppose x is any object. If $x \in A$, then $x \in B$, since $A \subseteq B$. If $x \in B$, then $x \in C$, since $B \subseteq C$. Therefore $x \in C$. Therefore $A \subseteq C$. ∎

(d) **Claim.** If $X = \{x \in \mathbf{N}: x^2 < 14\}$ and $Y = \{1, 2, 3\}$, then $X = Y$.
   "Proof."   Since $1^2 = 1 < 14$, $2^2 = 4 < 14$, and $3^2 = 9 < 14$, $X = Y$. ∎

(e) **Claim.** If A, B, and C are sets, and $A = B$, and $B = C$, then $A = C$.
   "Proof." $x \in A$
   $\qquad\quad$ iff $x \in B$
   $\qquad\quad$ iff $x \in C$.

   Thus $A = C$. ∎

(f) **Claim.** If $\mathcal{P}(A) = \mathcal{P}(B)$, then $A = B$.
   "Proof."   Suppose $A \neq B$. Then either $A \not\subseteq B$ or $B \not\subseteq A$. Without loss of generality, assume $A \not\subseteq B$. Then by the contrapositive of Theorem 2.5, $\mathcal{P}(A) \not\subseteq \mathcal{P}(B)$. Thus $\mathcal{P}(A) \neq \mathcal{P}(B)$. Hence $\mathcal{P}(A) = \mathcal{P}(B)$ implies $A = B$. ∎

2.2

Set Operations

In this section we present the most common ways to combine two sets to produce a third. These **binary operations** on sets (that is, operations on exactly two sets) will be generalized in the next section.

Definition.   Let A and B be sets. The **union** of A and B is defined by

$$A \cup B = \{x : x \in A \text{ or } x \in B\}.$$

The **intersection** of A and B is defined by

$$A \cap B = \{x : x \in A \text{ and } x \in B\}.$$

The **difference** of A and B is defined by

$$A - B = \{x : x \in A \text{ and } x \notin B\}.$$

$A \cup B$ can be thought of as a new set formed from A and B by choosing as elements the objects that appear in at least one of A or B; $A \cap B$ consists of objects that appear in both; $A - B$ contains those elements of A that are not in B (figures 2.1, 2.2, and 2.3). In case $A \cap B = \varnothing$, we say A and B are **disjoint.**

    Example.   For $A = \{1, 2, 4, 5, 7\}$ and $B = \{1, 3, 5, 9\}$,

$$A \cup B = \{1, 2, 3, 4, 5, 7, 9\}$$
$$A \cap B = \{1, 5\}$$
$$A - B = \{2, 4, 7\}$$
$$B - A = \{3, 9\}$$

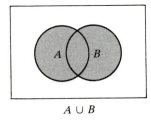

$A \cup B$

Figure 2.1

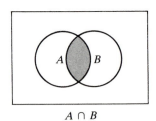

$A \cap B$

Figure 2.2

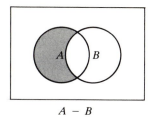

$$A - B$$

Figure 2.3

Example.   For $C = [1, 4)$ and $D = (2, 6]$,

$$C \cup D = [1, 6]$$
$$C \cap D = (2, 4)$$
$$C - D = [1, 2]$$

The sets $[1, 2]$ and $(2, 4)$ are disjoint.

The set operations obey certain rules that at times allow us to simplify our work, in the same manner that the distributive law for real numbers can be used to simplify $12(2 + \frac{1}{3} + \frac{1}{4} + \frac{1}{6}) = 33$. Several such laws have been gathered together in the next theorem. Those not proved are left as exercise 6.

Theorem 2.7   Let A, B, and C be sets. Then

(a) $A \subseteq A \cup B$.
(b) $A \cap B \subseteq A$.
(c) $A \cap \emptyset = \emptyset$.
(d) $A \cup \emptyset = A$.
(e) $A \cap A = A$.
(f) $A \cup A = A$.
(g) $A \cup B = B \cup A$.
(h) $A \cap B = B \cap A$.
(i) $A - \emptyset = A$.
(j) $\emptyset - A = \emptyset$.
(k) $A \cup (B \cup C) = (A \cup B) \cup C$.
(l) $A \cap (B \cap C) = (A \cap B) \cap C$.
(m) $A \cap (B \cup C) = (A \cap B) \cup (A \cap C)$.
(n) $A \cup (B \cap C) = (A \cup B) \cap (A \cup C)$.
(o) $A \subseteq B$ iff $A \cup B = B$.
(p) $A \subseteq B$ iff $A \cap B = A$.
(q) If $A \subseteq B$, then $A \cup C \subseteq B \cup C$.
(r) If $A \subseteq B$, then $A \cap C \subseteq B \cap C$.

Proof.

(b)   We must show that, if $x \in A \cap B$, then $x \in A$. Suppose $x \in A \cap B$. Then $x \in A$ and $x \in B$. Therefore $\langle$*because* $P \wedge Q \Rightarrow P\rangle\, x \in A$.

(f)   We must show that $x \in A \cup A$ iff $x \in A$. By the definition of union, $x \in A \cup A$ iff $x \in A$ or $x \in A$. This is equivalent to $x \in A$. Therefore $A \cup A = A$.

(h) $\langle$*This iff proof uses the definition of intersection and the equivalence of $P \wedge Q$ and $Q \wedge P$.*$\rangle$

$x \in A \cap B$
iff $x \in A$ and $x \in B$
iff $x \in B$ and $x \in A$
iff $x \in B \cap A$.

(m) $\langle$*As you read this proof, watch for the steps in which the definitions of union and intersection are used (two for each). Watch also for the use of the equivalence from Theorem 1.2(f).*$\rangle$

$x \in A \cap (B \cup C)$
iff $x \in A$ and $x \in B \cup C$
iff $x \in A$ and $(x \in B$ or $x \in C)$
iff $(x \in A$ and $x \in B)$ or $(x \in A$ and $x \in C)$
iff $x \in A \cap B$ or $x \in A \cap C$
iff $x \in (A \cap B) \cup (A \cap C)$.

Therefore $A \cap (B \cup C) = (A \cap B) \cup (A \cap C)$.

(p) $\langle$*The statement $A \subseteq B$ iff $A \cap B = A$ requires separate proofs for each implication. We make use of earlier parts of this theorem.*$\rangle$ First, assume that $A \subseteq B$. We must show that $A \cap B = A$. If $x \in A$, then from the hypothesis $A \subseteq B$, we have $x \in B$. Therefore $x \in A$ implies $x \in A$ and $x \in B$, so $x \in A \cap B$. This shows that $A \subseteq A \cap B$, which, combined with $A \cap B \subseteq A$ from part (b) of this theorem, gives $A \cap B = A$.

Second, assume that $A \cap B = A$. We must show that $A \subseteq B$. By parts (b) and (h) of this theorem, we have $B \cap A \subseteq B$ and $B \cap A = A \cap B$. Therefore $A \cap B \subseteq B$. By hypothesis, $A \cap B = A$, so $A \subseteq B$.   ∎

Recall that the universe of discourse is a collection of objects understood from the context or specified at the outset of a discussion and that all objects under consideration must belong to the universe.

Definition.   If U is the universe and $B \subseteq U$, then we define the **complement** of B to be the set $\widetilde{B} = U - B$.

Thus $\widetilde{B}$ is to be interpreted as all those elements of the universe that are not in B (figure 2.4).

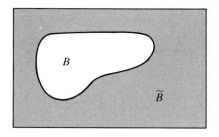

Figure 2.4

For the universe **R**, if $B = (0, \infty)$, then $\widetilde{B} = (-\infty, 0]$. If $A = \{x: x$ is rational$\}$, then $\widetilde{A} = \{x: x$ is irrational$\}$.

The complement is nothing more than a short way to write set difference in the case where the first set is the universe. The complement operation obeys several rules.

Theorem 2.8   Let U be the universe, and let A and B be subsets of U. Then

(a) $\widetilde{\widetilde{A}} = A$.

(b) $A \cup \widetilde{A} = U$.

(c) $A \cap \widetilde{A} = \varnothing$.

(d) $A - B = A \cap \widetilde{B}$.

(e) $A \subseteq B$ iff $\widetilde{B} \subseteq \widetilde{A}$.

(f) $\widetilde{A \cup B} = \widetilde{A} \cap \widetilde{B}$.   }

(g) $\widetilde{A \cap B} = \widetilde{A} \cup \widetilde{B}$.   }   De Morgan's Laws

(h) $A \cap B = \varnothing$ iff $A \subseteq \widetilde{B}$.

Proof.

(a)   By definition of the complement $x \in \widetilde{\widetilde{A}}$ iff $x \notin \widetilde{A}$ iff $x \in A$. Therefore $\widetilde{\widetilde{A}} = A$.

(e)   ⟨*For this part of the theorem we give two separate proofs. The first proof has two parts, and its first part is used to prove its second part.*⟩
      First proof of (e). We prove that if $A \subseteq B$ then $\widetilde{B} \subseteq \widetilde{A}$. Assume that $A \subseteq B$. Suppose $x \in \widetilde{B}$. Then $x \notin B$. Since $A \subseteq B$ and $x \notin B$, we have $x \notin A$. ⟨*This is the contrapositive of $A \subseteq B$.*⟩ Thus $x \in \widetilde{A}$. Therefore $\widetilde{B} \subseteq \widetilde{A}$. For the second part of this proof, we show that $\widetilde{B} \subseteq \widetilde{A}$ implies $A \subseteq B$. Assume that $\widetilde{B} \subseteq \widetilde{A}$. Then by the

first part of this proof, $\widetilde{\widetilde{A}} \subseteq \widetilde{\widetilde{B}}$. Therefore using part (a), $A \subseteq B$. Combining the two parts of this proof,

$A \subseteq B$ iff $\widetilde{B} \subseteq \widetilde{A}$.

      Second proof of (e). ⟨*This proof makes use of the fact that a conditional sentence is equivalent to its contrapositive.*⟩

$A \subseteq B$
    iff $(\forall x)(x \in A \Rightarrow x \in B)$
    iff $(\forall x)(x \notin B \Rightarrow x \notin A)$
    iff $(\forall x)(x \in \widetilde{B} \Rightarrow x \in \widetilde{A})$
    iff $\widetilde{B} \subseteq \widetilde{A}$.

(f)   The object x is a member of $\widetilde{A \cup B}$
        iff x is not a member of $A \cup B$
        iff it is not the case that $x \in A$ or $x \in B$
        iff $x \notin A$ and $x \notin B$ ⟨*See Theorem 1.2 (c).*⟩
        iff $x \in \widetilde{A}$ and $x \in \widetilde{B}$.
        iff $x \in \widetilde{A} \cap \widetilde{B}$.

The proofs of the remaining parts are left as exercise 7.   ■

Exercises 2.2

1.   Let $A = \{1, 3, 5, 7, 9\}$, $B = \{0, 2, 4, 6, 8\}$, $C = \{1, 2, 4, 5, 7, 8\}$, and $D = \{1, 2, 3, 5, 6, 7, 8, 9, 10\}$. Find

  ★ (a) $A \cup B$ (b) $A \cap B$
  ★ (c) $A - B$ (d) $A - (B - C)$
  ★ (e) $(A - B) - C$ (f) $A \cap (C \cap D)$
  ★ (g) $(A \cap C) \cap D$ (h) $A \cap (B \cup C)$
  ★ (i) $(A \cap B) \cup (A \cap C)$ (j) $(A \cup B) - (C \cap D)$

2.   Let U be the set of all integers. Let E, D, P, and N be the sets of all even, odd, positive, and negative integers, respectively. Find

  ★ (a) $E - P$ (b) $P - E$ ★ (c) $D - E$
      (d) $P - N$ ★ (e) $\widetilde{P}$ (f) $\widetilde{N}$
  ★ (g) $\widetilde{E}$ (h) $\widetilde{D}$ ★ (i) $E - N$
      (j) $U - P$ ★ (k) $\widetilde{U}$ (l) $\widetilde{\varnothing}$

3.   Let $U = \{1, 2, 3\}$ be the universe for the sets $A = \{1, 2\}$ and $B = \{2, 3\}$. Find

  ★ (a) $\mathscr{P}(A)$ (b) $\mathscr{P}(\widetilde{A})$
      (c) $\mathscr{P}(A) \cap \mathscr{P}(B)$ ★ (d) $\mathscr{P}(\widetilde{A}) \cup \mathscr{P}(\widetilde{B})$
      (e) $\mathscr{P}(A) - \mathscr{P}(B)$ ★ (f) $\mathscr{P}(\widetilde{A}) - \mathscr{P}(\widetilde{B})$

★ 4.   Let A, B, C, and D be as in exercise 1. What pairs of sets are disjoint?

5.   Let U, E, D, P, and N be as in exercise 2. What pairs of sets are disjoint?

6.   Prove the remaining parts of Theorem 2.7.

(d) 7.   Prove the remaining parts of Theorem 2.8.
(g)
8.   Let A, B, and C be sets.
  ☆ (a)   Prove that $A \subseteq B$ iff $A - B = \varnothing$.
  — (b)   Prove that if $A \subseteq B \cup C$ and $A \cap B = \varnothing$, then $A \subseteq C$.
     (c)   Prove that $C \subseteq A \cap B$ iff $C \subseteq A$ and $C \subseteq B$.
     (d)   Prove that if $A \subseteq B$ then $A - C \subseteq B - C$.
     (e)   Prove that $(A - B) - C = (A - C) - (B - C)$.
     (f)   Prove that if $A \subseteq C$ and $B \subseteq C$, then $A \cup B \subseteq C$.

9.   Let A, B, C, and D be sets with $C \subseteq A$ and $D \subseteq B$.
     (a)   Prove that $C \cap D \subseteq A \cap B$.
     (b)   Prove that $C \cup D \subseteq A \cup B$.
  ★ (c)   Prove that if A and B are disjoint then C and D are disjoint.
     (d)   Prove that $D - A \subseteq B - C$.

10. Let A, B, C, and D be sets. Prove that if $A \cup B = C \cup D$, $A \cap B = \varnothing$, and $C \subseteq A$, then $B \subseteq D$.

11. Let A and B be sets.
  ★ (a)   Prove that $\mathscr{P}(A \cap B) = \mathscr{P}(A) \cap \mathscr{P}(B)$. You may use exercise 8(c).
  — (b)   Prove that $\mathscr{P}(A) \cup \mathscr{P}(B) \subseteq \mathscr{P}(A \cup B)$.
  ★ (c)   Show that there are no sets A and B such that $\mathscr{P}(A - B) = \mathscr{P}(A) - \mathscr{P}(B)$.

12. Provide examples of sets to show that each of the following is false.
  ★ (a)   If $A \cup C \subseteq B \cup C$, then $A \subseteq B$.
  — (b)   If $A \cap C \subseteq B \cap C$, then $A \subseteq B$.
  ★ (c) $\mathscr{P}(A \cup B) = \mathscr{P}(A) \cup \mathscr{P}(B)$.
     (d) $\mathscr{P}(A) - \mathscr{P}(B) \subseteq \mathscr{P}(A - B)$.
  ★ (e) $A - (B - C) = (A - B) - (A - C)$.

13. Define the symmetric difference operation Δ on sets by:
     $A \Delta B = (A - B) \cup (B - A)$.
     Prove that:
     (a) $A \Delta B = B \Delta A$.
     (b) $A \Delta A = \varnothing$.
     (c) $A \Delta \varnothing = A$.
     (d)   Show that $A \Delta B = (A \cup B) - (A \cap B)$.

14. **Proofs to Grade.**
  ★ (a)   **Claim.** $A \subseteq B \Leftrightarrow A \cap B = A$.
           "Proof."   Assume that $A \subseteq B$. Suppose $x \in A \cap B$. Then $x \in A$ and $x \in B$, so $A \cap B = A$. Now assume that $A \cap B = A$. Suppose $x \in A$. Then $x \in A \cap B$, since $A = A \cap B$; and therefore $x \in B$. This shows that $x \in A$ implies $x \in B$, and so $A \subseteq B$.   ■
  ★ (b)   **Claim.** $A \cap \varnothing = A$.
           "Proof."   We know that $x \in A \cap \varnothing$ iff $x \in A$ and $x \in \varnothing$. Since $x \in \varnothing$ is false, $x \in A$ and $x \in \varnothing$ iff $x \in A$. Therefore $x \in A \cap \varnothing$ iff $x \in A$; that is, $A \cap \varnothing = A$.   ■
     (c)   **Claim.** $\mathscr{P}(A - B) - \{\varnothing\} \subseteq \mathscr{P}(A) - \mathscr{P}(B)$.
           "Proof."   Suppose $x \in \mathscr{P}(A - B) - \{\varnothing\}$. Then $x \in \mathscr{P}(A) - \mathscr{P}(B)$. Therefore $\mathscr{P}(A - B) \subseteq \mathscr{P}(A) - \mathscr{P}(B)$ ■

★ (d) **Claim.** $A \Delta A = \varnothing$ [exercise 13(b)].
"*Proof.*" By part (d) of exercise 13, $A \Delta A = (A \cup A) - (A \cap A)$. By Theorem 2.7, parts (e) and (f), $(A \cup A) - (A \cap A) = A - A$. But $x \in A - A$ iff both $x \in A$ and $x \not\in A$ iff $x \in \varnothing$. ⟨*Each is false.*⟩ Therefore $A \Delta A = \varnothing$. ∎

(e) **Claim.** $C \subseteq A \cup B$ iff $C \subseteq A$ or $C \subseteq B$.
"*Proof.*"   If $x \in C$, then $x \in A \cup B$
⇔ If $x \in C$, then $x \in A$ or $x \in B$
⇔ If $x \in C$, then $x \in A$ or, if $x \in C$, then $x \in B$
⇔ $C \subseteq A$ or $C \subseteq B$. ∎

★ (f) **Claim.** If $A \subseteq B$, then $A \cup B = B$.
"*Proof.*"   Let $A \subseteq B$. Then A and B are related as in this figure.

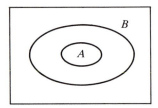

Since $A \cup B$ is the set of elements in either of the sets A or B, $A \cup B$ is the shaded area in this figure.

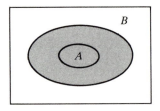

Since this is B, $A \cup B = B$. ∎

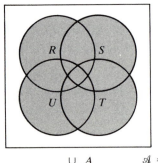

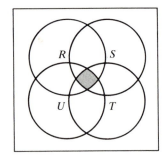

$$\underset{A \in \mathscr{A}}{\cup} A \qquad \mathscr{A} = \{R, S, T, U\} \qquad \underset{A \in \mathscr{A}}{\cap} A$$

Figure 2.5

2.3

Extended Set Operations and Indexed Families of Sets

A set of sets is often called a **family** of sets. In this section we extend the definition of union and intersection to families of sets, so that the union and intersection of two sets will be a special case. If $\mathscr{A}$ is a family of sets, the intersection over $\mathscr{A}$ will be the set of all elements common to all sets in $\mathscr{A}$, while the union over $\mathscr{A}$ will consist of those objects appearing in at least one of the sets in $\mathscr{A}$ (figure 2.5).

Definition.   Let $\mathscr{A}$ be a family of sets. The **union over** $\mathscr{A}$ is defined by

$$\bigcup_{A \in \mathscr{A}} A = \{x: (\exists A \in \mathscr{A})(x \in A)\}$$

and the **intersection over** $\mathscr{A}$ is

$$\bigcap_{A \in \mathscr{A}} A = \{x: (\forall A \in \mathscr{A})(x \in A)\}.$$

Let $\mathscr{A} = \{\{a, b, c\}, \{b, c, d\}, \{c, d, e, f\}\}$ be a family of three sets. Then $\bigcup_{A \in \mathscr{A}} A = \{a, b, c, d, e, f\}$, since each of the elements a, b, c, d, e, f, appears in at least one of the three sets in $\mathscr{A}$. Also, $\bigcap_{A \in \mathscr{A}} A = \{c\}$, since c is the only object appearing in every set in $\mathscr{A}$.

Theorem 2.9   For every set B in a family $\mathscr{A}$ of sets,

$$\bigcap_{A \in \mathscr{A}} A \subseteq B \text{ and } B \subseteq \bigcup_{A \in \mathscr{A}} A.$$

Proof.   Let $\mathscr{A}$ be a family of sets and $B \in \mathscr{A}$. If $x \in \bigcap_{A \in \mathscr{A}} A$, then $x \in A$ for every $A \in \mathscr{A}$. ⟨*Notice that the set A in the last sentence is a dummy symbol. It stands for any set in the family. The set B is in the family.*⟩ In particular, $x \in B$. Therefore $\bigcap_{A \in \mathscr{A}} A \subseteq B$. The proof that B is a subset of the union over $\mathscr{A}$ is left as exercise 3.   ∎

Example. Let $\mathscr{A}$ be the set of all subsets of the real line of the form $(-a, a)$ for some positive $a \in \mathbf{R}$. For example, $(-1, 1) \in \mathscr{A}$ and $(-\sqrt{2}, \sqrt{2}) \in \mathscr{A}$. Then $\bigcup\limits_{A \in \mathscr{A}} A = \mathbf{R}$ and $\bigcap\limits_{A \in \mathscr{A}} A = \{0\}$ (figure 2.6).

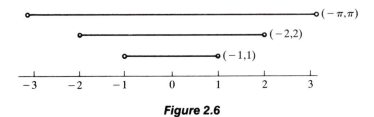

Figure 2.6

Example. If $\mathscr{B}$ is the family of all sets of the form $(-5, n]$, where n is a positive even integer, then $\bigcup\limits_{A \in \mathscr{B}} A = (-5, \infty)$ and $\bigcap\limits_{A \in \mathscr{B}} A = (-5, 2]$. Some members of the family $\mathscr{B}$ are $A_2 = (-5, 2]$, $A_4 = (-5, 4]$, and $A_{28} = (-5, 28]$.

It is often possible, as in the previous example, to associate a kind of identifying tag, or index, with each set in a family. For the family $\mathscr{B}$ and set $(-5, 2]$, the index used above is 2, and 28 is the index for the set $(-5, 28]$. With this method of indexing, the set of indices for $\mathscr{B}$ is the set of all positive even integers.

Definition. Let Δ be a nonempty set. Suppose for each $\alpha \in \Delta$, there is a corresponding set A_α. Then the family of sets $\mathscr{A} = \{A_\alpha : \alpha \in \Delta\}$ is an **indexed family of sets.** Each $\alpha \in \Delta$ is called an **index,** and Δ is called an **indexing set.**

Example. Let $\Delta = \mathbf{N}$. For all $n \in \mathbf{N}$, let $A_n = \{n, n + 1, 2n\}$. Then $A_1 = \{1, 2\}$, $A_2 = \{2, 3, 4\}$, $A_3 = \{3, 4, 6\}$, and so forth. The set with index 10 is $A_{10} = \{10, 11, 20\}$.

There is no real difference between a family of sets and an indexed family. Every family of sets could be indexed by finding a large enough set of indices to label each set in the family. An indexing set may be finite or infinite, as well as ordered (like the integers) or unordered (like the complex numbers).

Example. For the sets $A_1 = \{1, 2, 4, 5\}$, $A_2 = \{2, 3, 5, 6\}$, and $A_3 = \{3, 4, 5, 6\}$, the index set has been chosen to be $\Delta = \{1, 2, 3\}$. The family $\mathscr{A}$ indexed by Δ is $\mathscr{A} = \{A_1, A_2, A_3\} =$

$\{A_i: i \in \Delta\}$. The family $\mathcal{A}$ could be indexed by another set. For instance, if $\Gamma = \{10, 21, \pi\}$, and $A_{10} = \{1, 2, 4, 5\}$, $A_{21} = \{2, 3, 5, 6\}$, and $A_\pi = \{3, 4, 5, 6\}$, then $\{A_i: i \in \Delta\} = \{A_i: i \in \Gamma\}$.

Example.   Let Δ be the set $\{0, 2, 5, 6, 10\}$. For each $\alpha \in \Delta$, let $A_\alpha = \{3, |\alpha - 4|, \alpha^2 - 8\alpha + 16\}$. Then $A_0 = \{3, 4, 16\}$, and the elements of the set with index 10 are 3, 6, and 36. Notice that A_5 has only two elements, 1 and 3. The family of sets indexed by Δ contains $\{2, 3, 4\}$, and this set is indexed both by 2 and by 6. Since $A_2 = A_6$, the family of sets indexed by Δ has only four elements.

Unions and intersections over indexed families of sets are denoted as follows. Note the relationship between $\cup$ and $\exists$, and that between $\cap$ and $\forall$. This occurs because an element of the union is a member of *at least one* set in the family, while an element of the intersection is in *every* set in the family.

Definition.   If $\mathcal{A} = \{A_\alpha: \alpha \in \Delta\}$ is an indexed family of sets, then

$$\bigcup_{\alpha \in \Delta} A_\alpha = \bigcup_{A \in \mathcal{A}} A = \{x: (\exists \alpha \in \Delta)(x \in A_\alpha)\}$$

and

$$\bigcap_{\alpha \in \Delta} A_\alpha = \bigcap_{A \in \mathcal{A}} A = \{x: (\forall \alpha \in \Delta)(x \in A\alpha)\}.$$

In the example above with index set $\mathbf{N}$ and each $A_n = \{n, n + 1, 2n\}$ we have $\bigcup_{n \in \mathbf{N}} A_n = \mathbf{N}$. To convince someone that $27 \in \bigcup_{n \in \mathbf{N}} A_n$, we need only point out some index n (26 or 27 will do) such that $27 \in A_n$. Since there does not exist x such that for all $n \in \mathbf{N}$, $x \in A_n$, $\bigcap_{n \in \mathbf{N}} A_n = \emptyset$.

Example.   Let $\Delta = \{1, 2, 3\}$, $A_1 = \{1, 2, 4, 5\}$, $A_2 = \{2, 3, 5, 6\}$, $A_3 = \{3, 4, 5, 6\}$, and $\mathcal{A} = \{A_i: i \in \Delta\}$. Then $\bigcap_{i \in \Delta} A_i = \{5\}$ and $\bigcup_{i \in \Delta} A_i = \{1, 2, 3, 4, 5, 6\}$.

Example.   Let Δ be the set of positive reals. For each $a \in \Delta$, let $H_a = (-a, a)$. Let $\mathcal{A} = \{H_a: a \in \Delta\}$. Then, as we have seen before, $\bigcap_{a \in \Delta} H_a = \{0\}$ and $\bigcup_{a \in \Delta} H_a = \mathbf{R}$.

Example.   For each real number x, define $B_x = [x^2, x^2 + 1]$. Then $B_{-1/2} = [\frac{1}{4}, \frac{5}{4}]$, $B_0 = [0, 1]$, and $B_{10} = [100, 101]$. In this example we have different indices representing the same set. For example, $B_{-2} = B_2 = [4, 5]$. Here the index set is $\mathbf{R}$, $\bigcap_{x \in \mathbf{R}} B_x = \emptyset$, and $\bigcup_{x \in \mathbf{R}} B_x = [0, \infty)$.

Example.   Let A be any nonempty set. For each $a \in A$, let $X_a = \{a\}$. Then $\bigcup\limits_{a \in A} X_a = A$. If A has more than one element,

then $\bigcap\limits_{a \in A} X_a = \emptyset$.

With index set **N**, the union over the family $\mathscr{A} = \{A_i : i \in \mathbf{N}\}$ is sometimes written $\bigcup\limits_{i=1}^{\infty} A_i$ rather than $\bigcup\limits_{i \in \mathbf{N}} A_i$ and the intersection over $\mathscr{A}$ is written $\bigcap\limits_{i=1}^{\infty} A_i$. This notation is useful if we want to work with such sets as $\bigcup\limits_{i=1}^{2} A_i = A_1 \cup A_2$, $\bigcap\limits_{i=3}^{5} A_i = A_3 \cap A_4 \cap A_5$, and so forth.

Families of sets, whether indexed or not, obey a form of De Morgan's Laws stated for two sets in Theorem 2.8. The next theorem gives a statement of De Morgan's Laws for indexed families, and also restates Theorem 2.9 for indexed families.

Theorem 2.10   Let $\mathscr{A} = \{A_\alpha : \alpha \in \Delta\}$ be an indexed collection of sets. Then

(a) $\bigcap\limits_{\alpha \in \Delta} A_\alpha \subseteq A_\beta$ for each $\beta \in \Delta$.

(b) $A_\beta \subseteq \bigcup\limits_{\alpha \in \Delta} A_\alpha$ for each $\beta \in \Delta$.

(c) $\overline{\bigcap\limits_{\alpha \in \Delta} A_\alpha} = \bigcup\limits_{\alpha \in \Delta} \widetilde{A_\alpha}$.

(d) $\overline{\bigcup\limits_{\alpha \in \Delta} A_\alpha} = \bigcap\limits_{\alpha \in \Delta} \widetilde{A_\alpha}$.

Proof.   The proofs of parts (a) and (b) are similar to those for Theorem 2.9 and are left for exercise 4(a).

(c) $x \in \overline{\bigcap\limits_{\alpha \in \Delta} A_\alpha}$

      iff $x \notin \bigcap\limits_{\alpha \in \Delta} A_\alpha$

      iff it is not the case that for every $\alpha \in \Delta$, $x \in A_\alpha$

      iff for some $\beta \in \Delta$, $x \notin A_\beta$

      iff for some $\beta \in \Delta$, $x \in \widetilde{A_\beta}$

      iff $x \in \bigcup\limits_{\alpha \in \Delta} \widetilde{A_\alpha}$.

    Therefore $\overline{\bigcap\limits_{\alpha \in \Delta} A_\alpha} = \bigcup\limits_{\alpha \in \Delta} \widetilde{A_\alpha}$.

(d) *⟨One proof of part (d) is very similar to that given for part (c) and is left as exercise 4(b). However, since part (c) has been proved, it is permissible to use it, as follows.⟩*

$$\widetilde{\bigcup_{\alpha \in \Delta} A_\alpha} = \bigcup_{\alpha \in \Delta} \widetilde{\widetilde{A_\alpha}} \quad \langle A_\alpha = \widetilde{\widetilde{A}} \rangle$$

$$= \overline{\overline{\bigcap_{\alpha \in \Delta} \widetilde{A_\alpha}}} \quad \langle by \ (c) \rangle$$

$$= \bigcap_{\alpha \in \Delta} \widetilde{A_\alpha}. \quad \langle \widetilde{\widetilde{A}} = A \rangle \quad ■$$

Definition.   The family $\mathscr{A} = \{A_\alpha : \alpha \in \Delta\}$ of sets is **pairwise disjoint** iff for all α and β in Δ, if $A_\alpha \neq A_\beta$, then $A_\alpha \cap A_\beta = \varnothing$.

Two questions are commonly asked about this definition. The first question is why we bother with such a definition when we could more easily talk about a disjoint family satisfying $\bigcap_{\alpha \in \Delta} A_\alpha = \varnothing$. The answer is that the two ways of talking about disjointness of families of sets are not the same, and the situation covered by the definition given is by far the more interesting. A pairwise disjoint family with more than one set always satisfies the condition that the intersection over the whole family is empty. One reason why the pairwise disjoint idea is important will appear in the study of partitions (chapter 3, section 3). The fact that the two ideas of disjointness are not the same can be seen from the following examples.

Let $\mathscr{C} = \{A_1, A_2, A_3\}$, where $A_1 = \{1, 2\}$, $A_2 = \{2, 3\}$, and $A_3 = \{4, 5\}$. Then $\bigcap_{A \in \mathscr{C}} A = \varnothing$, but the family is not pairwise disjoint because $A_1 \cap A_2 \neq \varnothing$.

For the family $\mathscr{A} = \{A_n : n \in \mathbf{N}\}$ where, for each n, $A_n = (n, \infty)$, the family is not pairwise disjoint, but $\bigcap_{n \in \mathbf{N}} A_n = \varnothing$. In this case not only are there some pairs of sets that are not disjoint, but for every pair A_n, A_m, with, say $n < m$, the intersection $A_n \cap A_m = (m, \infty) = A_m$.

The second common question about the definition of pairwise disjoint has to do with the need for including the phrase ". . . whenever $A_\alpha \neq A_\beta$" in the definition. This is necessary because the family $\{A_1, A_2, A_3, A_4\}$ with $A_1 = \{6, 7\}$, $A_2 = \{5, 9\}$, $A_3 = \{6, 7\}$, and $A_4 = \{2, 8\}$ is the same family as $\{A_1, A_2, A_4\}$ and is pairwise disjoint. We do not require that two sets with different indices be different.

Exercises 2.3

1. Find the union and intersection of each of the following families or indexed collections.
   ★ (a) Let $\mathscr{A} = \{\{1, 2, 3, 4, 5\}, \{2, 3, 4, 5, 6\}, \{3, 4, 5, 6, 7\}, \{4, 5, 6, 7, 8\}\}$.
     (b) Let $\mathscr{A} = \{\{1, 3, 5\}, \{2, 4, 6\}, \{7, 9, 11, 13\}, \{8, 10, 12\}\}$.

★ (c) For each natural number n, let $A_n = \{1, 2, 3, \ldots, n\}$, and let $\mathscr{A} = \{A_n : n \in \mathbf{N}\}$.

(d) For each natural number n, let $B_n = \mathbf{N} - \{1, 2, 3, \ldots, n\}$, and let $\mathscr{B} = \{B_n : n \in \mathbf{N}\}$.

★ (e) Let $\mathscr{A}$ be the set of all sets of integers that contain 10.

(f) Let $A_1 = \{1\}$, $A_2 = \{2, 3\}$, $A_3 = \{3, 4, 5\}$, $\ldots$, $A_{10} = \{10, 11, \ldots, 19\}$, and let $\mathscr{A} = \{A_n : n \in \{1, 2, 3, \ldots, 10\}\}$.

★ (g) For each natural number, let $A_n = (0, 1/n)$, and let $\mathscr{A} = \{A_n : n \in \mathbf{N}\}$.

(h) Let $\mathbf{R}^+ = (0, \infty)$. For $r \in \mathbf{R}^+$, let $A_r = [-\pi, r)$, and let $\mathscr{A} = \{A_r : r \in \mathbf{R}^+\}$.

★ (i) For each real number r, let $A_r = [|r|, 2|r| + 1]$, and let $\mathscr{A} = \{A_r : r \in \mathbf{R}\}$.

(j) For each $n \in \mathbf{N}$, let $M_n = \{\ldots, -3n, -2n, -n, 0, n, 2n, 3n, \ldots\}$, and let $\mathscr{M} = \{M_n : n \in \mathbf{N}\}$.

(k) For each natural number $n \geq 3$, let $A_n = \left[\dfrac{1}{n}, 2 + \dfrac{1}{n}\right)$ and $\mathscr{A} = \{A_n : n \geq 3\}$.

(l) For each $n \in \mathbf{Z}$, let $C_n = [n, n + 1)$ and let $\mathscr{C} = \{C_n : n \in \mathbf{Z}\}$.

☆ 2. Which families in exercise 1 are pairwise disjoint?

3. Prove the remaining part of Theorem 2.9. That is, prove that if $\mathscr{A}$ is a family of sets and $B \in \mathscr{A}$, then $B \subseteq \underset{A \in \mathscr{A}}{\cup} A$.

☆ 4. (a) Prove parts (a) and (b) of Theorem 2.10.

(b) Give a direct proof of part (d) of Theorem 2.10 that does not use part (c).

5. Let $\mathscr{A} = \{A_\alpha : \alpha \in \Delta\}$ be an arbitrary family of sets and let B be a set. Prove that

★ (a) $B \cap \underset{\alpha \in \Delta}{\cup} A_\alpha = \underset{\alpha \in \Delta}{\cup} (B \cap A_\alpha)$.

(b) $B \cup \underset{\alpha \in \Delta}{\cap} A_\alpha = \underset{\alpha \in \Delta}{\cap} (B \cup A_\alpha)$.

6. Let $\mathscr{A} = \{A_\alpha : \alpha \in \Delta\}$ and $\mathscr{B} = \{B_\beta : \beta \in \Gamma\}$. Use exercise 5 to write

★ (a) $(\underset{\alpha \in \Delta}{\cup} A_\alpha) \cap (\underset{\beta \in \Gamma}{\cup} B_\beta)$ as a union of intersections.

(b) $(\underset{\alpha \in \Delta}{\cap} A_\alpha) \cup (\underset{\beta \in \Gamma}{\cap} B_\beta)$ as an intersection of unions.

7. Let $\mathscr{A}$ be a family of sets and suppose $\varnothing \in \mathscr{A}$. Prove that $\underset{A \in \mathscr{A}}{\cap} A = \varnothing$.

8. If $\mathscr{A} = \{A_\alpha : \alpha \in \Delta\}$ is a family of sets and if $\Gamma \subseteq \Delta$, prove that

★ (a) $\underset{\alpha \in \Gamma}{\cup} A_\alpha \subseteq \underset{\alpha \in \Delta}{\cup} A_\alpha$.

(b) $\underset{\alpha \in \Delta}{\cap} A_\alpha \subseteq \underset{\alpha \in \Gamma}{\cap} A_\alpha$.

9. Let $\mathscr{A}$ be a family of sets and suppose $B \subseteq A$ for every $A \in \mathscr{A}$. Prove that $B \subseteq \underset{A \in \mathscr{A}}{\cap} A$.

10. Let $\mathscr{A}$ be a family of sets and suppose that $A \subseteq B$ for every $A \in \mathscr{A}$. Prove that $\underset{A \in \mathscr{A}}{\cup} A \subseteq B$.

11. Prove that, if $\mathscr{A} = \{A_\alpha : \alpha \in \Delta\}$ is pairwise disjoint and Δ contains more than one element, then $\underset{\alpha \in \Delta}{\cap} A_\alpha = \varnothing$.

★ 12. Let $\mathscr{A} = \{A_\alpha : \alpha \in \Delta\}$ be an indexed family of sets. By Theorem 2.10 we know that $\bigcap_{\alpha \in \Delta} A_\alpha \subseteq A_\beta$ for every $\beta \in \Delta$. Prove that $\bigcap_{\alpha \in \Delta} A_\alpha$ is the largest such set by proving that if B is any other set such that $B \subseteq A_\beta$ for all $\beta \in \Delta$, then $B \subseteq \bigcap_{\alpha \in \Delta} A_\alpha$.

13. Let $\mathscr{A} = \{A_\alpha : \alpha \in \Delta\}$ be a family of sets. By Theorem 2.10 we know that $A_\beta \subseteq \bigcup_{\alpha \in \Delta} A_\alpha$ for every $\beta \in \Delta$. Prove that $\bigcup_{\alpha \in \Delta} A_\alpha$ is the smallest such set by proving that if B is any other set such that $A_\beta \subseteq B$ for all $\beta \in \Delta$, then $\bigcup_{\alpha \in \Delta} A_\alpha \subseteq B$.

★ 14. Give another example of an indexed collection of sets $\{A_\alpha : \alpha \in \Delta\}$ such that, for all α and $\beta \in \Delta$, $A_\alpha \cap A_\beta \neq \varnothing$ but $\bigcap_{\alpha \in \Delta} A_\alpha = \varnothing$.

15. Let $\mathscr{A} = \{A_i : i \in \mathbf{N}\}$ be a family of sets and k, m be natural numbers with $k \leq m$. Prove that

   (a) $\bigcup_{i=1}^{k+1} A_i = \bigcup_{i=1}^{k} A_i \cup A_{k+1}$.

   (b) $\bigcap_{i=1}^{k+1} A_i = \bigcap_{i=1}^{k} A_i \cap A_{k+1}$.

   (c) $\bigcup_{i=k}^{m} A_i \subseteq \bigcup_{i=1}^{\infty} A_i$.

   (d) $\bigcap_{i=1}^{\infty} A_i \subseteq \bigcap_{i=k}^{m} A_i$.

16. **Proofs to Grade.**

   ★ (a) **Claim.**   For any indexed family $\{A_\alpha : \alpha \in \Delta\}$, $\bigcap_{\alpha \in \Delta} A_\alpha \subseteq \bigcup_{\alpha \in \Delta} A_\alpha$.

   "Proof."   Choose any $A_\beta \in \{A_\alpha : \alpha \in \Delta\}$. Then $\bigcap_{\alpha \in \Delta} A_\alpha \subseteq A_\beta$ and $A_\beta \subseteq \bigcup_{\alpha \in \Delta} A_\alpha$. Therefore by transitivity of set inclusion,

   $\bigcap_{\alpha \in \Delta} A_\alpha \subseteq \bigcup_{\alpha \in \Delta} A_\alpha$.   ∎

   ★ (b) **Claim.**   If $A_\alpha \subseteq B$ for all $\alpha \in \Delta$, then $\bigcup_{\alpha \in \Delta} A_\alpha \subseteq B$.

   "Proof."   Suppose $x \in \bigcup_{\alpha \in \Delta} A_\alpha$. Then, since $A_\alpha \subseteq B$ for all $\alpha \in \Delta$, $x \in B$. Therefore $\bigcup_{\alpha \in \Delta} A \subseteq B$.   ∎

   (c) **Claim.**   For any indexed family $\{A_\alpha : \alpha \in \Delta\}$, $\bigcup_{\alpha \in \Delta} A_\alpha \subseteq \bigcap_{\alpha \in \Delta} A_\alpha$.

   "Proof."   Suppose $x \in \bigcup_{\alpha \in \Delta} A_\alpha$. Then $x \in A_\alpha$ for every $\alpha \in \Delta$, which implies $x \in A_\alpha$ for at least one $\alpha \in \Delta$ since $\Delta \neq \varnothing$. Therefore $x \in \bigcap_{\alpha \in \Delta} A_\alpha$.   ∎

   (d) **Claim.**   For any indexed family $\{A_\alpha : \alpha \in \Delta\}$, $\bigcap_{\alpha \in \Delta} A_\alpha \subseteq \bigcup_{\alpha \in \Delta} A_\alpha$.

   "Proof."   Assume $\bigcap_{\alpha \in \Delta} A_\alpha \not\subseteq \bigcup_{\alpha \in \Delta} A_\alpha$. Then for some $x \in \bigcap_{\alpha \in \Delta} A_\alpha$, $x \notin \bigcup_{\alpha \in \Delta} A_\alpha$. Since $x \notin \bigcup_{\alpha \in \Delta} A_\alpha$, it is not the case that $x \in A_\alpha$ for some $\alpha \in \Delta$. Therefore $x \notin A_\alpha$ for every $\alpha \in \Delta$. But since $x \in \bigcap_{\alpha \in \Delta} A_\alpha$, $x \in A_\alpha$ for every $\alpha \in \Delta$. This is a contradiction, so we conclude $\bigcap_{\alpha \in \Delta} A_\alpha \subseteq \bigcup_{\alpha \in \Delta} A_\alpha$.   ∎

2.4

Induction

The most fundamental number system is the system of natural numbers. You are already familiar with the set $\mathbf{N} = \{1, 2, 3, 4, \ldots\}$ and with the properties of addition and multiplication on $\mathbf{N}$. These properties are listed below.

1. *Successor properties.* The number 1 is not the successor of any natural number; every natural number except 1 is the successor of exactly one natural number; every natural number x has a unique successor $x + 1$.
2. *Closure properties.* For each pair x and y of natural numbers, the sum $x + y$ and the product xy are natural numbers.
3. *Associativity properties.* For all $x, y, z \in \mathbf{N}$,
$$x + (y + z) = (x + y) + z \quad \text{and} \quad x(yz) = (xy)z.$$
4. *Commutativity properties.* For all x and y in $\mathbf{N}$,
$$x + y = y + x \qquad \text{and} \qquad xy = yx.$$
5. *Distributivity properties.* For all $x, y, z \in \mathbf{N}$,
$$x(y + z) = xy + xz \quad \text{and} \quad (y + z)x = yx + zx.$$
6. *Cancellation properties.* For all $x, y, z \in \mathbf{N}$
$$\text{if } x + z = y + z \text{ or } xz = yz, \text{ then } x = y.$$
7. *Order properties.* For all $x, y, z \in \mathbf{N}$:
$x < y$ iff there is $w \in \mathbf{N}$ such that $x + w = y$
$x \leq y$ iff $x < y$ or $x = y$
$x < y$ and $y < z$ implies $x < z$
$x \leq y$ and $y \leq x$ implies $x = y$
if $x < y$, then $x + z < y + z$ and $xz < yz$.

These properties will be assumed as we focus on the remaining property of the natural number system, the *induction property*.

Principle of Mathematical Induction (PMI)

Let S be a subset of $\mathbf{N}$ with the properties

(i) $1 \in S$.
(ii)   For all $n \in \mathbf{N}$, $n \in S$ implies $n + 1 \in S$.

Then $S = \mathbf{N}$.

A set of natural numbers with the property that $n \in S$ implies $n + 1 \in S$ is called an **inductive set.** The set $\{5, 6, 7, 8, \ldots\}$ is inductive but does not contain 1. The set $\{1, 3, 5, 7, 9, \ldots\}$ contains 1 but is not inductive because, for example, 7 is a member but 8 is not. Many sets of natural numbers have the inductive property or contain 1, but only one set contains 1 and is also inductive. By the PMI, that set is **N**.

We shall see that there are three forms of induction, each useful in certain situations. The PMI in the form given is basic to our understanding of the natural numbers: the natural numbers consist of 1, then $1 + 1 = 2$, then $2 + 1 = 3, \ldots$, and so on. The PMI allows us to do two important things: first, to make inductive definitions; second, to prove that some properties are shared by all natural numbers.

An inductive definition is a means to define an infinite set of objects that can be indexed by the natural numbers; that is, there is a first object, a second object, a third, and so on. Inductive definitions follow the form and spirit of the PMI. The first object is defined. Then the $n + 1$st object is defined in terms of the nth object. The PMI ensures that the set of all n for which the corresponding object is defined is **N**.

Example.   The factorial of a natural number may be defined inductively. You have probably seen the noninductive definition

$$n! = n \cdot (n - 1) \cdot \cdots \cdot 3 \cdot 2 \cdot 1$$

The inductive definition of $n!$ proceeds as follows:

    (i)   Define $1! = 1$.
    (ii) For $n \in$ **N**, define $(n + 1)! = (n + 1)n!$

Let S be the set of n for which $n!$ is defined. First, $1 \in S$ because of part (i). Second, S is inductive because if $n \in S$, then $n!$ is defined and hence, by part (ii), $(n + 1)!$ is also defined. Thus $(n + 1) \in S$. By the PMI, $S =$ **N**. In other words, the set of numbers for which the factorial is defined is **N**, so $n!$ has been defined for all natural numbers. The inductive definition makes clear the relationship between the factorial of a number and the factorial of the next number; if you happen to know that $11! = 39{,}916{,}800$, then you compute $12! = 12 \cdot 11! = 479{,}001{,}600$.

Example.   Sets may also be defined inductively. Suppose we let O be the collection of natural numbers defined by

    (i) $1 \in O$.
    (ii) If $n \in O$, then $n + 2 \in O$.

The set O is, of course, the set of odd natural numbers which has the familiar noninductive definition

$$O = \{n : n = 2k + 1 \text{ for some } k \in \mathbf{N}\}.$$

You have probably used mathematical induction to prove statements like the one in the following example.

Example. We will prove that for every natural number n, $1 + 3 + 5 + \cdots + (2n - 1) = n^2$. The correctness of this statement can be verified for, say, $n = 6$, by adding $1 + 3 + 5 + 7 + 9 + 11 = 36$. A proof by induction that the property holds for all n proceeds as follows:

Proof. Let $S = \{n \in \mathbf{N} : 1 + 3 + 5 + \cdots + (2n - 1) = n^2\}$. ⟨*Our aim is to show that the statement is true for every natural number by showing that $S = \mathbf{N}$.*⟩

(i) $1 = 1^2$, so $1 \in S$.
(ii) Let n be a natural number such that $n \in S$. Then
      $1 + 3 + 5 + \cdots + (2n - 1) = n^2$. ⟨*We have* not *assumed what is to be proved. We assume only that* some n *is in S to show that $n + 1 \in S$ follows.*⟩ Therefore

$$1 + 3 + 5 + \cdots + (2n - 1) + [2(n + 1) - 1]$$
$$= n^2 + 2(n + 1) - 1$$
      ⟨*Compare the statements for n and for $n + 1$.*⟩
$$= n^2 + 2n + 1$$
$$= (n + 1)^2.$$

This shows that $n + 1 \in S$.
(iii) By the PMI, $S = \mathbf{N}$. That is, for every natural number n,
      $1 + 3 + 5 + \cdots + (2n - 1) = n^2$. ∎

Recall that the distributive law that says that for any numbers m, x_1, and x_2, $m(x_1 + x_2) = mx_1 + mx_2$. It is also true that no matter how many numbers are involved, the numbers $m(x_1 + x_2 + \cdots + x_n)$ and $mx_1 + mx_2 + \cdots + mx_n$ are equal. Although we would otherwise accept this as being obviously true, we will, as an example, give a proof of this generalization of the distributive law. Since $+$ is defined only for two numbers, we must be certain to make a sum $x_1 + x_2 + \cdots + x_n$ of n numbers meaningful by thinking of it as a sum of two numbers. We use the following inductive definition:

$$\text{For } n \geq 3, \; x_1 + x_2 + \cdots + x_n = (x_1 + x_2 + \cdots + x_{n-1}) + x_n.$$

This definition assures us that a sum of n numbers has been unambiguously defined for every natural number n. The sum of one number is understood to be the number; we know all about the sum of two numbers; $x_1 + x_2 + x_3$ is the sum of $x_1 + x_2$ and x_3; $x_1 + x_2 + x_3 + x_4$ is the sum of $x_1 + x_2 + x_3$ and x_4; and so on. We are now prepared to prove the generalized distributive law.

Example. For every list $x_1, x_2, \ldots, x_n$ of n numbers, and every number m, $m(x_1 + x_2 + \cdots + x_n) = mx_1 + mx_2 + \cdots + mx_n$.

Proof. Let S be the set of natural numbers with the desired property. That is,

$S = \{n \in \mathbf{N}$: for every list $x_1, x_2, \ldots, x_n$ of n numbers,
$$m(x_1 + x_2 + \cdots + x_n) = mx_1 + mx_2 + \cdots + mx_n\}.$$

(i) $m(x_1) = mx_1$ for every list of 1 number, so $1 \in S$.
(ii)   Suppose $n \in S$, and consider the situation for a list $x_1, x_2, \ldots, x_{n+1}$ of $n + 1$ numbers. Then
$m(x_1 + x_2 + \cdots + x_{n+1})$
$= m[(x_1 + x_2 + \cdots + x_n) + x_{n+1}]$ 　　〈*by definition*〉
$= m(x_1 + x_2 + \cdots + x_n) + mx_{n+1}$
　　　　　　　　　　　　　　　〈*by the distributive law*〉
$= (mx_1 + mx_2 + \cdots + mx_n) + mx_{n+1}$
　　　　　　　　　　　　　〈*by the assumption that $n \in S$*〉
$= mx_1 + mx_2 + \cdots + mx_n + mx_{n+1}$. 　〈*by definition*〉

Therefore $n + 1 \in S$.
(iii)   By the PMI, $S = \mathbf{N}$. That is, the generalized distributive law holds for the sum of n numbers, for every $n \in \mathbf{N}$. ■

In actual practice most people, when they write induction proofs, do not define the set S of all numbers that satisfy the property in question. If the property is expressed with the open sentence $P(n)$, the induction proof takes the form:

(i)   〈*Basis step*〉 Show that the proposition $P(1)$ is true.
(ii)   〈*Induction step*〉 Show that for all natural numbers n, if $P(n)$ is true, then $P(n + 1)$ is true.
(iii)   〈*Conclusion*〉 By steps (i) and (ii) and the PMI, $P(n)$ is true for all $n \in \mathbf{N}$.

The assumption in part (ii), that the property P is true for n, is called the **hypothesis of induction.** Every good proof by induction will use the hypothesis of induction to show that the property P is true for $n + 1$.

Example.   For all $n \in \mathbf{N}$, $n + 3 < 5n^2$.

Proof.

(i) $1 + 3 < 5 \cdot 1^2$, so the statement is true for 1.
(ii)   Assume that for some n, $n + 3 < 5n^2$. Then

$$(n + 1) + 3 = n + 3 + 1 < 5n^2 + 1$$
$$\langle by\ the\ hypothesis\ of\ induction \rangle$$

and

$$5n^2 + 1 < 5n^2 + 10n + 5 = 5(n + 1)^2$$

so

$$(n + 1) + 3 < 5(n + 1)^2.$$

Thus the property holds for $n + 1$.
(iii) By the PMI, $n + 3 < 5n^2$ for every $n \in \mathbf{N}$.   ∎

Example.   It is easy to see that the polynomial $x - y$ divides the polynomials $x^2 - y^2$ and $x^3 - y^3$ because $x^2 - y^2 = (x - y)(x + y)$ and $x^3 - y^3 = (x - y)(x^2 + xy + y^2)$. This suggests the possibility that for every natural number n, $x - y$ divides $x^n - y^n$. We prove this by induction.

Proof.

(i) $x - y$ obviously divides $x^1 - y^1 = x - y$, so the statement holds for $n = 1$.
(ii)   Assume that $x - y$ divides $x^n - y^n$ for some n. We must show that $x - y$ divides $x^{n+1} - y^{n+1}$. We write

$$x^{n+1} - y^{n+1} = xx^n - yy^n = xx^n - yx^n + yx^n - yy^n$$
$$= (x - y)x^n + y(x^n - y^n).$$

Now $x - y$ divides the first term because it divides $x - y$ and the second term because it divides $x^n - y^n$ $\langle by\ the\ hypothesis\ of\ induction \rangle$. Therefore $x - y$ divides the sum; so $x - y$ divides $x^{n+1} - y^{n+1}$.
(iii) By the PMI, $x - y$ divides $x^n - y^n$ for every natural number n.   ∎

From our examples you should not jump to the conclusion that all inductive steps in a proof by induction are proved by cleverly adding some amount to one or both sides of an expression in the hypothesis of induction. To help convince you this is so, we present another example of a proof using the PMI.

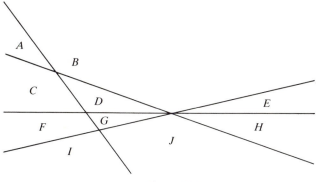

Figure 2.7

Example. Consider any "map" formed by drawing straight lines in a plane to represent boundaries. Figure 2.7 shows ten countries labeled A through J, formed by drawing four lines in the plane. The problem is to color the countries so that adjoining countries (those with a line segment as a common border) have different colors. This has been done in figure 2.8 using only the two colors white and grey. In fact, every map formed by drawing n straight lines can be colored using exactly two colors, and this can be proved by induction on the number of lines.

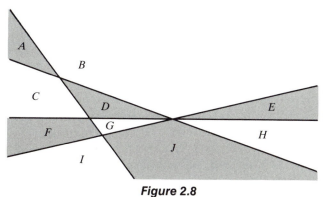

Figure 2.8

Proof.

   (i)   If a map is made by drawing one straight line, then there are only two countries. Thus every map formed with one line can be colored with two colors.

  (ii) Assume that for some n, *every* map formed by drawing n lines can be colored with exactly two colors. Now consider a map with $n + 1$ lines. Before coloring this map, choose

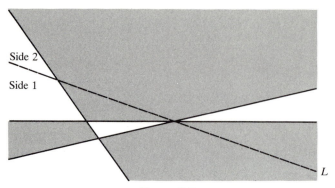

Figure 2.9

any one of the lines and label it *L*. Now color the map as though the line *L* were not there, using exactly two colors by the hypothesis of induction. ⟨*Such a coloring is shown in figure 2.9, with the line L shown as a dashed line. Of course, only part of the plane can be shown.*⟩ To color the map *with* line *L*, proceed as follows: Call one half-plane determined by *L* side 1, and the other half-plane side 2. Leave all colors on side 1 exactly as they were but change every color on side 2 to the other color. This gives a coloring to every country in the map with line *L*. (See figure 2.10.) It remains to verify that adjacent countries in this map have different colors.

First, if two countries have a segment of line *L* as a border, then they had the same color because they were parts of the same country in the original coloring. But now the country on side 2 is the other color and thus is colored differently from the country on side 1.

If two countries do not have line *L* as a border, then they were originally colored differently. If the countries lie on side 1, they are colored differently as before; if the countries lie on side 2, the colors are switched, but still different.

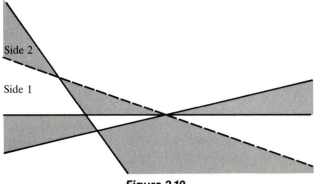

Figure 2.10

(iii)   By (i), (ii), and the PMI, every map formed by drawing n straight lines in a plane can be colored with exactly two colors.   ∎

As one more example of the use of the PMI, we will prove a small but useful result, needed in a later theorem, about the comparative sizes of natural numbers. Suppose a and b are any two natural numbers. Even if b is smaller than a, the result says that a can eventually be surpassed by taking natural number multiples of b. One reason for considering this theorem is that it happens to be one of those facts about natural numbers that can be proved without the PMI. We give two proofs, neither of which is "more correct" than the other. Notice once again that the key to an induction proof is in the inductive step where we use an assumption about a natural number n to draw a conclusion about $n + 1$.

Theorem 2.11   For all natural numbers a and b there exists a natural number s such that $a < sb$.

Proof 1.   Let a and b be two natural numbers. Then $a < a + 1$ and since $1 \le b$, $a + 1 \le (a + 1)b$. Therefore by choosing s to be the natural number $a + 1$, we have that $a < sb$.   ∎

Proof 2.   By induction on b.

(i)   If $b = 1$, choose s to be $a + 1$. Then $a < a + 1 = sb$.

(ii)   Suppose the statement is true when $b = n$, for some natural number n. Then there is some natural number s such that $a < sn$. Choosing the same s, we have $a < sn < s(n + 1)$, so the statement is true when $b = n + 1$.

(iii)   By parts (i) and (ii) and the PMI, the statement is true for all natural numbers a and b.   ∎

In some cases using the PMI is not the most convenient form of inductive proof. To use the PMI successfully, one must be able to show that if n has a certain property, then $n + 1$ has the property. In some cases there is no apparent connection between the property for n and for $n + 1$. An alternate form of induction, called **complete** (or strong) **induction**, is needed.

Theorem 2.12   (The Principle of Complete Induction (PCI))   Let S be a subset of $\mathbf{N}$ with the following property:

For all natural numbers m, if $\{1, 2, \ldots, m - 1\} \subseteq S$, then $m \in S$.

Then $S = \mathbf{N}$.

Proof. Let S be a subset of the natural numbers such that for all m, if $\{1, 2, \ldots, m - 1\} \subseteq S$, then $m \in S$. We use the PMI to prove that for every n, $\{1, 2, \ldots, n\} \subseteq S$.

(i) $1 \in S$ because every natural number less than 1 is in S. ⟨*If $m = 1$, we take the set $\{1, 2, \ldots, m - 1\}$ to be the empty set. Since $\emptyset \subseteq S$ is true, the property ensures that $1 \in S$.*⟩ Thus $\{1\} \subseteq S$.

(ii) Assume now that $\{1, 2, \ldots, n\} \subseteq S$. Then by the condition on S, we have from $\{1, 2, \ldots, n\} \subseteq S$ that $n + 1 \in S$. Therefore $\{1, 2, \ldots, n + 1\} \subseteq S$.

(iii) By the PMI, $\{1, 2, \ldots, n\} \subseteq S$ for every $n \in \mathbf{N}$. It follows that $\mathbf{N} \subseteq S$, and since $S \subseteq \mathbf{N}$, we conclude $S = \mathbf{N}$. ∎

The PCI could have been accepted rather than the PMI as the fundamental property of $\mathbf{N}$ because the two are equivalent. Notice that the PCI does not require the separate step of showing that $1 \in S$ because the PCI property reduces to $\emptyset \subseteq S$ implies $1 \in S$ in the case when $m = 1$. In practice many PCI proofs do include a separate proof of $1 \in S$. In fact, to prove that for all m, $\{1, 2, \ldots, m - 1\} \subseteq S$ implies $m \in S$, special considerations may be necessary when $m = 1$ or $m = 2$. This caution may apply to the PMI, too, when $m = 2$ [see exercise 12(a)].

Both the PMI and PCI have more general forms that can be used to prove that some property holds for all natural numbers greater than, say, 4. The basis step in such a PMI proof is to show that the property holds for 5. A PCI proof would use the statement

for all natural numbers $m > 4$, if $\{5, 6, \ldots, m - 1\} \subseteq S$, then $m \in S$.

The next example uses such a generalization for $n > 1$. It is also a result that is difficult to prove by using the PMI.

Example. Every natural number $n > 1$ has a prime factor.

Proof. Recall that a number p is prime iff $p > 1$ and its only factors in $\mathbf{N}$ are 1 and p. Let S be $\{n \in \mathbf{N}: n > 1$ and n has a prime factor$\}$. Notice that 1 is not in S, but 2 is in S. Let m be a natural number greater than 1. Assume that for all $k \in \{2, \ldots, m - 1\}$, $k \in S$. We must show that $m \in S$. If m has no factors other than 1 and m, then m is prime, and so m has a prime factor—*itself*. If m has a factor x other than 1 and m, then $1 < x < m$ so $x \in S$. Therefore x has a prime factor, which must also be a prime factor of m. In either case $m \in S$. Therefore $S = \{n \in \mathbf{N}: n > 1\}$, and every natural number greater than 1 has a prime factor. ∎

Another property characteristic of **N** is the well-ordering principle (WOP). The WOP is equivalent to the PMI and the PCI.

Theorem 2.13   (The Well-Ordering Principle (WOP))   Every nonempty subset of **N** has a smallest element.

Proof.   Let T be a nonempty subset of **N** and suppose T has no smallest element. ⟨*This is a proof by contradiction.*⟩ Let $S =$ **N** $- T$.

(i)   Since 1 is the smallest element of **N** and T has no smallest element, $1 \notin T$. Therefore 1 must be in S.

(ii)   Suppose $n \in S$. Observe that no numbers less than n belong to T, for if some number less than n were in T, then one of those numbers would be the smallest element in T. Now that we know $1, 2, \ldots, n$ are not in T, $n + 1$ cannot be in T, for otherwise it would be the smallest element of T. Therefore $(n + 1) \in S$. By the PMI, $S =$ **N**. Thus if T has no smallest elements, we arrive at the contradiction $T = \varnothing$. Therefore if T is a nonempty subset of **N**, T has a smallest element.   ■

Theorems 2.12 and 2.13 show that the PMI implies both the PCI and the WOP. In exercise 5 you are asked to show that the PCI implies both the PMI and WOP. Similarly, exercise 6 will show that the WOP implies each of the other properties. The theorems together with the exercises establish that all three properties are equivalent. Any one of the PMI, PCI, or WOP may be used as a basic characteristic of the natural numbers. *They are the first proof methods to turn to whenever it must be shown that a property holds for all natural numbers.* In some cases one of the principles is more appropriate than others, but sometimes more than one are suitable.

Proofs using the WOP frequently take the form of assuming that some desired property does not hold for all natural numbers. This produces a nonempty set of natural numbers that do not have the property. By working with the *smallest* such number, one can often find a contradiction. As an example of this method, we will prove that for every $n \in$ **N**, $n + 1 = 1 + n$. The purpose of this example is not to prove that $n + 1 = 1 + n$, but to see how a proof using the WOP is done. So imagine for a moment that we did not know that addition is commutative, and we will show how the statement can be proved (from the associative property) by using the WOP.

Example.   For every natural number n, $n + 1 = 1 + n$.

Proof.   Suppose there exist natural numbers n such that $n + 1 \neq 1 + n$. Let b be the smallest such number. Obviously, $1 + 1 = 1 + 1$, so $b \neq 1$. Thus b must be of the form $b =$

$c + 1$ for some $c \in \mathbf{N}$. ⟨*See the successor properties for* $\mathbf{N}$.⟩ Then $(c + 1) + 1 \neq 1 + (c + 1)$. By the associative property, $(c + 1) + 1 \neq (1 + c) + 1$. Subtracting 1 ⟨*from the right side*⟩ of each expression, we have $c + 1 \neq 1 + c$. But this is a contradiction, because $c < b$ and b is the smallest natural number with the property. We conclude that $n + 1 = 1 + n$ for all natural numbers n. ■

As a more interesting application of the WOP, we will prove the division algorithm for natural numbers. An extension of this algorithm to the integers is presented as exercise 10.

Theorem 2.14   (*The Division Algorithm for* $\mathbf{N}$.)   Let a and $b \in \mathbf{N}$, with $b \leq a$. Then there exist $q \in \mathbf{N}$ and $r \in \mathbf{N} \cup \{0\}$ such that $a = qb + r$, where $0 \leq r < b$. The numbers q and r are the quotient and remainder, respectively.

Proof.   Let $T = \{s \in \mathbf{N}: a < sb\}$. By Theorem 2.11, this set is nonempty. Therefore by the WOP, T contains a smallest element w. Let $q = w - 1$ and $r = a - qb$. Since $a \geq b$, $w > 1$ and $q \in \mathbf{N}$. Since $q < w$, $a \geq qb$ and $r \geq 0$. By definition of r, $a = qb + r$. Now suppose $r \geq b$. Then $a - (w - 1)b \geq b$, so $a \geq wb$. This contradicts the fact that $a < wb$. Therefore $r < b$. ■

Exercises 2.4

★ 1. Which of these sets have the inductive property?
    (a) $\varnothing$ (b) $\{2, 4, 6, 8, 10, \ldots\}$
    (c) $\{1, 2, 4, 5, 6, 7, \ldots\}$ (d) $\{17\}$

2. Suppose S is inductive. Which of the following must be true?
    (a) If $n + 1 \in S$, then $n \in S$.   ★ (b) If $n \in S$, then $n + 2 \in S$.
    (c) If $n + 1 \notin S$, then $n \notin S$.   (d) If $6 \in S$, then $11 \in S$.
    ★ (e) $6 \in S$ and $11 \in S$.

3. (a) For natural numbers n, the value of $n!$ (n factorial) is defined inductively as follows:

$$1! = 1 \qquad \text{and} \qquad (n + 1)! = (n + 1) \cdot n! \text{ for } n \geq 1.$$

    Find $4!$, $7!$, and $(n + 2)!/n!$.
    ★ (b) For natural numbers n, the nth Fibonacci number is defined inductively by

$$f_1 = 1, f_2 = 1 \qquad \text{and} \qquad f_{n+2} = f_{n+1} + f_n \text{ for } n \geq 1$$

    Find f_4, f_7, and $f_{n+3} - f_{n+1}$.

4. Use the PMI to prove the following for all natural numbers n, or for the natural numbers indicated.
   (a) $1 + 4 + 7 + \cdots + (3n - 2) = \frac{1}{2}n(3n - 1)$.
   (b) $3 + 11 + 19 + \cdots + (8n - 5) = 4n^2 - n$.
   (c) $2^1 + 2^2 + 2^3 + \cdots + 2^n = 2^{n+1} - 2$.
   (d) $1 \cdot 1! + 2 \cdot 2! + 3 \cdot 3! + \cdots + n \cdot n! = (n + 1)! - 1$.
   (e) $\dfrac{n^3}{3} + \dfrac{n^5}{5} + \dfrac{7n}{15}$ is an integer.
   (f) $2^n \geqslant 1 + n$.
   (g) $n^3 + 5n + 6$ is divisible by 3.
   (h) $4^n - 1$ is divisible by 3.
   (i) For all $n > 2$, the sum of angle measures of the interior angles of a convex polygon of n sides is $(n - 2) \cdot 180°$.
   (j) $3^n > 1 + 2^n$, for all $n > 1$.
   (k) $(n + 1)! > 2^{n+3}$, for $n \geqslant 5$.
   (l) $4^{n+4} > (n + 4)^4$.
   (m) If a set A has n elements, then $\mathcal{P}(A)$ has 2^n elements.
   (n) $10^n + 3 \cdot 4^{n+2} + 5$ is divisible by 9.
   (o) $f_1 + f_2 + \cdots + f_n = f_{n+2} - 1$ where f_i is the ith Fibonacci number. [See exercise 3(b).]

5. (a) Use the PCI to prove the PMI.
   (b) Use the PCI to prove the WOP.

6. ★ (a) Use the WOP to prove the PMI.
   (b) Use the WOP to prove the PCI.

7. In this exercise, as in some examples, you are to prove some well-known facts about numbers as a way of demonstrating use of the WOP. Use the WOP to prove the following:
   (a) If $a > 0$, then for every natural number n, $a^n > 0$.
   (b) For all positive integers a and b, $b \neq a + b$. (*Hint:* Suppose for some a there is b such that $b = a + b$. By the WOP, there is a smallest a_0 such that, for some b, $b = a_0 + b$. Now apply the WOP again.)

8. Use the PCI to prove that the nth Fibonacci number is an integer. [See exercise 3(b).]

9. Prove Theorem 2.11 by induction on the natural number a.

10. **The Division Algorithm for Integers:** If a and b are integers, then there exist integers q and r such that $a = bq + r$, where $0 \leqslant r < |b|$.
    (a) Find q and r when $a = 7, b = 2$; $a = -6, b = -2$; $a = -5, b = 3$; $a = -5, b = -3$; $a = 1, b = -4$; $a = 0, b = 3$.
    (b) Use the WOP to prove the Division Algorithm for Integers.

11. A puzzle called the Towers of Hanoi consists of a board with 3 pegs and several disks of differing diameters that fit over the pegs. In the starting position all the disks are placed on one peg, with the largest at bottom, and the others with smaller and smaller diameters up to the top disk (figure 2.11). A move is made by lifting the top disk off a peg and placing it on another peg so that there is no smaller disk beneath it. The object of the puzzle is to transfer all the disks from one peg to another.

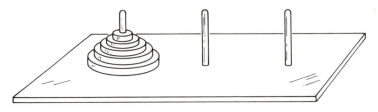

Figure 2.11

With a little practice, perhaps using coins of various sizes, you should convince yourself that if there are 3 disks, the puzzle can be solved in 7 moves. With 4 disks, 15 moves are required. Use the PMI to prove that with n disks, the puzzle can be solved in $2^n - 1$ moves.

12. In a certain kind of tournament, every player plays every other player exactly once and either wins or loses. There are no ties. Define a *top* player to be a player who, for every other player x, either beats x or beats a player y who beats x.
    (a) Show that there can be more than one top player.
    ☆ (b) Use the WOP to show that in every such tournament with n players ($n \in \mathbf{N}$), there is at least one top player.
    (c) Use the PMI to show that every n-player tournament has a top player.

13. **Proofs to Grade.**
    ★ (a) **Claim.** For all $n \in \mathbf{N}$, in every set of n horses, all horses have the same color.
        "Proof." Clearly in every set containing exactly 1 horse, all horses have the same color. Now suppose all horses in every set of n horses have the same color. Consider a set of $n + 1$ horses. If we remove one horse, the horses in the remaining set of n horses all have the same color. Now consider a set of n horses obtained by removing some other horse. All horses in this set have the same color. Therefore, all horses in the set of $n + 1$ horses have the same color. By the PMI, the statement is true for every $n \in \mathbf{N}$. ∎

    (b) **Claim.** The WOP implies the PCI.
        "Proof." By exercise 5(a), the PCI implies the PMI. By exercise 6(a), the WOP implies the PMI. Therefore, the WOP implies the PCI. ∎

    (c) **Claim.** For every natural number n, $n^2 + n$ is odd.
        "Proof." The number $n = 1$ is odd. Suppose $n \in \mathbf{N}$ and $n^2 + n$ is odd. Then

        $$(n + 1)^2 + (n + 1) = n^2 + 2n + 1 + n + 1$$
        $$= (n^2 + n) + (2n + 2)$$

        is the sum of an odd and an even number. Therefore $(n + 1)^2 + (n + 1)$ is odd. By the PMI, the property that $n^2 + n$ is odd is true for all natural numbers n. ∎

    ★ (d) **Claim.** Every natural number n greater than 1 has a prime factor.
        "Proof." Suppose that, whenever $1 < m < n$, m has a prime factor. If n is prime, then n is a prime factor of n. If n is composite, then $n = rs$ where r and s are natural numbers less than n. By the hypothesis of induction, r

has a prime factor p. Since p divides r and r divides n, n has a prime factor. In either case n has a prime factor. By the PCI, every natural number greater than 1 has a prime factor.   ■

2.5

Principles of Counting

This section introduces some basic techniques for counting the elements of finite sets and examines several applications. Our approach is somewhat in-formal, since the underlying concepts will be studied very carefully in Chap-ter 5. We say that the set A is **finite** iff it has n elements for some $n \in \mathbf{N}$. The number of elements in A is denoted either $\#(A)$ or, more briefly, $\#A$. For example,

$\# \{a, b\} = 2$.
$\# \{5, 7, 2, 7\} = 3$.
$\# (\varnothing) = 0$.

The first counting principle we accept for now as being completely obvious.

Theorem 2.15 *(The Sum Rule)*   If A and B are disjoint fi-nite sets with $\#A = m$ and $\#B = n$, then $\#(A \cup B) = m + n$.

The Sum Rule is more powerful than it looks. It is proved rigorously (Theorem 5.6) as part of the careful study in Chapter 5. It can also be ex-tended to the union of finitely many finite sets. We use the sigma notation for summation; $\sum_{i=1}^{k} a_i$ is an abbreviation for $a_1 + a_2 + \cdots + a_n$.

Theorem 2.16   If $\mathcal{A} = \{A_1, A_2, \ldots, A_k\}$ is a family of pair-wise disjoint sets and $\#A_i = a_i$ for $1 \le i \le k$, then

$$\# \bigcup_{i=1}^{k} A_i = \sum_{i=1}^{k} a_i.$$

Proof.   The proof is by induction on the number k of sets in the family $\mathcal{A}$. If $k = 1$, then $\cup \mathcal{A} = A_1$, so $\#(\cup \mathcal{A}) = a_1$.

Suppose now that the statement is true for some natural number k. Let $\mathcal{A}$ be a family $\{A_1, A_2, \ldots, A_k, A_{k+1}\}$ of $k + 1$ sets.

Then $\#(\cup\mathcal{A}) = \#(\overset{k}{\underset{i=1}{\cup}} A_i \cup A_{k+1})$ ⟨*by exercise 15 of section*

2.3⟩. But $\overset{k}{\underset{i=1}{\cup}} A_i$ and A_{k+1} are disjoint so this number, by the Sum

Rule, is $\#(\overset{k}{\underset{i=1}{\cup}} A_i) + \#A_k$. The family $\{A_i : i = 1, \ldots, k\}$ has k

sets so by the hypothesis of induction this number is

$$\sum_{i=1}^{k} a_i + \#A_{k+1} = \sum_{i=1}^{k} a_i + a_{k+1} = \sum_{i=1}^{k+1} a_i.$$

By the PMI, the statement is true for every family of k sets, for all
$k \in \mathbf{N}$. ∎

If A and B are not disjoint (see figure 2.12), then $\#A + \#B$ overcounts
$\#(A \cup B)$ by counting twice each element of $A \cap B$. Theorem 2.17 states that
this overcounting can be corrected by subtracting $\#(A \cap B)$. This theorem
will also be proved from a more formal point of view in Chapter 5.

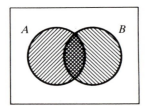

Figure 2.12

Theorem 2.17   For finite sets A and B,

$$\#(A \cup B) = \#A + \#B - \#(A \cap B).$$

Example.   During one week a total of 46 patients were treated
by Dr. Medical for either a broken leg or a sore throat. Of these,
32 had a broken leg and 20 had a sore throat. How many did she
treat for both ailments? Letting B be the set of patients with broken legs and S the set of patients with sore throats, the solution is

$$\#(B \cap S) = \#B + \#S - \#(B \cup S) = 32 + 20 - 46 = 6.$$

As is easily seen from the Venn diagram in figure 2.13, we could as well
solve for the number of patients with a broken leg but no sore throat, or with
a sore throat but no broken leg.

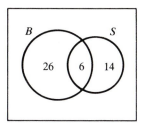

Figure 2.13

Example.   It is standard practice in manufacturing quality control to inspect a random sample of items produced to estimate the percentage of defective products. A manufacturer of plastic cups tested 100 items for two types of defects. It was found that 8 had defect A, 10 had defect B, and 3 had both defects. How many inspected cups had neither defect? The solution is to find the number of cups with at least one defect and subtract this number from the total number of cups inspected. Thus the number of cups without defects is

$$100 - (8 + 10 - 3) = 100 - 15 = 85.$$

Theorem 2.17 can be extended to three or more sets by the **principle of inclusion and exclusion.** The idea is that in counting the number of elements in the union of several sets by counting the number of elements in each set, we have *included* too many elements more than once; so some need to be *excluded*, or subtracted, from the total. When more than two sets are involved, this first attempt at exclusion will subtract too many elements, so that some need to be added back or *included* again, and so forth. For three sets A, B, C, the principle of inclusion and exclusion states that

$$\#(A \cup B \cup C) = [\#A + \#B + \#C] - [\#(A \cap B) + \#(A \cap C) + \#(B \cap C)] + \#(A \cap B \cap C).$$

The principle of inclusion and exclusion is often applied to determine the number of elements not in any of several sets (see exercise 5).

The next principle is as simple to state as the Sum Rule yet has a vast range of applications.

Theorem 2.18   *(The Product Rule)*   If two independent tasks T_1 and T_2 are to be performed, and T_1 can be performed in m ways and T_2 in n ways, then the two tasks can be performed in mn ways.

Example. A two-symbol variable name in the programming language BASIC must be one letter of the alphabet followed by one decimal digit. Thus the number of two-symbol variable names in BASIC is 260, because there are 26 ways to choose a letter (task number one) and 10 ways to choose a decimal digit (task number two).

Example. If there are 7 children visiting a pet store with 31 puppies, and each child is allowed to select one or more puppies, then there are $7 \cdot 31 = 217$ ways ownership certificates might be filled out listing an owner's name and a dog's tag number.

The Product Rule, like the Sum Rule, can be extended by induction. The proof is exercise 7.

Theorem 2.19 If k independent tasks $T_1, T_2, \ldots, T_k$ are to be performed and the number of ways task T_i can be performed is n_i, then the number of ways the tasks can be performed in sequence is $\prod\limits_{i=1}^{k} n_i = n_1 n_2 \ldots n_k$.

Example. If in the pet store example each child, from youngest to oldest, is asked to point out his or her choice for the cutest puppy, the number of different ways this could be accomplished is 31^7. On the other hand, if each child in turn is allowed to select and keep exactly one puppy, the total number of ways this could possibly be done is $31 \cdot 30 \cdot 29 \cdot 28 \cdot 27 \cdot 26 \cdot 25$. The factors decrease because the tasks of choosing puppies to keep are not independent; no child may select a puppy already taken.

Example. If the set A has n elements, then forming a subset of A amounts to carrying out n independent tasks, where each task is to decide whether or not to place the element in the subset. Since each task has two outcomes there are 2^n ways this process can be carried out, so $\mathscr{P}(A)$ has 2^n elements. This argument is simply a restatement of Theorem 2.4.

Example. Suppose that a standard license plate has three letters followed by three digits. The letters I, O and Q are not used on standard plates. The number of possible standard plates is $12,167,000 = 23 \cdot 23 \cdot 23 \cdot 10 \cdot 10 \cdot 10$.

Personalized plates may have from 1 to 7 letters or digits, and there may be blank spaces between symbols. To determine the number of possible personalized plates (we include in our count the prohibited plates, such as POLICE or those containing obscenities), we must determine the number of plates with each number of symbols. By the product rule the number of plates with one symbol is then 36, the number of plates with two symbols is 36^2, the number of plates with 3 symbols is $36^2 \cdot 37$ (because the middle symbol could be a blank space), and so forth. By the sum rule, the total number of possible personalized plates is

$$36 + 36^2 + 36^2 \cdot 37 + \cdots + 36^2 37^5 = 92{,}366{,}150{,}744.$$

Definition. A **permutation** of a set with n elements is an arrangement of the elements of the set in some order.

Since there are n ways to choose the first element, $n - 1$ ways to choose the second, and so forth, the number of permutations of a set with n elements is $n! = n(n - 1)(n - 2) \cdots 3 \cdot 2 \cdot 1$. For example, the number of different possible orders in which a radio disk jockey can play the tapes of 10 golden oldies is $10! = 3{,}628{,}000$.

Example. Four rodeo contestants entered a bull-riding contest in which the first rider to last 30 seconds would get the $1000 prize. List all possible orders in which the riders A, B, C, and D might be arranged. There are $4! = 24$ such permutations.

ABCD	BACD	CABD	DABC
ABDC	BADC	CADB	DACB
ACBD	BCAD	CBAD	DBAC
ACDB	BCDA	CBDA	DBCA
ADBC	BDAC	CDAB	DCAB
ADCB	BDCA	CDBA	DCBA

Definition. An r-element subset of objects selected from an n-element set is called a **combination of n elements taken r at a time.** The number of ways such a selection can be made is the number of subsets having r elements that are included in a set of size n. This number is called a **binomial coefficient** and is denoted by $\binom{n}{r}$. The symbol $\binom{n}{r}$ is read "n choose r" or "n binomial r."

There is only one 0-element subset of $\varnothing$, so we take $\begin{pmatrix} 0 \\ 0 \end{pmatrix}$ to be 1.

Example.   The four subsets of $\{a, b, c, d\}$ having three elements are $\{a, b, c\}$, $\{a, b, d\}$, $\{b, c, d\}$, so $\begin{pmatrix} 4 \\ 3 \end{pmatrix} = 4$. The set $\{a, b, c, d\}$ has 6 subsets with two elements, so $\begin{pmatrix} 4 \\ 2 \end{pmatrix} = 6$. The 6 subsets are $\{a, b\}$, $\{a, c\}$, $\{a, d\}$, $\{b, c\}$, $\{b, d\}$, and $\{c, d\}$.

Theorem 2.20   Let $n, r \in \mathbf{Z}$ and $0 \leqslant r \leqslant n$. The number of r-element subsets of an n-element set is $\begin{pmatrix} n \\ r \end{pmatrix} = \dfrac{n!}{r!(n - r)!}$

Proof.   The elements of an n-element set can be arranged in $n!$ ways. The n elements may also be arranged as follows. We select r objects, which can be done in $\begin{pmatrix} n \\ r \end{pmatrix}$ ways; then order these selected elements, which can be done in $r!$ ways; and finally order the remaining $n - r$ elements, which can be done in $(n - r)!$ ways. Thus the number of ways to order n objects is $n! = \begin{pmatrix} n \\ r \end{pmatrix} \cdot r! \cdot (n - r)!$, so

$$\begin{pmatrix} n \\ r \end{pmatrix} = \frac{n!}{r!(n - r)!} \quad \blacksquare$$

For example, in a company with 15 employees, the number of ways 5 employees could be selected for bonus pay is

$$\begin{pmatrix} 15 \\ 5 \end{pmatrix} = \frac{15!}{5!10!} = \frac{15 \cdot 14 \cdot 13 \cdot 12 \cdot 11}{5 \cdot 4 \cdot 3 \cdot 2 \cdot 1} = 3003.$$

We assume all 5 employees are to be given the same bonus so there is no ordering in this situation; think of the selection of the 5 employees as being made simultaneously. If the 5 employees were to be given 5 different bonus amounts, we would use the extended product rule to conclude that 5 elements chosen from a set of 15 elements can be ordered in

$$15 \cdot 14 \cdot 13 \cdot 12 \cdot 11 = 360{,}360$$

ways. We might also think of giving the different bonus amounts by arranging 5 selected employees in order. The number of ways to give five different bonuses is then the number of combinations times the number of permutations within each combination, or $3003 \cdot 5!$ which is again 360,360.

Note that selecting a set of 5 employees to receive a bonus amounts to selecting a set of 10 employees who will *not* receive a bonus, so that $\binom{15}{10}$ should be equal to $\binom{15}{5}$. The fact that this statement is true in general can be seen from the formula for $\binom{n}{r}$.

Corollary 2.21   For $n, r \in \mathbf{Z}$ with $0 \leqslant r \leqslant n$,

$$\binom{n}{r} = \binom{n}{n-r}.$$

Example.   In a group of 20 automobiles, 7 fail a safety inspection. How many ways can 5 cars be selected from the 20 so that a) all fail the safety inspection? b) two fail the inspection?

a)   If all 5 fail the inspection, then they must all be selected from the 7 unsafe cars. This can be done in $\binom{7}{5} = 21$ ways.

b)   The two that fail the inspection can be selected from the 7 unsafe vehicles in $\binom{7}{2} = 21$ ways. The remaining 3 of the set of 5 must be selected from the 13 approved cars. This can be done in $\binom{13}{3} = 286$ ways. Thus the number of ways to select 5 cars among which two fail the inspection is $21 \cdot 286 = 6006$.

Several relationships exist among binomial coefficients (see exercise 15). Some of these relationships can be proved either algebraically, from the formulas, or combinatorially. A combinatorial proof is one that is based on the meaning of the binomial coefficients.

Theorem 2.22   For $n, r \in \mathbf{N}$ with $1 \leqslant r \leqslant n$,

$$\binom{n}{r} = \binom{n-1}{r} + \binom{n-1}{r-1}.$$

Proof.   ⟨*We leave the algebraic proof of this theorem to exercise 15 and give a combinatorial argument.*⟩ Choose one particular element x from a set A of n elements. The number of subsets of A

with r elements is $\binom{n}{r}$. The collection of r-element subsets may be divided into two disjoint collections: those subsets containing x and those subsets not containing x. We count the number of subsets in each collection and add the results. First, there are $\binom{n-1}{r}$ r-element subsets of A that do not contain x, since each is a subset of $A - \{x\}$. Second, there are $\binom{n-1}{r-1}$ r-element subsets of A that do contain x, because each of these corresponds to the $(r-1)$-element subset of $A - \{x\}$ obtained by removing x from the subset. Thus the sum of the number of subsets in the two collections is

$$\binom{n-1}{r} + \binom{n-1}{r-1} = \binom{n}{r}. \quad \blacksquare$$

Our next result explains why $\binom{n}{r}$ is called a binomial coefficient.

Theorem 2.23 *(Binomial Expansion)* For $n \geqslant 0$,

$$(a + b)^n = \sum_{r=0}^{n} \binom{n}{r} a^r b^{n-r}.$$

Proof. Since $(a + b)^n = \underbrace{(a + b)(a + b) \cdots (a + b)}_{n \text{ times}}$, each term of the product contains one term from each of the n factors $(a + b)$. Since each has a total of n a's and b's, the terms of the product are of the form $a^r b^{n-r}$. The coefficient of $a^r b^{n-r}$ is the number of times this term is obtained in multiplying out. Since $a^n b^{n-r}$ is obtained exactly by choosing the term a from r of the factors $(a + b)$, the coefficient is $\binom{n}{r}$. $\blacksquare$

Perhaps you are familiar with Pascal's triangle, named for the great seventeenth-century mathematician and scientist, Blaise Pascal (1623–1662). The triangle, shown in Figure 2.14, is a simple means for computing the binomial coefficients and was actually used hundreds of years before Pascal studied it.

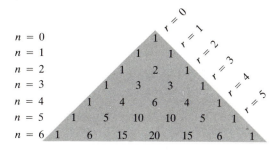

Figure 2.14

For example, we read from the row for $n = 4$ that

$$(a + b)^4 = \underline{1}a^4 + \underline{4}a^3b + \underline{6}a^2b^2 + \underline{4}ab^3 + \underline{1}b^4.$$

The triangle is constructed by beginning with the configuration

$$1$$
$$1 \qquad 1$$

and constructing a new row by putting 1's on the far left and right. All other numbers in each row are found by adding the two numbers directly to the left and right in the preceding row. Thus the first 10 in row 5 is the sum of 4 and 6 from row 4. The triangle has the property that the entry in the nth row and rth "column" has the value $\binom{n}{r}$. Thus the triangle exhibits the property of binomial coefficients in Theorem 2.22.

Exercises 2.5

1. Find
   (a) $\#\{n \in \mathbf{Z}: n^2 < 41\}$
   (b) $\#\{2, 6, 2, 6, 2\}$
   (c) $\#\{x \in \mathbf{R}: x^2 = -1\}$
   (d) $\#\{n \in \mathbf{N}: n + 1 = 4n - 10\}$

2. Suppose $\#A = 24$, $\#B = 21$, $\#(A \cup B) = 37$, $\#(A \cap C) = 11$, $\#(B - C) = 10$, and $\#(C - B) = 12$. Find
   (a) $\#(A \cap B)$ ★ (b) $\#(A - B)$
   (c) $\#(B - A)$ (d) $\#(B \cup C)$
   (e) $\#C$ (f) $\#(A \cup C)$

★ 3. How many natural numbers less than one million are either squares or cubes of natural numbers?

4. Eight male and five female astronauts are in training for a Mars landing mission.
   (a) If exactly one astronaut will fly the first mission, what basic counting rule may be used to compute the number of ways the selection may be made?
   (b) If one male and one female are to fly the mission, what counting rule may be used to compute the number of ways the selection may be made?

5. Among the 40 first-time campers at Camp Forlorn one week, 14 fell into the lake during the week, 13 suffered from poison ivy, and 16 got lost trying to find the dining hall. Three of these campers had poison ivy rash and fell into the lake, five fell into the lake and got lost, 8 had poison ivy and got lost, and 2 experienced all three misfortunes. How many first-time campers got through the week without any of these mishaps?

6. (a) If you have 10 left shoes and 9 right shoes and don't care whether they match, how many "pairs" of shoes can you select?
   (b) A cafeteria has 3 meat selections, 2 vegetable selections and 4 dessert selections for a given meal. If a meal consists of one meat, one vegetable, and one dessert, how many different meals could be constructed?
   (c) There are 3 roads from Abbottville to Bakerstown, 4 roads from Bakerstown to Cadez, and 5 roads from Cadez to Detour Village. How many different routes are there from Abbottville through Bakerstown and then Cadez to Detour City?

7. Prove Theorem 2.19 by induction on the number of tasks.

8. Find the number of ways 7 school children can line up to board a school bus.

9. Suppose that the seven children of Exercise 8 are three girls and four boys. Find the number of ways they could line up subject to these conditions.
   ★ (a) The three girls are first in line.
   (b) The three girls are together in line.
   (c) The four boys are together in line.
   (d) No two boys are together.

10. The number of 4-digit numbers that can be formed using exactly the digits 1, 3, 3, 7 is less than 4!, because the two 3's are indistinguishable. Prove that the number of permutations of n objects, m of which are alike, is $n!/m!$. Generalize to the case when m_1 are alike, m_2 others are alike, and so forth.

11. Among ten lottery finalists, four will be selected to win individual amounts of $1000, $2000, $5000, and $10,000. In how many ways may the money be distributed?

12. From a second grade class of 11 boys and 8 girls, 3 are selected for flag duty.
    (a) How many selections are possible?
    (b) How many of these selections have exactly two boys?
    (c) Exactly one boy?

13. Among 14 radios, 5 have broken dials. If 4 radios are selected, how many such selections are possible in which 2 radios have broken dials?

14. Find
    (a) $(a + b)^6$ (b) $(a + 2b)^4$
    (c) the coefficient of a^3b^{10} in the expansion of $(a + b)^{13}$
    (d) the coefficient of a^2b^{10} in the expansion of $(a + 2b)^{12}$

15. (a) Prove algebraically from the formulas that for $n, r \in \mathbf{Z}$ with $1 \leqslant r \leqslant n$,

$$\binom{n}{r} = \binom{n-1}{r} + \binom{n-1}{r-1}.$$

(b) Prove combinatorially that if n is odd, then the number of ways to select an even number of objects from n is equal to the number of ways to select an odd number of objects.

(c) Using the fact that an n-element set has 2^n subsets, give a combinatorial proof that for $n \geqslant 0$, $\sum_{r=0}^{n} \binom{n}{r} = 2^n$.

(d) Using the fact that $\binom{n+1}{r} = \binom{n}{r} + \binom{n}{r-1}$ from part (a), prove by induction that for $n \geqslant 0$, $\sum_{r=0}^{n} \binom{n}{r} = 2^n$.

☆ (e) Give a combinatorial proof of Vandermonde's identity: for $n, m, r \in \mathbf{Z}$ with $0 \leqslant r \leqslant n + m$,

$$\binom{n+m}{r} = \binom{n}{0}\binom{m}{r} + \binom{n}{1}\binom{m}{r-1}$$
$$+ \binom{n}{2}\binom{m}{r-2} + \cdots + \binom{n}{r}\binom{m}{0}.$$

(f) Prove that $\binom{2n}{n} + \binom{2n}{n+1} = \frac{1}{2}\binom{2n+2}{n+1}$

16. **Proofs to Grade**

(a) **Claim.**   For $n \geqslant 0$, $\sum_{r=0}^{n} \binom{n}{r} = 2^n$.

   "Proof."   Since $2^n = (1 + 1)^n$, we have by the binomial expansion that
$$2^n = \sum_{r=0}^{n} \binom{n}{r} 1^r 1^{n-r} = \sum_{r=0}^{n} \binom{n}{r}.$$

(b) **Claim.**   For $n \geqslant 0$,

$$\binom{n}{0} - \binom{n}{1} + \binom{n}{2} - \cdots + (-1)^k \binom{n}{k} + \cdots + (-1)^n \binom{n}{n} = 0.$$

   "Proof." $0 = (-1 + 1)^n = \sum_{k=0}^{n} \binom{n}{k}(-1)^k(1)^{n-k} =$
$$\binom{n}{0} - \binom{n}{1} + \binom{n}{2} - \cdots + (-1)^k \binom{n}{k} + \cdots + (-1)^n \binom{n}{n}.$$

(c) **Claim.**   For $n \geqslant 0$, the number of ways to select an even number of objects from n is equal to the number of ways to select an odd number.
   "Proof."   From part (b) of this exercise ⟨*the claim made there is correct*⟩ we have that

$$\binom{n}{0} + \binom{n}{2} + \cdots = \binom{n}{1} + \binom{n}{3} + \cdots$$

The left side of this equality gives the number of ways to select an even number of objects from n and the right side is the number of ways to select an odd number.

3

Relations

Given a set of objects, we may want to say that certain objects are related in some way. For example, we may say that two people are related if they have the same citizenship or the same blood type, or if they like the same kinds of food. If a and b are integers, we might say that a is related to b when a divides b. In this chapter we will study the idea of "is related to" by making precise the notion of a relation and then concentrating on certain relations called equivalence relations.

3.1

Cartesian Products and Relations

The study of relations begins with the concept of an **ordered pair,** symbolized as (a, b), which has the property that if either of the **coordinates** a or b is changed, the ordered pair changes. Thus two ordered pairs (a, b) and (x, y) are *equal* iff $a = x$ and $b = y$.

The adjective "ordered" is used to describe (a, b) because, for example, $(3, 7) \neq (7, 3)$ although the set $\{3, 7\} = \{7, 3\}$. Even though the same symbol $(3, 7)$ could be used to represent the open interval—i.e., the set of all real numbers strictly between 3 and 7—it will always be clear from the context whether the interval or the pair is meant. A more rigorous definition of an ordered pair as a set is given as exercise 11.

Generalizing, we can say that the **ordered *n*-tuples** $(a_1, a_2, \ldots, a_n)$ and $(x_1, x_2, \ldots, x_n)$ are equal iff $a_i = x_i$ for $i = 1, 2, \ldots, n$. Thus the 5-tuples $(3, 7, 1, 3, 6)$, $(3, 7, 1, 3, 8)$, and $(7, 3, 1, 3, 6)$ are all different. An ordered 2-tuple is just an ordered pair; an ordered 3-tuple is usually called an **ordered triple.**

Definition.   Let A and B be sets. The set of all ordered pairs having first coordinate in A and second coordinate in B is called the **Cartesian product** (or **cross product**) of A and B and is written $A \times B$. Thus

$$A \times B = \{(a, b): a \in A \text{ and } b \in B\}.$$

If $A = \{1, 2\}$ and $B = \{2, 3, 4\}$, then

$$A \times B = \{(1, 2), (1, 3), (1, 4), (2, 2), (2, 3), (2, 4)\}$$

whereas

$$B \times A = \{(2, 1), (2, 2), (3, 1), (3, 2), (4, 1) \ (4, 2)\}$$

which in this example is not equal to $A \times B$.

If $A = \{a, \{x\}, 2\}$ and $B = \{\{a\}, 1\}$, then

$$A \times B = \{(a, \{a\}), (a, 1), (\{x\}, \{a\}), (\{x\}, 1), (2, \{a\}), (2, 1)\}.$$

The Cartesian product of three or more sets is defined similarly. For example, if A, B, and C are sets, then $A \times B \times C = \{(a, b, c): a \in A \text{ and } b \in B \text{ and } c \in C\}$. The sets $A \times B \times C$ and $A \times (B \times C)$ are different. The first is a set of ordered triples but the second is a set of ordered pairs. In practice this distinction is often glossed over.

By the Product Rule (Theorem 2.18), if A has m elements and B has n elements, then $A \times B$ has mn elements (see exercise 14) and $A \times A$ has m^2 elements. In addition, if C has k elements, then both $A \times B \times C$ and $(A \times B) \times C$ have mnk elements.

Some useful relationships between the Cartesian product of sets and the other set operations are gathered in the next theorem.

Theorem 3.1   If A, B, C, and D are sets, then

(a) $A \times (B \cup C) = (A \times B) \cup (A \times C).$
(b) $A \times (B \cap C) = (A \times B) \cap (A \times C).$
(c) $A \times \varnothing = \varnothing.$
(d) $(A \times B) \cap (C \times D) = (A \cap C) \times (B \cap D).$
(e) $(A \times B) \cup (C \times D) \subseteq (A \cup C) \times (B \cup D).$

Proof.

(a) ⟨*Since both $A \times (B \cup C)$ and $(A \times B) \cup (A \times C)$ are sets of ordered pairs, objects of the form (x, y) will be used to show each set is a subset of the other. We use an "iff-argument."*⟩
The ordered pair $(x, y) \in A \times (B \cup C)$
iff $x \in A$ and $y \in B \cup C$
iff $x \in A$ and $(y \in B$ or $y \in C)$
iff $(x \in A$ and $y \in B)$ or $(x \in A$ and $y \in C)$
iff $(x, y) \in A \times B$ or $(x, y) \in A \times C$
iff $(x, y) \in (A \times B) \cup (A \times C)$.
Therefore $A \times (B \cup C) = (A \times B) \cup (A \times C)$.

(e) If $(x, y) \in (A \times B) \cup (C \times D)$, then $(x, y) \in A \times B$ or $(x, y) \in C \times D$. If $(x, y) \in A \times B$, then $x \in A$ and $y \in B$. Thus $x \in A \cup C$ and $y \in B \cup D$. ⟨$A \subseteq A \cup C$, $B \subseteq B \cup D$.⟩ Therefore $(x, y) \in (A \cup C) \times (B \cup D)$. If $(x, y) \in C \times D$, a similar argument shows $(x, y) \in (A \cup C) \times (B \cup D)$. This shows that $(A \times B) \cup (C \times D) \subseteq (A \cup C) \times (B \cup D)$.

Parts (b), (c), and (d) are given as exercise 2.   ∎

Definition.   Let A and B be sets. A **relation from A to B** is a subset of $A \times B$. Subsets of $A \times A$ are called **relations on A**. If R is a relation from A to B and sets A and B are understood, we simply say that R is a relation.

If $A = \{1, 2, 3\}$ and $B = \{a, 5, \{b\}, c\}$, then $A \times B$ and the set $\{(1, a), (2, \{b\}), (2, c)\}$ are relations from A to B. On the other hand, $\{(a, 1), (\{b\}, 2), (c, 2)\}$ and $\{(5, 1), (5, 2)\}$, are relations from B to A.

Since $\varnothing \subseteq A \times B$ for any sets A and B, $\varnothing$ is a relation from A to B. Some care must be taken in mathematical arguments not to overlook this relation.

Since every subset of $A \times B$ is a relation from A to B, there are many different relations. In fact, if A has m elements and B has n elements then $A \times B$ has mn elements; so there are 2^{mn} relations from A to B.

If R is a relation and if $(a, b) \in R$, we write $a \, R \, b$, and read this as "*a is R-related to b.*" Likewise $a \, \cancel{R} \, b$ means $(a, b) \notin R$.

Let P be the set of all people. Let $L = \{(a, b) \in P \times P: a$ has the same last name as $b\}$. Then L is a relation on P. We observe that Sally Brown L Charlie Brown while Buddy Holly $\cancel{L}$ Clyde McPhatter.

Let LTE $= \{(x, y) \in \mathbf{R} \times \mathbf{R}: x \leq y\}$. Then $(2, 5) \in$ LTE, so we write 2 LTE 5. Indeed, $(x, y) \in$ LTE iff $x \leq y$, so LTE is the "less than or equal to" relation on $\mathbf{R}$. Thus 2 LTE 5 is consistent with the notation $2 \leq 5$.

Consider the relation S on the set $\mathbf{N} \times \mathbf{N}$ given by $(m, n)\ S\ (k, j)$ iff $m + n = k + j$. Then $(3, 17)\ S\ (12, 8)$ but $(5, 4)$ is not S-related to $(6, 15)$. Notice that S is a relation from $\mathbf{N} \times \mathbf{N}$ to $\mathbf{N} \times \mathbf{N}$ and that the description above is somewhat simpler than defining S by writing

$$S = \{((m, n), (k, j)) \in (\mathbf{N} \times \mathbf{N}) \times (\mathbf{N} \times \mathbf{N}): m + n = k + j\},$$

Definition. The **domain** of the relation R from A to B is the set

$$\mathrm{Dom}(R) = \{x \in A: \text{there exists } y \in B \text{ such that } x\ R\ y\}.$$

The **range** of the relation R is the set

$$\mathrm{Rng}(R) = \{y \in B: \text{there exists } x \in A \text{ such that } x\ R\ y\}.$$

Thus the domain of R is the *set of all first coordinates of ordered pairs* in R, and $\mathrm{Rng}(R)$ is the *set of all second coordinates*. By definition, $\mathrm{Dom}(R) \subseteq A$ and $\mathrm{Rng}(R) \subseteq B$.

For $A = \{s, p, q, \{r\}\}$ and $B = \{a, b, c, d\}$, let R be the relation $\{(p, a), (q, b), (q, c), (\{r\}, b)\}$. Then $\mathrm{Dom}(R) = \{p, q, \{r\}\}$ and $\mathrm{Rng}(R) = \{a, b, c\}$.

Every set of ordered pairs is a relation. If M is any set of ordered pairs, then M is a relation from A to B, where A and B are any sets for which $\mathrm{Dom}(M) \subseteq A$ and $\mathrm{Rng}(M) \subseteq B$.

You are already familiar with the idea of a **graph** of a relation R as a pictorial or geometric representation of the ordered pairs in R. We use the **Cartesian** (or rectangular) **coordinate system** in the examples below.

Example. The graph of the relation $R = \{(p, a), (q, b), (q, c), (\{r\}, b)\}$ is shown in figure 3.1.

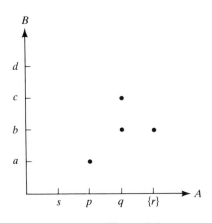

Figure 3.1

Example.   Let $S = \{(x, y) \in \mathbf{R} \times \mathbf{R}: x^2/324 + y^2/64 \leq 1\}$. The graph of S is given in figure 3.2.

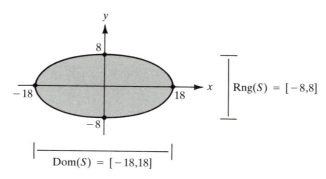

Figure 3.2

Let A be any set. The set $I_A = \{(x, x): x \in A\}$ is called the **identity relation** on A. For $A = \{1, 2\}$, $I_A = \{(1, 1), (2, 2)\}$. Obviously, $\text{Dom}(I_A) = \text{Rng}(I_A) = A$.

Example.   For the identity relation on $[-2, \infty)$, we can picture only a portion of the relation (figure 3.3).

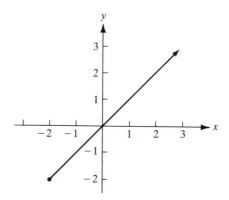

Figure 3.3

Let $A = [2, 4]$ and $B = (1, 3) \cup \{4\}$. Let S be the relation on the reals given by $x \, S \, y$ iff $x \in A$. Let T be the relation given by $x \, T \, y$ iff $y \in B$. The graphs of S and T are shown in figures 3.4(a) and (b). Note that $S = A \times \mathbf{R}$ and $T = \mathbf{R} \times B$. Figure 3.4(c) shows the graph of $S \cap T = (A \times \mathbf{R}) \cap (\mathbf{R} \times B) = (A \cap \mathbf{R}) \times (B \cap \mathbf{R}) = A \times B$. (Note the use of Theorem 3.1(d) in this computation.) The graph of $S \cup T$ is given in figure 3.4(d).

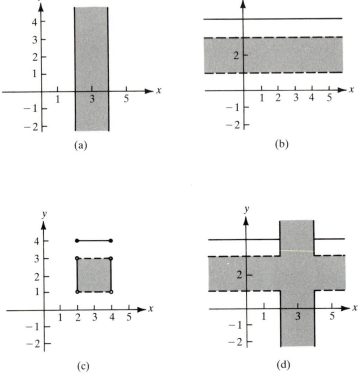

Figure 3.4

Another important kind of graph may be used to represent a relation. To form the **directed graph** or **digraph** representing the relation R on a set A, we think of the objects in A as vertices and think of R as telling us which vertices are connected by edges. The edges are directed from one vertex to another, like arrows. There is an edge from x to y exactly when $(x, y) \in R$.

Example.   The digraph corresponding to the relation $S = \{(6, 12), (2, 6), (2, 12), (6, 6), (12, 2)\}$ on the set $V = \{2, 5, 6, 12\}$ is shown in figure 3.5. Figure 3.6 shows that the digraph for the relation "divides" on the set $\{3, 6, 9, 12\}$.

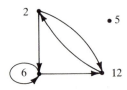

Figure 3.5

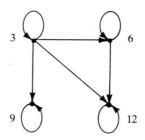

Figure 3.6

It is easy to construct the digraph corresponding to a relation, and vice versa. Consider the digraph in figure 3.7 showing required courses for a mathematics major at Someplace College. Here the arrowheads are omitted, but an edge drawn down from one vertex (course) to another means that the course closer to the top of the graph is a prerequisite for the other. As a further simplification, we have not drawn an edge when there is a path, or sequence of edges, between the courses. For example, the edge from MATH 121 to MATH 350 is not shown, although the graph shows that MATH 121 is a prerequisite for MATH 350.

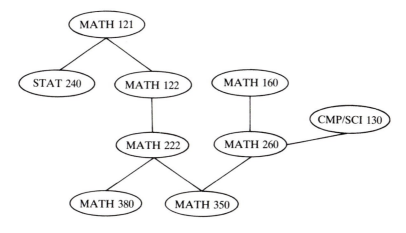

Figure 3.7

The relation P corresponding to this digraph contains, for example, (MATH 222, MATH 350), and (MATH 122, MATH 380), but not (STAT 240, MATH 380).

One reason for the importance of relations is their central role in the database theory, which is the study of means of storing, manipulating, and retrieving large collections of information in computing machinery. In a relational database, relations among domains of data are represented in tables. For example, a portion of the table PREREQ might look like this:

COURSE	COURSE
MATH 121	MATH 222
MATH 122	MATH 380
MATH 122	MATH 350
MATH 160	MATH 260
CMP SCI 130	MATH 260
⋮	⋮

The table PREREQ represents a relation, with each row of the table being an ordered pair in the relation. In this example PREREQ contains pairs from the relation P of the digraph above, but it might also represent the relation of all prerequisite courses at Someplace College.

It will soon be clear that another reason for the importance of relations is that functions are defined (in chapter 4) to be relations with a special property. The remainder of this section is devoted to methods of constructing new relations from given relations. These ideas are important in the study of digraphs, but they are particularly important to us in the study of functions.

The two most fundamental methods for constructing new relations from given ones are **inversion** and **composition.** Inversion is a matter of switching the order of each pair in a relation.

Definition.   If R is a relation from A to B, then the **inverse** of R is

$$R^{-1} = \{(y, x): (x, y) \in R\}.$$

The inverse of the relation $\{(1, b), (1, c), (2, c)\}$ is $\{(b, 1), (c, 1), (c, 2)\}$. For any set A, $I_A^{-1} = I_A$. The inverse of LTE on **R** is the relation "greater than or equal to" since x LTE$^{-1} y$ iff y LTE x iff $y \leq x$ iff $x \geq y$.

The digraph of the inverse of a relation on a set differs from the digraph of the relation only in that the directions of the arrows are reversed. Figure 3.8 shows the digraphs of R and R^{-1}, where R is the relation $\subseteq$ on the set $\{\varnothing, \{1\}, \{3\}, \{1, 2\}\}$.

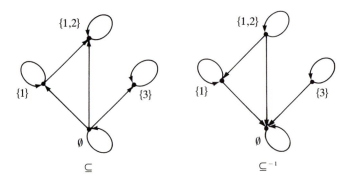

Figure 3.8

Theorem 3.2   Let R be a relation from A to B.

(a) R^{-1} is a relation from B to A.
(b) $\text{Dom}(R^{-1}) = \text{Rng}(R)$.
(c) $\text{Rng}(R^{-1}) = \text{Dom}(R)$.

Proof.

(a)   Suppose $(x, y) \in R^{-1}$. ⟨*We show* $(x, y) \in B \times A$.⟩ Then
       $(y, x) \in R$. Since R is a relation from A to B, $R \subseteq A \times B$.
       Thus $y \in A$ and $x \in B$. Therefore $(x, y) \in B \times A$. This
       proves $R^{-1} \subseteq B \times A$.
(b) $x \in \text{Dom}(R^{-1})$ iff there exists $y \in A$ such that
       $(x, y) \in R^{-1}$ iff there exists $y \in A$ such that $(y, x) \in R$ iff
       $x \in \text{Rng}(R)$.
(c)   This is similar to part (b).   ■

Given a relation from A to B and another from B to C, composition is
a method of constructing a relation from A to C.

Definition.   Let R be a relation from A to B, and let S be a relation
from B to C. The **composite** of R and S is $S \circ R =$

$$\{(a, c): \text{there exists } b \in B \text{ such that } (a, b) \in R \text{ and } (b, c) \in S\}.$$

Since $S \circ R \subseteq A \times C$, $S \circ R$ is a relation from A to C. It is always true
that $\text{Dom}(S \circ R) \subseteq \text{Dom}(R)$, but it is not always true that $\text{Dom}(S \circ R) =$
$\text{Dom}(R)$. We have adopted the right-to-left notation for $S \circ R$ used in analysis. To determine $S \circ R$, remember that the relation R is from the first set to
the second, and S is from the second to the third.
    Let $A = \{1, 2, 3, 4\}$, $B = \{p, q, r, s\}$, and $C = \{x, y, z\}$. Let $R =$
$\{(1, p), (1, q), (2, q), (3, r), (4, s)\}$ be a relation from A to B, and let $S =$
$\{(p, x), (q, x), (q, y), (s, y)\}$ be a relation from B to C. These relations are
illustrated in figure 3.9. An element a from A is related to an element c of C
under $S \circ R$ if there is at least one "intermediate" element b of B—intermediate in the sense that $(a, b) \in R$ and $(b, c) \in S$. For example, since

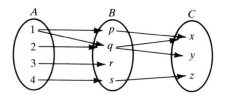

Figure 3.9

$(1, p) \in R$ and $(p, x) \in S$, then $(1, x) \in S \circ R$. By determining intermediates, which is the same idea as following arrows from A through B to C in figure 3.9, we have $S \circ R = \{(1, x), (1, y), (2, x), (2, y), (4, y)\}$.

If R is a relation from A to B, and S is a relation from B to A, then $R \circ S$ and $S \circ R$ are both defined but, owing to the asymmetry in the definition of composition, you should not expect that $R \circ S = S \circ R$. Even when R and S are relations on the same set, it may happen that $R \circ S \neq S \circ R$.

Example.   Let $R = \{(x, y) \in \mathbf{R} \times \mathbf{R}: y = x + 1\}$ and let $S = \{(x, y) \in \mathbf{R} \times \mathbf{R}: y = x^2\}$. Then

$$
\begin{aligned}
R \circ S &= \{(x, y): (x, z) \in S \text{ and } (z, y) \in R \text{ for some } z \in \mathbf{R}\} \\
&= \{(x, y): z = x^2 \text{ and } y = z + 1 \text{ for some } z \in \mathbf{R}\} \\
&= \{(x, y): y = x^2 + 1\}.
\end{aligned}
$$

$$
\begin{aligned}
S \circ R &= \{(x, y): (x, z) \in R \text{ and } (z, y) \in S \text{ for some } z \in \mathbf{R}\} \\
&= \{(x, y): z = x + 1 \text{ and } y = z^2 \text{ for some } z \in \mathbf{R}\} \\
&= \{(x, y): y = (x + 1)^2\}.
\end{aligned}
$$

Clearly, $S \circ R \neq R \circ S$, since $x^2 + 1$ is seldom equal to $(x + 1)^2$.

The last theorem of this section collects several results about inversion, composition, and the identity relation. We prove only part (b) and the first part of (c), leaving the rest as exercise 10.

Theorem 3.3   Suppose A, B, C, and D are sets. Let R be a relation from A to B, S be a relation from B to C, and T be a relation from C to D.

(a) $(R^{-1})^{-1} = R$.
(b) $T \circ (S \circ R) = (T \circ S) \circ R$; that is, composition is associative.
(c) $I_B \circ R = R$ and $R \circ I_A = R$.
(d) $(S \circ R)^{-1} = R^{-1} \circ S^{-1}$.

Proof.

(b) The pair $(x, w) \in T \circ (S \circ R)$ for some $x \in A$ and $w \in D$
  iff $(\exists z \in C)((x, z) \in S \circ R$ and $(z, w) \in T)$
  iff $(\exists z \in C)((\exists y \in B)((x, y) \in R$ and $(y, z) \in S)$ and $(z, w) \in T)$
  iff $(\exists z \in C)(\exists y \in B)((x, y) \in R$ and $(y, z) \in S$ and $(z, w) \in T)$
  iff $(\exists y \in B)(\exists z \in C)((x, y) \in R$ and $(y, z) \in S$ and $(z, w) \in T)$

iff $(\exists y \in B)((x, y) \in R$ and $(\exists z \in C)((y, z) \in S$ and
$(z, w) \in T))$
iff $(\exists y \in B)((x, y) \in R$ and $(y, w) \in T \circ S)$
iff $(x, w) \in (T \circ S) \circ R$.
Therefore $T \circ (S \circ R) = (T \circ S) \circ R$.

(c) ⟨*We first show that* $I_B \circ R \subseteq R$.⟩ Suppose $(x, y) \in I_B \circ R$.
Then there exists $z \in B$ such that $(x, z) \in R$ and
$(z, y) \in I_B$. Since $(z, y) \in I_B$, $z = y$. Thus $(x, y) \in R$
⟨*since* $(x, y) = (x, z) \in R$⟩. Conversely, suppose
$(p, q) \in R$. Then $(q, q) \in I_B$ and thus $(p, q) \in I_B \circ R$.
Thus $I_B \circ R = R$.   ∎

Exercises 3.1

1.   Let A and B be nonempty sets. Prove that $A \times B = B \times A$ iff $A = B$.

2.   Complete the proof of Theorem 3.1 by proving
★ (b) $A \times (B \cap C) = (A \times B) \cap (A \times C)$.
    (c) $A \times \varnothing = \varnothing$.
    (d) $(A \times B) \cap (C \times D) = (A \cap C) \times (B \cap D)$.

3.   Give an example of sets A, B, and C such that $(C \times C) - (A \times B) \neq (C - A) \times (C - B)$.

4.   Let T be the relation $\{(3, 1), (2, 3), (3, 5), (2, 2), (1, 6), (2, 6), (1, 2)\}$. Find
    (a) $\text{Dom}(T)$.   (b) $\text{Rng}(T)$.
    (c) T^{-1}.   (d) $(T^{-1})^{-1}$.

5.   Find the domain and range for the relation W on $\mathbf{R}$ given by $x\,W\,y$ iff
★ (a) $y = 2x + 1$.   (b) $y = x^2 + 3$.
★ (c) $y = \sqrt{x} - 1$.   (d) $y = 1/x^2$.
★ (e) $y < x^2$.   (f) $|x| < 2$ and $y = 3$.
    (g) $|x| < 2$ or $y = 3$.   (h) $y \neq x$.

6.   The inverse of $R = \{(x, y) \in \mathbf{R} \times \mathbf{R}: y = 2x + 1\}$ is the relation $R^{-1} = \{(x, y) \in \mathbf{R} \times \mathbf{R}: y = (x - 1)/2\}$. Use this form to give the inverses of the following relations. In (i), (j), and (k), P is the set of all people.
★ (a) $R_1 = \{(x, y) \in \mathbf{R} \times \mathbf{R}: y = x\}$
    (b) $R_2 = \{(x, y) \in \mathbf{R} \times \mathbf{R}: y = -5x + 2\}$
★ (c) $R_3 = \{(x, y) \in \mathbf{R} \times \mathbf{R}: y = 7x - 10\}$
    (d) $R_4 = \{(x, y) \in \mathbf{R} \times \mathbf{R}: y = x^2 + 2\}$
★ (e) $R_5 = \{(x, y) \in \mathbf{R} \times \mathbf{R}: y = -4x^2 + 5\}$
    (f) $R_6 = \{(x, y) \in \mathbf{R} \times \mathbf{R}: y < x + 1\}$
★ (g) $R_7 = \{(x, y) \in \mathbf{R} \times \mathbf{R}: y > 3x - 4\}$
    (h) $R_8 = \{(x, y) \in \mathbf{R} \times \mathbf{R}: y = 2x/(x - 2)\}$
★ (i) $R_9 = \{(x, y) \in P \times P: y$ is the father of $x\}$
    (j) $R_{10} = \{(x, y) \in P \times P: y$ is a sibling of $x\}$
    (k) $R_{11} = \{(x, y) \in P \times P: y$ loves $x\}$

7. Find these composites for the relations defined in exercise 6.
   ★ (a) $R_1 \circ R_1$ (b) $R_1 \circ R_2$
      (c) $R_2 \circ R_2$ ★ (d) $R_2 \circ R_3$
      (e) $R_2 \circ R_4$ (f) $R_4 \circ R_2$
   ★ (g) $R_4 \circ R_5$ (h) $R_6 \circ R_2$
      (i) $R_6 \circ R_4$ ★ (j) $R_6 \circ R_6$
      (k) $R_5 \circ R_5$ (l) $R_8 \circ R_8$
   ★ (m) $R_3 \circ R_8$ (n) $R_8 \circ R_3$
   ☆ (o) $R_9 \circ R_9$ (p) $R_{10} \circ R_9$
      (q) $R_{11} \circ R_9$

8. Give the digraphs for these relations on the set $\{1, 2, 3\}$.
   (a) $=$ (b) $S = \{(1, 3), (2, 1)\}$
   (c) $\leqslant$ (d) S^{-1}, where $S = \{(1, 3), (2, 1)\}$
   (e) $\neq$ (f) $S \circ S$, where $S = \{(1, 3), (2, 1)\}$

9. Let $A = \{a, b, c, d\}$. Give an example of relations R and S on A such that
   (a) $R \circ S \neq S \circ R$.   (b) $(S \circ R)^{-1} \neq S^{-1} \circ R^{-1}$.

10. Complete the proof of Theorem 3.3.

11. One way to define an ordered pair in terms of sets is to say $(a, b) = \{\{a\}, \{a, b\}\}$. Using this definition, prove that $(a, b) = (x, y)$ iff $a = x$ and $b = y$.

12. Assume ordered pairs have been defined as in exercise 11. Show by example that $(A \times B) \times C = A \times (B \times C)$ is not always true.

13. We may define ordered triples in terms of ordered pairs by saying that $(a, b, c) = ((a, b), c)$. Use this definition to prove that $(a, b, c) = (x, y, z)$ iff $a = x$ and $b = y$ and $c = z$.

14. Prove that $\#(A \times B) = (\#A)(\#B)$.

15. **Proofs to Grade.**
    ★ (a) **Claim.** $(A \times B) \cup C = (A \times C) \cup (B \times C)$.
       "Proof." $x \in (A \times B) \cup C$
          iff $x \in A \times B$ or $x \in C$
          iff $x \in A$ and $x \in B$ or $x \in C$
          iff $x \in A \times C$ or $x \in B \times C$
          iff $x \in (A \times C) \cup (B \times C)$. ∎
    ★ (b) **Claim.** If $A \subseteq B$ and $C \subseteq D$, then $A \times C \subseteq B \times D$.
       "Proof." Suppose $A \times C \not\subseteq B \times D$. Then there exists a pair $(a, c) \in A \times C$ with $(a, c) \notin B \times D$. But $(a, c) \in A \times C$ implies that $a \in A$ and $c \in C$, whereas $(a, c) \notin B \times D$ implies that $a \notin B$ and $c \notin D$. However, $A \subseteq B$ and $C \subseteq D$, so $a \in B$ and $c \in D$. This is a contradiction. Therefore $A \times C \subseteq B \times D$. ∎
       (c) **Claim.** If $A \times B = A \times C$ and $A \neq \varnothing$, then $B = C$.
       "Proof." Suppose $A \times B = A \times C$. Then

$$\frac{A \times B}{A} = \frac{A \times C}{A} \quad \text{so } B = C. \quad \blacksquare$$

    ★ (d) **Claim.** If $A \times B = A \times C$ and $A \neq \varnothing$, then $B = C$.

*"**Proof**."*   To show $B = C$, suppose $b \in B$. Choose any $a \in A$. Then $(a, b) \in A \times B$. But since $A \times B = A \times C$, $(a, b) \in A \times C$. Thus $b \in C$. This proves $B \subseteq C$. A proof of $C \subseteq B$ is similar. Therefore $B = C$.   ■

(e)   **Claim.**   Let R and S be relations from A to B and from B to C, respectively. Then $S \circ R = (R \circ S)^{-1}$.
*"**Proof**."*   The ordered pair $(x, y) \in S \circ R$ iff $(y, x) \in R \circ S$ iff $(x, y) \in (R \circ S)^{-1}$. Therefore $S \circ R = (R \circ S)^{-1}$.   ■

(f)   **Claim.**   Let R be a relation from A to B. Then $I_A \subseteq R^{-1} \circ R$.
*"**Proof**."*   Suppose $(x, x) \in I_A$. Choose any $y \in B$ such that $(x, y) \in R$. Then $(y, x) \in R^{-1}$. Thus $(x, x) \in R^{-1} \circ R$. Therefore $I_A \subseteq R^{-1} \circ R$.   ■

(g)   **Claim.**   Suppose R is a relation from A to B. Then $R^{-1} \circ R \subseteq I_A$.
*"**Proof**."*   Let $(x, y) \in R^{-1} \circ R$. Then for some $z \in B$, $(x, z) \in R$ and $(z, y) \in R^{-1}$. Thus $(y, z) \in R$. Since $(x, z) \in R$ and $(y, z) \in R$, $x = y$. Thus $(x, y) = (x, x)$ and $x \in A$, so $(x, y) \in I_A$.   ■

3.2

Equivalence Relations

Each of the three properties set forth in the next definition is important in its own right. Relations that possess all three properties are particularly valuable.

Definition.   Let A be a set and R be a relation on A.
R is **reflexive on** A iff for all $x \in A$, $x \mathrel{R} x$.
R is **symmetric** iff for all x and $y \in A$, if $x \mathrel{R} y$, then $y \mathrel{R} x$.
R is **transitive** iff for all x, y, and $z \in A$, if $x \mathrel{R} y$ and $y \mathrel{R} z$, then $x \mathrel{R} z$.

Consider these properties for a relation R on a nonempty set A. Only the reflexive property actually asserts that some ordered pairs belong to R. A proof that R is reflexive must then show $x \mathrel{R} x$ for *all* $x \in A$. Since the identity relation on A is the set $I_A = \{(x, x): x \in A\}$, R is reflexive on A iff $I_A \subseteq R$. Of course, if A is empty, then R is trivially reflexive on A because $x \mathrel{R} x$ for all $x \in A$.

Symmetry and transitivity are defined by conditional sentences, so most proofs involving these properties are direct proofs. Neither property requires that R contain any ordered pairs. In fact, the empty relation $\varnothing$ is a symmetric and transitive relation on any set A.

For the set $B = \{2, 5, 6, 7\}$, let $S_1 = \{(2, 5), (5, 6), (2, 6)\}$ and $S_2 = \{(2, 5), (2, 2)\}$. Both S_1 and S_2 are transitive relations on B. They are not reflexive on B and not symmetric.

Let R be the relation "is a subset of" on $\mathscr{P}(\mathbf{Z})$, the power set of $\mathbf{Z}$. R is reflexive on $\mathscr{P}(\mathbf{Z})$ since every set is a subset of itself. R is transitive by

Theorem 2.3. Notice that $\{1, 2\} \subseteq \{1, 2, 3\}$ but $\{1, 2, 3\} \not\subseteq \{1, 2\}$. Therefore R is not symmetric.

Let STNR designate the relation $\{(x, y) \in \mathbf{Z} \times \mathbf{Z}: xy > 0\}$ on $\mathbf{Z}$. In this example x STNR x for all x in $\mathbf{Z}$ except the integer 0; hence the relation STNR is not reflexive on $\mathbf{Z}$. STNR is symmetric since, if x and y are integers and $xy > 0$, then $yx > 0$. STNR is also transitive. To verify this, we assume that x STNR y and y STNR z. Then $xy > 0$ and $yz > 0$. If y is positive, then both x and z are positive; so $xz > 0$. If y is negative, then both x and z are negative; so $xz > 0$. Thus in either case, x STNR z. Therefore STNR is symmetric, transitive, and not reflexive on A; whence the acronym that names the relation.

The identity relation I_A on any set A has all three properties. It is, in fact, the relation "equals," because $x \, I_A \, y$ iff x equals y. Equality is a way of comparing objects according to whether they are the same. Equivalence relations, defined later, are a method of grouping objects according to whether they share a common trait. For example, if T is the set of all triangles, we might say two triangles are alike (equivalent) if they are congruent. This generates the relation $R = \{(x, y) \in T \times T: x \text{ is congruent to } y\}$ on T, which is reflexive on T, symmetric, and transitive. The notion of equivalence, then, is embodied in these three properties.

The properties of reflexivity, symmetry, and transitivity can be nicely characterized by properties of the digraph of the relation. A relation is reflexive, for example, exactly when every vertex has a **loop** (an edge from the vertex to itself). This can be seen in figure 3.10, which shows the digraphs of three relations on $A = \{2, 3, 6\}$. Figure 3.10(a) is the digraph of the relation "divides," which is reflexive on A because every integer divides itself. Figure 3.10(b) is the digraph of "$\geqslant$," which is also reflexive on A. Figure 3.10(c) is the digraph of the relation S, where $x \, S \, y$ iff $x + y > 7$. Since $2 + 2 = 4$, $(2, 2) \not\in S$, S is not reflexive, and there is no loop at 2.

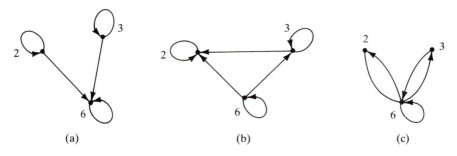

(a)   (b)   (c)

Figure 3.10

In the digraph of a symmetric relation, if there is an edge from vertex x to y, then there must be an edge from y to x. Thus a relation is symmetric exactly when between any two vertices of the digraph there are either no edges or both edges. In figure 3.10 only the digraph of S satisfies this condition; S is a symmetric relation.

For the digraph of a transitive relation, if there is an edge from x to y and an edge from y to z, there must be an edge from x to z. The graphs in figure 3.10(a) and (b) have this property, and both "divides" and "$\geq$" are transitive relations. The relation S is not transitive because $2\ S\ 6$ and $6\ S\ 3$, but 2 is not S-related to 3.

Definitions.   A relation R on a set A is an **equivalence relation on A** iff R *is reflexive on A, symmetric, and transitive.* For $x \in A$, the **equivalence class of x** determined by R is the set $x/R = \{y \in A: x\ R\ y\}$. This is read "the class of x modulo R" or "x mod R." The set of all equivalence classes is called A **modulo R** and is denoted
$A/R = \{x/R: x \in A\}$.

The relation $H = \{(1, 1), (2, 2), (3, 3), (1, 2), (2, 1)\}$ is an equivalence relation on the set $A = \{1, 2, 3\}$. Here $1/H = 2/H = \{1, 2\}$ and $3/H = \{3\}$. Thus $A/H = \{\{1, 2\}, \{3\}\}$.

The relation $\square$ on **R** given by $x \square y$ iff $x^2 = y^2$ is an equivalence relation on **R**. In this example $2/\square = \{2, -2\}$. Notice that $-\pi/\square = \{-\pi, \pi\}$, and $0/\square = \{0\}$. For any $x \in$ **R**, $x/\square = \{x, -x\}$.

For the set P of all people, let L be the relation on P given by $x\ L\ y$ iff x and y have the same last name. Then L is an equivalence relation on P (if we make the assumption that everyone has a last name). The equivalence class of Charlie Brown modulo L is the set of all people whose last name is Brown. Thus, in addition to Charlie, Charlie Brown/L contains Sally Brown, Buster Brown, Leroy Brown, and all other people who are like Charlie Brown in the sense that they have Brown as a last name. Furthermore, Buster Brown/L = Charlie Brown/L.

Two integers have the same **parity** iff they are either both even or both odd. Let $R = \{(x, y) \in$ **Z** $\times$ **Z**$: x$ and y have the same parity$\}$. R is an equivalence relation on **Z** with two equivalence classes, the even integers E and the odd integers D. If x is odd, $x/R = D$, while if x is even, $x/R = E$. Thus **Z**$/R = \{E, D\}$.

The digraph of an equivalence relation has a striking property. Consider the relation S on the set $A = \{21, 64, 82, 113, 247, 1042\}$ given by $x\ S\ y$ iff x and y have the same number of digits. The digraph of this equivalence relation (figure 3.11) looks like three separate digraphs. A subset of the vertex set of a graph, together with all the edges connecting vertices in the subset, is called a **subdigraph.** In this case the three subdigraphs on the vertex sets $\{21, 64, 82\}$, $\{113, 247\}$, and $\{1042\}$ can be described as follows. Each subdigraph has all possible edges connecting its vertices. (A digraph with all possible edges is called **complete.**) Furthermore, no edge connects a vertex

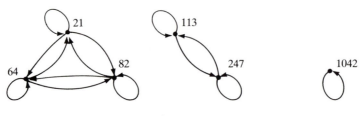

Figure 3.11

in one of the three subsets with a vertex in another subset. The digraph in figure 3.11 thus is a union of three components, each of which is a complete subdigraph. The three components are the three equivalence classes, {21, 64, 82}, {113, 247}, and {1042}.

What we have seen is that the equivalence relation S gives us a way of walling off, or partitioning, the set A into disjoint subsets. We will return to this idea in the next section.

The last example we give of an equivalence relation has many applications. Recall that an integer m divides another integer n iff there exists an integer k such that $n = km$. For a fixed integer $m \neq 0$, let $\equiv_m$ be the relation on **Z** given by

$$x \equiv_m y \text{ iff } m \text{ divides } (x - y).$$

We shall see in Theorem 3.4 that $\equiv_m$ is an equivalence relation on **Z**.

Another notation for $x \equiv_m y$ is $x \equiv y \pmod{m}$, which is read "x is congruent to y modulo m." We read $x/\equiv_m$ as "the equivalence class of x modulo m" or simply "the class of x mod m." The set of equivalence classes for $\equiv_m$ is denoted $\mathbf{Z}_m$, called "**Z** mod m."

As an example, let's concentrate on congruence modulo 3. We see that $4 \equiv 1 \pmod 3$ because 3 divides $4 - 1$. Also $10 \equiv 16 \pmod 3$, since 3 divides $10 - 16$; that is, $10 - 16 = 3k$ for the integer $k = -2$. However $5 \not\equiv -6 \pmod 3$ since 3 fails to divide 11.

Next let us calculate all the equivalence classes of $\mathbf{Z}_3$. For $x \in \mathbf{Z}$, the equivalence class of x modulo 3 is $\{y \in \mathbf{Z} : x \equiv y \pmod 3\}$. For 0, we see $0 \equiv 0 \pmod 3$, $3 \equiv 0 \pmod 3$, and $-6 \equiv 0 \pmod 3$. It can be shown that $0/\equiv_3 = \{\ldots, -12, -9, -6, -3, 0, 3, 6, 9, 12, \ldots\}$, the set of all multiples of 3. Similarly $1/\equiv_3 = \{\ldots, -8, -5, -2, 1, 4, 7, 10, \ldots\}$ and $2/\equiv_3 = \{\ldots, -10, -7, -4, -1, 2, 5, 8, 11, 14, \ldots\}$. There are no other equivalence classes. Thus $\mathbf{Z}_3 = \{0/\equiv_3, 1/\equiv_3, 2/\equiv_3\}$.

The notation $x/\equiv_m$ we have used is rather cumbersome. Usually the equivalence class of x modulo m is written $\bar{x}$ or $[x]$. This notation works well when the modulus m remains unchanged throughout a discussion. The disadvantage is that if we are dealing with, say, classes modulo 6 as well as with classes modulo 3, the symbol $\bar{1}$ could have two different meanings. As an equivalence class modulo 6,

$$\bar{1} = \{\ldots, -5, 1, 7, 13, \ldots\}$$

but as an equivalence class modulo 3,

$$\bar{1} = \{\ldots, -5, -2, 1, 4, 7, 10, 13, \ldots\}$$

Theorem 3.4 The relation $\equiv_m$ is an equivalence relation on the integers. Furthermore, $\mathbf{Z}_m$ has m distinct elements.

Proof. It is clear that $\equiv_m$ is a set of ordered pairs of integers and, hence, is a relation on $\mathbf{Z}$. We shall first show that $\equiv_m$ is an equivalence relation.

(i)   To show reflexivity on $\mathbf{Z}$, let x be an integer. We show that $x \equiv x \pmod{m}$. Since $m \cdot 0 = 0 = x - x$, m divides $x - x$. Thus $\equiv_m$ is reflexive on $\mathbf{Z}$.

(ii) For symmetry, suppose $x \equiv y \pmod{m}$. Then m divides $x - y$. Thus there is an integer k so that $x - y = km$. But this means that $-(x - y) = -(km)$, or that $y - x = (-k)m$. Therefore m divides $(y - x)$, so that $y \equiv x \pmod{m}$.

(iii) Suppose $x \equiv y \pmod{m}$ and $y \equiv z \pmod{m}$. Thus m divides both $x - y$ and $y - z$. Therefore there exist integers h and k such that $x - y = hm$ and $y - z = km$. But then $h + k$ is an integer, and $x - z = (x - y) + (y - z) = hm + km = (h + k)m$. Thus m divides $x - z$, so $x \equiv z \pmod{m}$. Therefore $\equiv_m$ is transitive.

Now that $\equiv_m$ is known to be an equivalence relation, it follows that $x \equiv_m y$ iff $\bar{x} = \bar{y}$ (exercise 6). The remainder of the proof shows there are exactly m distinct equivalence classes. We claim

$$\mathbf{Z}_m = \{\bar{0}, \bar{1}, \bar{2}, \ldots, \overline{(m-1)}\}.$$

First, each $\bar{k}$ for $k = 0, 1, 2, \ldots, (m-1)$ is an equivalence class and hence in $\mathbf{Z}_m$. Now suppose $\bar{x} \in \mathbf{Z}_m$. By the Division Algorithm, there exist integers q and r such that $x = mq + r$ with $0 \leq r \leq m - 1$. Thus $mq = x - r$. Therefore $x \equiv r \pmod{m}$ and, by Theorem 3.4, it follows that $\bar{x} = \bar{r}$. Thus $\bar{x} \in \{\bar{0}, \bar{1}, \bar{2}, \ldots, \overline{(m-1)}\}$, and the claim is verified.

We will know that $\mathbf{Z}_m$ has exactly m elements when we show that $\bar{0}, \bar{1}, \bar{2}, \ldots, \overline{(m-1)}$ are all distinct. Suppose $\bar{k} = \bar{r}$ where $0 \leq r \leq k \leq (m-1)$. Then $k \equiv r \pmod{m}$, and thus m divides $k - r$. But $0 \leq k - r \leq m - 1$, so $k - r = 0$. Therefore $k = r$, and the m equivalence classes are distinct. ∎

Exercises 3.2

1. Indicate which of the following relations on the given sets are reflexive, which are symmetric, and which are transitive.
   ★ (a) $\{(1, 2)\}$ on the set $A = \{1, 2\}$ (b) $\leqslant$ on **N**
      (c) $=$ on **N**   (d) $<$ on **N**
   ★ (e) $\geqslant$ on **N**   (f) $\neq$ on **N**
      (g)   "divides" on **N**   (h) $\{(x, y) \in \mathbf{Z} \times \mathbf{Z}: x + y = 10\}$
      (i) $\perp = \{(\ell, m): \ell \text{ and } m \text{ are lines and } \ell \text{ is perpendicular to } m\}$
      (j) R, where $(x, y) R (z, w)$ iff $x + z \leqslant y + w$, on the set $\mathbf{R} \times \mathbf{R}$
   ★ (k) S, where $x S y$ iff x is a sibling of y, on the set P of all people
      (l) T, where $(x, y) T (z, w)$ iff $x + y \leqslant z + w$, on the set $\mathbf{R} \times \mathbf{R}$

2. Let A be the set $\{1, 2, 3,\}$. List the ordered pairs in a relation on A which is
   ★ (a)   not reflexive, not symmetric, and not transitive.
      (b)   reflexive, not symmetric, and not transitive.
      (c)   not reflexive, symmetric, and not transitive.
   ★ (d)   reflexive, symmetric, and not transitive.
      (e)   not reflexive, not symmetric, and transitive.
      (f)   reflexive, not symmetric, and transitive.
      (g)   not reflexive, symmetric, and transitive.
      (h)   reflexive, symmetric, and transitive.

✭ 3. Repeat exercise 2 for relations on **R** by sketching graphs of relations with the desired properties.

4. For each of the following, verify that the relation is an equivalence relation. Then give information about the equivalence classes as specified.
   ☆ (a)   On **N**, the relation R given by $a R b$ iff the prime factorizations of a and b have the same number of 2's. Name three elements in each of these classes: $1/R$, $4/R$, $72/R$.
      (b)   The relation R on **Z** given by $x R y$ iff $x^2 = y^2$. Give the equivalence class of 0; of 4; of -72.
      (c)   The relation R on **N** given by $m R n$ iff m and n have the same digit in the tens places. Name an element of $106/R$ that is less than 50; between 150 and 300; greater than 1,000. Repeat the problem for the class $635/R$.
   ☆ (d)   The relation R on $\mathbf{N} \times \mathbf{N}$ given by $(x, y) R (z, w)$ iff $xw = yz$. Find an element (a, b) of $(2, 3)/R$ such that $a = 6$; $a = 10$; $a = 50$. Describe all ordered pairs in the equivalence class of $(2, 3)$.
      (e)   The relation R on the set of all ordered 3-tuples from the set $\{1, 2, 3, 4\}$ given by $(x, y, z) R (a, b, c)$ iff $y = b$. List five elements of $(4, 2, 1)/R$. How many elements are in the equivalence class of $(1, 1, 1)$?

5. Calculate the equivalence classes for the relation of
   ★ (a)   congruence modulo 5.   (b)   congruence modulo 8.
      (c)   congruence modulo 1.   (d)   congruence modulo 7.

6. For the equivalence relation $\equiv_m$, prove that
   (a)   if $x \equiv_m y$, then $\bar{x} = \bar{y}$.
   (b)   if $\bar{x} = \bar{y}$, then $x \equiv_m y$.
   (c)   if $\bar{x} \cap \bar{y} \neq \varnothing$, then $\bar{x} = \bar{y}$.

7. The properties of reflexivity, symmetry, and transitivity are related to the identity relation and the operations of inversion and composition. Prove that
   (a) R is reflexive iff $I_A \subseteq R$.   ★ (b) R is symmetric iff $R = R^{-1}$.
   (c) R is transitive iff $R \circ R \subseteq R$.

8. Prove that if R is a symmetric, transitive relation on A and the domain of R is A, then R is reflexive on A.

9. The **complement** of a digraph has the same vertex set as the original digraph, and an edge from x to y exactly when the original digraph does not have an edge from x to y. The two digraphs D and $\widetilde{D}$ shown below are complementary. Call a digraph symmetric (transitive) iff its relation is symmetric (transitive).
   (a)   Show that the complement $\widetilde{D}$ of a symmetric digraph D is symmetric.
   (b)   Show by example that the complement of a transitive digraph need not be transitive.

☆ 10. Let L be a relation on a set A that is reflexive and transitive (but not necessarily symmetric). Let R be the relation defined on A by $x\,R\,y$ iff $x\,L\,y$ and $y\,L\,x$. Prove R is an equivalence relation.

11. A relation R on a set A is called
    irreflexive iff for all $x \in A$, $(x, x) \notin R$.
    asymmetric iff for all $x, y \in A$, $(x, y) \in R$ implies $(y, x) \notin R$.
    antisymmetric iff for all $x, y \in A$, $(x, y) \in R$ and $(y, x) \in R$ implies $x = y$.
    (a)   Give a mathematical example of a set and a relation which is
       (i)   reflexive and asymmetric.
       (ii)   antisymmetric.
    ★ (b)   Prove that if a relation is asymmetric, then it is antisymmetric.

12. **Proofs to Grade.**
    (a)   **Claim.**   If the relation R is symmetric and transitive, it is also reflexive.
       "Proof."   Since R is symmetric, if $(x, y) \in R$, then $(y, x) \in R$. Thus $(x, y) \in R$ and $(y, x) \in R$, and since R is transitive, $(x, x) \in R$. Therefore R is reflexive.   ∎
    (b)   **Claim.**   If the relations R and S are symmetric, then $R \cap S$ is symmetric.
       "Proof."   Suppose $(x, y) \in R \cap S$. Then $(x, y) \in R$ and $(x, y) \in S$. Since R and S are symmetric, $(y, x) \in R$ and $(y, x) \in S$. Therefore $(y, x) \in R \cap S$.   ∎
    ★ (c)   **Claim.**   If the relations R and S are transitive, then $R \cap S$ is transitive.
       "Proof."   Suppose $(x, y) \in R \cap S$ and $(y, z) \in R \cap S$. Then $(x, y) \in R$ and $(y, z) \in S$. Therefore $(x, z) \in R \cap S$.   ∎

3.3

Partitions

If R is an equivalence relation on a nonempty set A and $x \in A$, then the set of elements in A that are R-related to x is the equivalence class of x. Looking back at examples from section 3.2, you may see that every equivalence class is nonempty, that every equivalence class is a subset of A, that the union of the equivalence classes is A, and that the classes of two elements x and y are either equal (when $x \ R \ y$) or have no elements in common (when $x \ \cancel{R} \ y$). These observations are summarized in the next theorem.

Theorem 3.5   Let R be an equivalence relation on a nonempty set A. Then

(a)   For all $x \in A$, $x \in x/R$. (Thus $x/R \neq \varnothing$.)
(b) $x/R \subseteq A$ for all $x \in A$.
(c) $A = \bigcup_{x \in A} x/R$.
(d) $x \ R \ y$ iff $x/R = y/R$.
(e) $x \ \cancel{R} \ y$ iff $x/R \cap y/R = \varnothing$.

Proof.

(a)   Since R is reflexive on A, $(x, x) \in R$. Thus $x \in x/R$.
(b)   This is from the definition of x/R.
(c)   First $\bigcup_{x \in A} x/R \subseteq A$ because each $x/R \subseteq A$. To prove $A \subseteq$
       $\bigcup_{x \in A} x/R$, suppose $t \in A$. By part (a), $t \in t/R \subseteq \bigcup_{x \in A} x/R$.
       Thus $A = \bigcup_{x \in A} x/R$.
(d)   (i)   Suppose $x \ R \ y$. To show $x/R = y/R$, we first show
               $x/R \subseteq y/R$. If $z \in x/R$, then $x \ R \ z$. From $x \ R \ y$, by
               symmetry, $y \ R \ x$. Then, by transitivity, $y \ R \ z$. Thus
               $z \in y/R$. The proof that $y/R \subseteq x/R$ is similar.
       (ii)   Suppose $x/R = y/R$. Since $y \in y/R$, $y \in x/R$. Thus
               $x \ R \ y$.
(e)   (i)   If $x/R \cap y/R = \varnothing$, then $\langle$since $y \in y/R \rangle$ $y \notin x/R$.
               Thus $x \ \cancel{R} \ y$.
       (ii)   Finally, we show $x \ \cancel{R} \ y$ implies $x/R \cap y/R = \varnothing$. $\langle We$
               $prove \ the \ contrapositive. \rangle$ Suppose $x/R \cap y/R \neq \varnothing$.
               Let $m \in x/R \cap y/R$. Then $x \ R \ m$ and $y \ R \ m$. There-
               fore $x \ R \ m$ and $m \ R \ y$. Thus $x \ R \ y$.   ∎

If R is an equivalence relation on a set A, Theorem 3.5 tells us that the set A may be thought of as a union of a collection of nonempty subsets that are pairwise disjoint. In other words, imposing an equivalence relation on A produces a set of equivalence classes, so that every element of A is in exactly one class and any two classes are either equal or disjoint. This leads to the concept of a partition of a set.

Definition.   Let A be a set and $\mathcal{A}$ be a collection of subsets of A. $\mathcal{A}$ is a **partition of A** iff

    (i)   If $X \in \mathcal{A}$, then $X \neq \varnothing$.
    (ii) If $X \in \mathcal{A}$ and $Y \in \mathcal{A}$, then $X = Y$ or $X \cap Y = \varnothing$.
    (iii) $\underset{X \in \mathcal{A}}{\cup}\, X = A$.

Examples.   The collections $\{\{0\}, \{1, -1\}, \{2, -2\}, \ldots\}$ and $\{E, D\}$, where E and D are the sets of even and odd integers, respectively, are two different partitions of $\mathbf{Z}$ (figures 3.12 and 3.13).

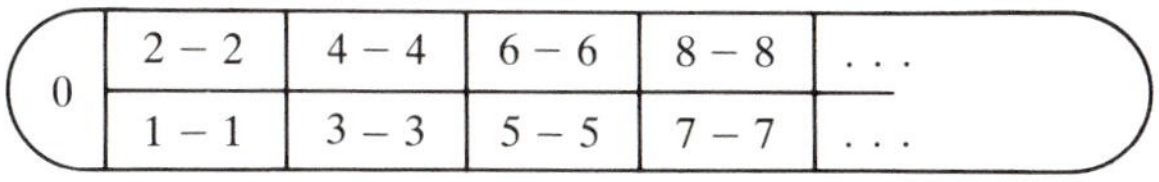

Figure 3.12:   A partition of $\mathbf{Z}$

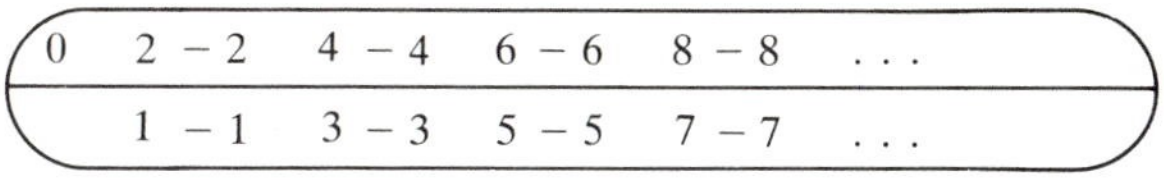

Figure 3.13:   A partition of $\mathbf{Z}$

    The collection $\{\{1\}, \{2\}, \{3\}, \ldots\}$ is a partition of $\mathbf{N}$. In fact, $\{\{x\}: x \in A\}$ is always a partition of a nonempty set A.
    For each $n \in \mathbf{Z}$, let $G_n = [n, n + 1)$. Then $\{G_n: n \in \mathbf{Z}\}$ is a partition of $\mathbf{R}$.

    Theorem 3.5 may be restated to say that every equivalence relation on a set determines a partition of that set. The relation $T = \{(5, 7), (7, 6), (6, 6), (5, 5), (7, 7), (7, 5), (6, 5), (5, 6), (6, 7), (4, 4)\}$ is an equivalence relation on the set $A = \{4, 5, 6, 7\}$ with equivalence classes $\{4\}$ and $\{5, 6, 7\}$, so the corresponding partition is $A = \{\{4\}, \{5, 6, 7\}\}$.
    We shall soon see that every partition in turn determines an equivalence relation. Thus each concept may be used to describe the other. This is

to our advantage, for we may use either concept, choosing the one that lends itself more readily to the situation at hand.

The method of producing an equivalence relation from a partition is based on the idea that two objects will be said to be equivalent iff they belong to the same member of the partition. For example, let A be $\{1, 2, 3, 4, 5\}$ and $\mathcal{B} = \{\{1, 2\}, \{3\}, \{4, 5\}\}$ be a partition of A. To make an equivalence relation Q on A, we note that since $\{1, 2\} \in \mathcal{B}$, 1 is related to 1 and 2, 2 is related to 1 and 2. Also 3 is related to 3, and so forth. Therefore $Q = \{(1, 1), (1, 2), (2, 1), (2, 2), (3, 3), (4, 4), (4, 5), (5, 4), (5, 5)\}$ is the equivalence relation associated with $\mathcal{B}$.

The next theorem asserts that this method of using a partition to define a relation always produces an equivalence relation and, furthermore, that the set of equivalence classes of the relation is the same as the original partition.

Before taking up the theorem, one more example may be instructive. The state of Missouri is divided (partitioned) into three telephone area codes (figure 3.14). That is, the set A of all telephones in Missouri is partitioned in three subsets, A_{314}, A_{417}, and A_{818}. How can we use this fact to define an equivalence relation on the set A? Obviously, we should say that two telephones in A are related iff they are in the same partition element, that is, iff they have the same area code.

Figure 3.14

Theorem 3.6   Let $\mathcal{B}$ be a partition of the set A. For x and $y \in A$, define $x \, Q \, y$ iff there exists $C \in \mathcal{B}$ such that $x \in C$ and $y \in C$. Then

(a) Q is an equivalence relation on A.
(b) $A/Q = \mathcal{B}$.

Proof.   As a special case, observe that if A is empty, then the only partition of A is the empty family $\mathcal{B}$. In this case Q will be the empty relation on A, and $A/Q = \varnothing = \mathcal{B}$, as promised. Assume now that A is nonempty.

(a)   We prove Q is transitive and leave the proofs of symmetry and reflexivity on A for exercise 6.

Let x, y, $z \in A$. Assume $x \, Q \, y$ and $y \, Q \, z$. Then there are sets C and D in $\mathcal{B}$ such that x, $y \in C$ and y, $z \in D$. Since $\mathcal{B}$ is a partition of A, the sets C and D are either identical or disjoint; but since y is an element of both sets, they cannot be disjoint. Hence there is a set $C \, \langle = D \rangle$ that contains both x and z, so that $x \, Q \, z$. Therefore Q is transitive.

(b)   We first show $A/Q \subseteq \mathcal{B}$. Let $x/Q \in A/Q$. Since $\mathcal{B}$ is a partition of A, choose $B \in \mathcal{B}$ such that $x \in B$. We claim $x/Q = B$. If $y \in x/Q$, then $x \, Q \, y$; there is some $C \in \mathcal{B}$ such that $x \in C$ and $y \in C$. Since either $C = B$ or $C \cap B = \varnothing$, and $x \in C \cap B$, $y \in B$. On the other hand, if $y \in B$, then $x \, Q \, y$, and so $y \in x/Q$. Therefore $x/Q = B$.

To show $\mathcal{B} \subseteq A/Q$, let $B \in \mathcal{B}$. As an element of a partition, $B \neq \varnothing$. Choose any $t \in B$; then we claim $B = t/Q$. If $s \in B$, then $t \, Q \, s$, so $s \in t/Q$. On the other hand, if $s \in t/Q$, then $t \, Q \, s$; so s and t are elements of the same member of $\mathcal{B}$, which must be B. ∎

Let $A = \{1, 2, 3, 4\}$ and $\mathcal{B} = \{\{1\}, \{2, 3\}, \{4\}\}$. The equivalence relation Q associated with the partition $\mathcal{B}$ is $\{(1, 1), (2, 2), (2, 3), (3, 2), (3, 3), (4, 4)\}$. The equivalence classes of this relation are $1/Q = \{1\}$, $2/Q = \{2, 3\} = 3/Q$, and $4/Q = \{4\}$, so the set of equivalence classes is precisely $\mathcal{B}$.

For $\mathbf{Z}$, let $\mathcal{A}$ be the partition $\{A_0, A_1, A_2, A_3\}$ where
$A_0 = \{\ldots, -8, -4, 0, 4, 8, \ldots\}$, $A_1 = \{\ldots, -7, -3, 1, 5, 9, \ldots\}$,
$A_2 = \{\ldots, -6, -2, 2, 6, 10, \ldots\}$, and $A_3 = \{\ldots, -5, -1, 3, 7, 11, \ldots\}$.
For integers x and y, we see x and y are in the same set A_i iff $x = 4k_1 + i$ and $y = 4k_2 + i$, for some integers k_1 and k_2 or, in other words, iff $x - y$ is a multiple of 4. Thus the equivalence relation associated with the partition $\mathcal{A}$ is our old friend, congruence modulo 4.

Exercises 3.3

1.   Let $C = \{i, -1, -i, 1\}$, where $i = \sqrt{-1}$. The relation R on C given by $x \, R \, y$ iff $xy = \pm 1$ is an equivalence relation on C. Give the partition of C associated with R.

☆ 2.   Let C be as in exercise 1. The relation S on $C \times C$ given by $(x, y) \, S \, (u, v)$ iff $xy = uv$ is an equivalence relation. Give the partition of $C \times C$ associated with S.

3. Describe the equivalence relation on **Z** determined by the partition $\{A, B\}$ where $A = \{x: x < 3\}$ and $B = \mathbf{Z} - A$.

4. List the ordered pairs in the equivalence relation on $A = \{1, 2, 3, 4, 5\}$ associated with these partitions:
    ★ (a) $\{\{1, 2\}, \{3, 4, 5\}\}$ (b) $\{\{1\}, \{2\}, \{3, 4\}, \{5\}\}$
       (c) $\{\{2, 3, 4, 5\}, \{1\}\}$

5. Partition the set $D = \{1, 2, 3, 4, 5, 6, 7\}$ into two subsets: those symbols made from straight line segments only (like 4), and those that are drawn with at least one curved segment (like 2). Describe or draw the digraph of the corresponding equivalence relation on D.

6. Complete the proof of Theorem 3.6 by proving that if $\mathscr{B}$ is a partition of A, and $x \, Q \, y$ iff there exists $C \in \mathscr{B}$ such that $x \in C$ and $y \in C$, then
    (a) Q is symmetric.   (b) Q is reflexive on A.

★ 7.   Let R be a relation on a set A which is reflexive and symmetric but not transitive. Let $R(x) = \{y: x \, R \, y\}$. [Note that $R(x)$ is the same as x/R except that R is not an equivalence relation in this exercise.] Does the set $\mathscr{A} = \{R(x): x \in A\}$ always form a partition of A? Prove that your answer is correct.

8. Repeat exercise 7, assuming R is reflexive and transitive but not symmetric.

9. Repeat exercise 7, assuming R is symmetric and transitive but not reflexive.

10. Let A be a set with at least three elements.
    (a)   Is there a partition of A with exactly one element?
    ★ (b)   If $\mathscr{B} = \{B_1, B_2\}$ is a partition of A with $B_1 \neq B_2$, is $\{\widetilde{B}_1, \widetilde{B}_2\}$ a partition of A? Explain. (Here $\widetilde{B}_i = A - B_i$.) What if $B_1 = B_2$?
    (c)   If $\mathscr{B} = \{B_1, B_2, B_3\}$ is a partition of A, is $\{\widetilde{B}_1, \widetilde{B}_2, \widetilde{B}_3\}$ a partition of A? Explain. Consider the possibility that two or more of the elements of $\mathscr{B}$ may be equal.
    (d)   If $\mathscr{B} = \{B_1, B_2\}$ is a partition of A, $\mathscr{C}_1$ is a partition of B_1, and $\mathscr{C}_2$ is a partition of B_2, and $B_1 \neq B_2$, prove that $\mathscr{C}_1 \cup \mathscr{C}_2$ is a partition of A.

11. **Proofs to Grade.**
    (a)   **Claim.**   If $\mathscr{A}$ is a partition of a set A and $\mathscr{B}$ is a partition of a set B, then $\mathscr{A} \cup \mathscr{B}$ is a partition of $A \cup B$.
          "Proof."
          (i)   If $X \in \mathscr{A} \cup \mathscr{B}$, then $X \in \mathscr{A}$ or $X \in \mathscr{B}$. In either case $X \neq \varnothing$.
          (ii) If $X \in \mathscr{A} \cup \mathscr{B}$ and $Y \in \mathscr{A} \cup \mathscr{B}$, then $X \in \mathscr{A}$ and $Y \in \mathscr{A}$, or $X \in \mathscr{A}$ and $Y \in \mathscr{B}$, or $X \in \mathscr{B}$ and $Y \in \mathscr{A}$, or $X \in \mathscr{B}$ and $Y \in \mathscr{B}$. Since both $\mathscr{A}$ and $\mathscr{B}$ are partitions, in each case either $X = Y$ or $X \cap Y = \varnothing$.
          (iii) Since $\bigcup_{X \in \mathscr{A}} X = A$ and $\bigcup_{X \in \mathscr{B}} X = B$, $\bigcup_{X \in \mathscr{A} \cup \mathscr{B}} X = A \cup B$.   ∎
    ★ (b)   **Claim.**   If $\mathscr{B}$ is a partition of A, and if $x \, Q \, y$ iff there exists $C \in \mathscr{B}$ such that $x \in C$ and $y \in C$, then the relation Q is symmetric.
          "Proof."   First, $x \, Q \, y$ iff there exists $C \in \mathscr{B}$ such that $x \in C$ and $y \in C$. Also, $y \, Q \, x$ iff there exists $C \in \mathscr{B}$ such that $y \in C$ and $x \in C$. Therefore $x \, Q \, y$ iff $y \, Q \, x$.   ∎

3.4

Graphs

We have seen several ways relations are represented—as graphs on Cartesian coordinates, in tables, and as directed graphs. Another representation for a relation is as another kind of graph. Informally, a **graph** is a collection of points, called **vertices,** and lines, called **edges,** connecting some of the vertices. These edges are not directed from one vertex to another, as they would be in the special case of a digraph (section 3.1).

The study of graphs has applications in computer science, statistics, operations research, linguistics, chemistry, genetics, electrical networks, and other areas. Graph theory is also a significant branch of mathematics in its own right. Our goal in this section is to present enough of the basic facts and terminology to introduce the subject as an extension of our study of relations.

We begin with an example representing conversations at a party. The five people at the party are Doc, Grumpy, Sneezy, Dopey, and Happy. Rather than listing the ordered pairs in the relation S on this set of people defined by $x \, S \, y$ iff x had a conversation with y, we describe the relation with a graph (figure 3.15).

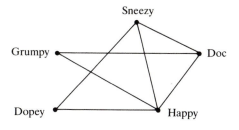

Figure 3.15

This graph has an edge connecting two vertices x and y exactly when x is S-related to y. It can be seen from the graph that Doc spoke with each of the others except Dopey and that Grumpy had a conversation only with Doc and Happy. The graph does not show anything about where the party-goers stood, how long they talked, or whether they had more than one conversation. Exactly the same information about whether two people did or did not have a conversation is conveyed by either of the graphs in figure 3.16. These three graphs are essentially the same. (After we define the word, we will call these graphs *isomorphic.*)

Possibly because there are so many applications for graph theory and so many people who work with graphs, there is a great deal of variation in

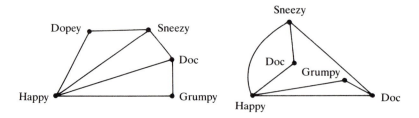

Figure 3.16

terminology and in types of graphs studied. One variation of some interest would be to allow multiple edges between vertices. Such a graph is called a **multigraph.** Our conversation graph example did not need to use more than one edge between vertices, but could have if we had wanted to represent separate conversations between two people. In a graph where vertices represent cities, multiple edges might represent alternate routes or transportation modes between cities.

Another kind of graph would allow an edge from a vertex to itself—a **loop.** Our example did not show that anyone at the party talked to himself, but it is easy to imagine relations for which the corresponding graph would have one or more loops. In this section we will deal only with **simple graphs**—those having no multiple edges and no loops.

Theoretically, a **graph** G is a pair (V, E) where V is a nonempty set and E is a set of unordered pairs of distinct elements of V. We call V the **vertex set** of G, and E is called the **edge set.** Two vertices u and v are said to be **adjacent** iff the edge $\{u, v\}$, which is often written simply as uv, is in E. The edge uv is said to be **incident** with u and v. Also, we say that uv **joins** u and v.

The **order** of a graph is the number of vertices in the graph. The **size** of a graph is the number of edges. For a vertex v, the **degree** of v is the number of edges incident with v.

The graphs shown in figure 3.17 both have order 6 and size 9. In each graph every vertex has degree 3. The vertex set of the graph in figure 3.17(a) contains 6 names. Names are adjacent iff they have two consecutive letters in common. For example, Dar*la* is adjacent to *La*uren. The vertices of the graph in Figure 3.17(b) are numbers; two numbers are adjacent iff they lie in

the same row or the same column of the array $\begin{bmatrix} 1 & 2 & 3 \\ 4 & 5 & 6 \end{bmatrix}$.

Two graphs (V, E) and (V', E') are called **isomorphic** iff there is a correspondence between the vertex sets V and V' that preserves adjacencies. What this means is that there is a way to relabel the vertices of V as vertices in V' so that there is an edge in E joining two vertices of V exactly when there is an edge in E' joining the corresponding vertices of V'.

The graph (V, E) of figure 3.17(b) above is isomorphic to the graph (V', E') shown in figure 3.18. The tables beside the graph show one possible edge-preserving correspondence.

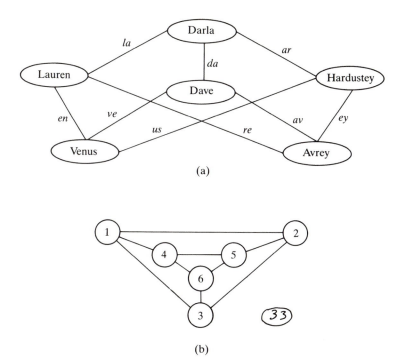

(a)

(b)

Figure 3.17

The three graphs shown in figures 3.15 and 3.16 are easily seen to be isomorphic; it is not even necessary to relabel the vertices. The two graphs in figure 3.17 are *not* isomorphic, even though the graphs have the same size and order, and every vertex in each graph has degree three. To see why this is true, look at the vertices 1, 2, 3 and the edges 12, 13, 23 of graph (b). If the graphs were isomorphic, there would have to be three corresponding vertices in graph (a), each related to the other two. However, there are no such vertices in graph (a) because that graph contains no triangles.

In general, it is very difficult to determine whether two graphs are isomorphic.

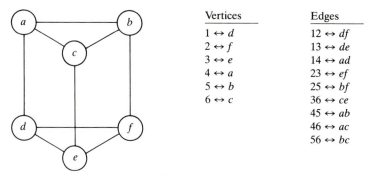

Vertices	Edges
1 ↔ d	12 ↔ df
2 ↔ f	13 ↔ de
3 ↔ e	14 ↔ ad
4 ↔ a	23 ↔ ef
5 ↔ b	25 ↔ bf
6 ↔ c	36 ↔ ce
	45 ↔ ab
	46 ↔ ac
	56 ↔ bc

Figure 3.18

If you look again at figure 3.15, you will see that the degrees of the vertices are 3, 3, 4, 2, and 2. The sum of the degrees is 14, which is even. For the graphs in figure 3.17 the sum of the degrees of the vertices is 18, which is also even. The explanation for the fact that the sum of the degrees of the vertices of a graph is even is the same as the explanation for the fact that if a group of people shake hands, the total number of hands shaken must be even.

Theorem 3.7   (a). (The Handshaking Lemma). For each graph G, the sum of the degrees of the vertices of G is even. (b) For every graph G, the number of vertices of G having odd degree is even.

Proof.

(a)   Each edge is incident with two vertices. Thus the sum of the degrees of the vertices is exactly twice the number of edges. Therefore the sum is even.

(b)   Obviously, the sum of the degrees of the vertices that have even degree is an even number. If there were an odd number of vertices with odd degree, then the sum of all the degrees would be odd.   ∎

A graph $G(V, E)$ for which the edge set is empty is called a **null graph.** In any graph a vertex with degree 0 is called an **isolated vertex.** Thus a graph is null exactly when every vertex is isolated. Clearly, any two null graphs with the same order are isomorphic. Figure 3.19(a) shows N_5, a null graph on 5 vertices.

A graph $G(V, E)$ in which every pair of distinct vertices are related is called a **complete graph.** If G has order n, then each vertex must be adjacent to $n - 1$ vertices. Thus the sum of the degrees of the vertices for a complete graph on n vertices is $n(n - 1)$ so the number of edges is $\frac{1}{2}n(n - 1)$. Any two complete graphs on n points are isomorphic. Figure 3.19(b) shows K_5, a complete graph on 5 vertices.

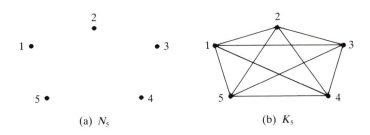

(a) N_5 (b) K_5

Figure 3.19

A **subgraph** (V_1, E_1) of the graph (V_2, E_2) is a graph such that $V_1 \subseteq V_2$ and $E_1 \subseteq E_2$. Since (V_1, E_1) is itself a graph, the edges in E_1 must join vertices of V_1. (An edge cannot terminate at a point that is not a vertex). Figure 3.20 shows three subgraphs of our graph of conversations.

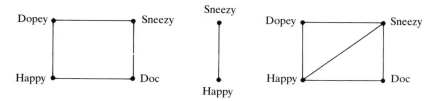

Figure 3.20

The **complement** $\widetilde{G}$ of a graph $G(V, E)$ is the graph with vertex set V in which two vertices are adjacent iff they are not adjacent in G. Thus the complement of $\widetilde{G}$ is G. Also, the complement of a complete graph is null, and vice versa. Figure 3.21 shows a graph and its complement. Notice that if (V, E_1) and (V, E_2) are complementary graphs, then $(V, E_1 \cup E_2)$ is the complete graph on the vertex set V.

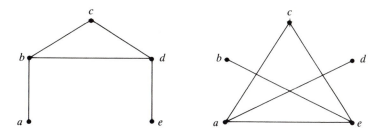

Figure 3.21

A **walk**† in a graph G is a finite‡ alternating sequence of vertices and incident edges, beginning and ending with a vertex. A walk can be described by listing the vertices and edges or more simply just by giving the sequence of adjacent vertices. A walk $v_0, v_1, v_2, \ldots, v_m$ is said to **traverse** the vertices in the sequence. We call v_0 the **initial vertex** of the walk and v_m the **terminal vertex**. The **length** of a walk is the number of edges in the walk. Listed below are some examples of walks in the graph shown in figure 3.22.

†What we call a walk is called by some authors a path, an edge-sequence, a route, a trail, or a chain.

‡By a finite sequence we mean that the objects listed could be labeled 1, 2, 3, . . ., n, for some natural number n.

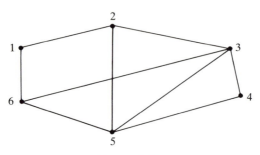

Figure 3.22

w_1: 6, 3, 5
w_2: 3, 5
w_3: 2, 5, 3, 4, 5
w_4: 1, 6, 3, 6, 5, 2, 1
w_5: 1, 2, 3, 4, 5, 6, 1

The walk w_3 originates at vertex 2, has length 4, and terminal vertex 5. The walk w_3 is also an example of a **trail,** which is a walk with distinct edges. Each of w_1, w_2, and w_5 is also a trail, but w_4 is not. A walk in which all vertices traversed are distinct is called a **path.** Every path must be a trail, but w_3 is an example of a trail that is not a path. Both w_1 and w_2 are paths.

A walk that originates and terminates at the same vertex is said to be **closed.** We allow **closed paths,** by permitting the initial and terminal vertex to be the only instance where a vertex is repeated. A closed walk with distinct edges is called a **cycle.** The walk w_4: 1, 6, 3, 6, 5, 2, 1 is closed but is not a cycle. The walk w_5: 1, 2, 3, 4, 5, 6, 1 is a cycle and also a closed path. The cycle w_6: 5, 3, 6, 1, 2, 3, 4, 5 is not a closed path.

If we think of a graph as representing a map of cities (vertices) and connecting roads (edges), then it would be important to a road inspector to find cycles in the graph, because the inspector would prefer to drive along roads only once. A salesman might look for closed paths, preferring to visit cities exactly once. It seems clear that if there is a way to drive from one city to another, then there ought to be a way that avoids going through the same city twice. The next theorem states this fact and tells us that the length of a path or closed path is limited.

Theorem 3.8 (a). If there is a walk originating at v and terminating at u in a graph G, then there is a path from v to u. (b) If G has order n, then the length of a path in G is at most $n - 1$. The length of a closed path is at most n.

Proof.

(a) Suppose $v, v_1, v_2, \ldots, u$ is a walk from v to u, with $v \neq u$. If the walk is not a path, then some vertex appears twice in the sequence. Let x be the first such vertex. Then the walk contains at least one closed walk of the form x, v_j, v_{j+1}, . . ., v_m, x.

Delete the vertices v_j, v_{j+1}, . . ., v_m, x from the sequence v, v_1, v_2, . . ., u. If the result is a path, we are done. Otherwise another such repeated vertex can be found and the deletion process repeated. Since we delete at least one vertex each time and there was a finite number of vertices in v, v_1, v_2, . . ., u, eventually no more vertices can be deleted, so this process must result in a path from v to u.

In the case $v = u$, the same process is applied to delete all repetitions of vertices except the initial and terminal vertex. The result is a closed path.

(b)   Consider a path in the graph G, where G has n vertices. If the path has length t, then there are $t + 1$ vertices traversed by the path. In a path that is not closed all vertices are distinct, so there are at most n vertices traversed. Thus $t + 1 \leq n$ and the path has length at most $n - 1$. In a closed path the initial and terminal vertices are the same, and there is no other repeated vertex. Thus if the closed path has length t, there are t distinct vertices. Therefore if G has n vertices, the length of a closed path is at most n.   ∎

Let u be a vertex in graph G. The vertex v is said to be **reachable** (or **accessible**) from u iff there is a path from u to v. If v is reachable from u, then the number of edges in a path of minimum length from u to v is called the **distance from u to v** and denoted by $d(u, v)$. For any vertex u, we say u is reachable from u and $d(u, u) = 0$.

In the graph with vertex set $V = \{a, b, c, q, e, f, g, h, i, j, k\}$ shown in figure 3.23, the vertex c is reachable only from the vertices c, q, g, h, j, and k. Also $d(q, j) = 3$, $d(c, k) = 2$, $d(k, q) = 2$, and $d(c, q) = 1$. If we think of a graph as representing cities and roads, then the graph of figure 3.23 might represent cities on three islands.

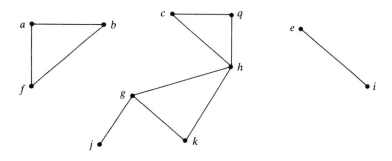

Figure 3.23

By our definition, every vertex of a graph is reachable from itself. Also, if v is reachable from u, then by reversing the order of the edges in a path from v to u, we see that u is reachable from v. Finally, if v is reachable from u and w is reachable from v, then by following the path from u to v and then the path from v to w, we see that there is a walk and therefore a path from u to w. Thus reachability is an equivalence relation on a graph.

A graph G is said to be **connected** if every vertex is reachable from every other vertex. Otherwise G is **disconnected.**

Examination of the disconnected graph of figure 3.23 suggests a way to describe a disconnected graph. For each vertex v in a graph G, let $C(v)$, the **component** of v, be the subgraph consisting of all vertices of the graph reachable from v, together with all the edges of G joining these vertices. We see in figure 3.23 three distinct components, namely $C(a)$ with vertex set $\{a, b, f\}$, $C(h)$ with vertex set $\{c, g, h, j, k, q\}$, and $C(e)$ with vertex set $\{e, i\}$. Of course, $C(b) = C(a)$, $C(h) = C(k)$, and so forth. Each of these components is a connected graph. Furthermore, if the vertex set of any component is a proper subset of the vertex set of any subgraph (V, E'), then (V, E') is disconnected. We can summarize all this by saying that a component is a **maximal connected subgraph.**

Theorem 3.9   For each vertex v in a graph G, let $C(v)$ be the component of v in G. That is, $C(v)$ is a subgraph of G whose vertex set is all vertices reachable from v and whose edges are all the edges of G joining these vertices. Then

(a) $C(v) = C(w)$ iff w is reachable from v.
(b)   If $C(v) \neq C(w)$, then no vertex is in both $C(v)$ and $C(w)$.
(c)   For each v, $C(v)$ is connected.
(d)   Each component $C(v)$ is a maximal connected subgraph of G.

Proof.   Note that the vertex set of each component $C(v)$ is the equivalence class of v under the reachability relation. We leave the proof of parts (a) and (b) as exercise 15.

(c)   Let x and y be any two vertices in $C(v)$. Then both x and y are reachable in G from v, so y is reachable from v, and v is reachable from x. Since all the edges needed to reach x and y from v are also in $C(v)$, y is reachable from x in $C(v)$. Therefore $C(v)$ is connected.

(d)   We must show that if $G' = (V', E')$ is a subgraph of G, and V' properly contains the vertices of any component $C(v)$, then G' is disconnected. Suppose u is a vertex in V' that is not in $C(v)$. Then u is not reachable from v in G, so there can be no path from v to u in G'. Thus G' is disconnected.   ■

Theorem 3.9 tells us that every vertex belongs to exactly one component, and that the collection of components is pairwise disjoint. Further, components are maximal connected subgraphs. It follows that every isolated point forms a component, and that a graph is connected iff it has exactly one component.

Exercises 3.4

1. List the degrees of the vertices of each of these graphs. Verify Theorem 3.7 in each case.

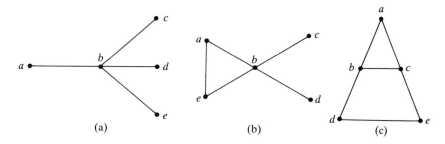

(a)   (b)   (c)

2. If possible, give an example of a graph with order 6 such that
    (a) the vertices have degrees 1, 1, 1, 1, 1, 5 (a star graph).
    (b) the vertices have degrees 1, 1, 1, 1, 1, 1.
    (c) the vertices have degrees 2, 2, 2, 2, 2, 2.
    (d) the vertices have degrees 1, 2, 2, 2, 3, 3.
    (e) exactly two vertices have even degree.
    (f) exactly two vertices have odd degree.

3. If possible, give an example of two nonisomorphic graphs
    (a) with order 6 and size 6.   ★ (b) with order 4 and size 6.

4. Show that the graphs shown are isomorphic by giving a correspondence between them that preserves edges. List also the corresponding edges.

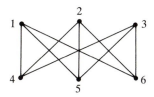
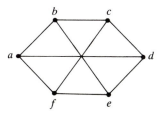

5. Repeat exercise 4 for these graphs.

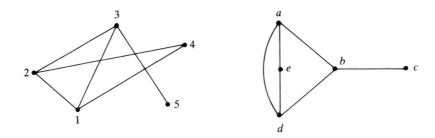

6. Counting the graph itself, the graph G 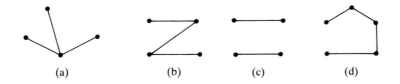 has ten different subgraphs with 4 vertices. We do not count isomorphic graphs as different. For example, among subgraphs with 3 edges, the graphs and and are the only three different subgraphs. List the other subgraphs.

7. Give the complements of these graphs.

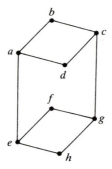

      (a)   (b)   (c)   (d)

8. Which graphs in exercise 7 are isomorphic to their complements?

9. A graph is **regular** iff all its vertices have the same degree. Prove that the complement of a regular graph is regular.

10. For the graph shown, give
    (a) all paths of length 4 and initial vertex g. (There are eight.)
    (b) all cycles of length 6 and initial vertex c. (There are eight.)
    (c) a path of length 7.
    (d) a trail of length 9 that is not a path.

11. Give an example of a graph with 6 vertices having degrees 1, 1, 2, 2, 2, 2 that is
    (a) connected.   (b) disconnected.

12. Give an example of a graph with 6 vertices having
    (a) one component.   (b) two components.
    (c) three components.   (d) six components.

☆ 13. Prove that in every simple graph of order $n \geq 2$ there are two vertices with the
    same degree.

14. Prove that every cycle has length ≥ 3.

15. Prove parts (a) and (b) of Theorem 3.9. Keep in mind that $C(v)$ and $C(w)$ are
    graphs, not just vertex sets; to be the same, they must have the same edges.

16. Show that for a simple graph G, either G or $\tilde{G}$ is connected. (*Hint:* Suppose G
    is disconnected and that $C(x), C(y), \ldots, C(z)$ are the components of G. Con-
    sider adjacencies in $\tilde{G}$.)

17. Give an example of a graph with order 6 such that
    (a) two vertices u and v have distance 5.
    (b) for any two vertices u and v, $d(u, v) \leq 2$.

18. Verify these properties for the distance between vertices in a graph:
    (a) $d(u, v) \geq 0$ (b) $d(u, v) = 0$ iff $u = v$
    (c) $d(u, w) \leq d(u, v) + d(v, w)$

19. Let u and r be vertices in a graph such that $d(u, r) \geq 2$. Show that there exists
    a vertex w such that $d(u, w) + d(w, r) = d(u, r)$.

20. An edge e of a connected graph is called a **bridge** iff, when e is removed from
    the edge set, the resulting subgraph is disconnected. For example, the bridges
    in the graph below are 12, 23, and 56. Give an example of a connected graph of
    order 7
    (a) with no bridges.   (b) with one bridge.
    (c) with 6 bridges.

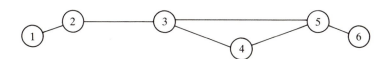

21. A **tree** is a connected graph with no cycles. The three graphs shown below are
    trees. In fact, a tree with one, two, or three vertices must be isomorphic, re-
    spectively, to these three graphs.
    (a) Give the two nonisomorphic trees of order 4.
    (b) Give the three nonisomorphic trees of order 5.

22. Prove that a graph is a tree iff every pair of vertices is connected by exactly one path.

23. Prove that in a tree, every edge is a bridge.

24. **Proofs to Grade.**
    ☆ (a) **Claim.** Every closed path is a cycle.
        "***Proof.***" Let $x, y_1, y_2, \ldots, y_n, x$ be a closed path in a graph. Since no vertex other than the initial and terminal vertex is repeated, all edges are distinct. Therefore $x, y_1, y_2, \ldots, y_n, x$ is a closed walk with distinct edges. That is, the closed path is a cycle. ■

    (b) **Claim.** If v and w are vertices in a graph such that $d(u, v) = 3$ and $d(u, w) = 4$, then $d(v, w) \leqslant 7$.
        "***Proof.***" Suppose $d(u, v) = 3$ and $d(u, w) = 4$. Then there is a path u, x_1, x_2, v from u to v with length 3 and a path u, y_1, y_2, y_3, w from u to w with length 4. Then $v, x_2, x_1, u, y_1, y_2, y_3, w$ is a path from v to w with length 7. The distance from v to w is the length of the shortest path from v to w and there is a path of length 7, so $d(v, w) \leqslant 7$. ■

4

Functions

The general notion of a function should be familiar to you through previous courses in algebra, trigonometry, or calculus. Intuitively, a function is thought of as a rule of correspondence between two sets that assigns to each object in the first set exactly one object from the second set. The statement "Distance is a function of time" means that there is a rule according to which the distance an object has traveled is associated with the time elapsed. The idea of a function as a rule is useful but not precise enough for a careful study. Unfortunately, the idea of a function as a rule gives the impression that a function must be given by a formula, which is not always the case.

The word "function" was first used by G. W. Leibniz in 1694. J. Bernoulli defined a function as "any expression involving variables and constants" in 1698. The familiar notation $f(x)$ was first used by Euler in 1734. It is only relatively recently that it has become standard practice to treat a function as a relation with special properties. This is possible because the rule that makes an object in one set correspond to an object from a second set may be thought of as producing ordered pairs.

The basic properties of functions, some operations on functions, and induced set functions will be presented in this chapter and used throughout the remainder of the book.

4.1

Functions as Relations

> **Definition.**   A **function f from A to B** is a relation from A to B that satisfies
>
>   (i) $\text{Dom}(f) = A$.
>   (ii) If $(x, y) \in f$ and $(x, z) \in f$, then $y = z$.
>
> In the case where $A = B$, we say f is a **function on A.**

A function f from A to B may also be called a **mapping** of A to B. We write $f: A \rightarrow B$, and this is read "f maps A to B" or "f is a function from A to B." As required by the definition, the domain of f is A. The set B is called the **codomain of f**. As with any relation, $\text{Rng}(f) = \{v: (\exists u)(u, v) \in f\}$. No restriction is placed on the sets A and B. They may be sets of numbers, ordered pairs, functions, or even sets of sets of functions.

Let $A = \{1, 2, 3\}$, $B = \{2, 5, 6\}$. The sets

$$r_1 = \{(1, 2), (2, 5), (3, 6), (2, 6)\}$$
$$r_2 = \{(1, 2), (2, 6), (3, 5)\}$$
$$r_3 = \{(1, 5), (2, 5), (3, 2)\}$$

are all relations from A to B. Since $(2, 5)$ and $(2, 6)$ are ordered pairs in r_1 with the same first coordinate, and $5 \neq 6$, r_1 is *not* a function from A to B. Both r_2 and r_3 satisfy the conditions (i) and (ii), so they are functions from A to B.

Now let us consider the set $r_4 = \{(1, 2), (3, 6)\}$. Is r_4 a function? Certainly, r_4 satisfies condition (ii) for functions, but our question is not well posed because we are not given a domain and codomain to consider. Since $\text{Dom}(r_4) = \{1, 3\} \neq A$, r_4 is *not* a function from A to B. However, r_4 is a function from $\{1, 3\}$ to B.

Let $G = \{(x, y) \in \mathbf{N} \times \mathbf{N}: y = x + 2\}$. Then G is a function from $\mathbf{N}$ to $\mathbf{N}$. Also, G is a function from $\mathbf{N}$ to $\mathbf{R}$. Indeed, since $\text{Rng}(G) = \{3, 4, 5, 6, \ldots\}$, G is a function from $\mathbf{N}$ to any set that includes $\{3, 4, 5, 6, \ldots\}$.

Let $H = \{(x, y) \in \mathbf{Z} \times \mathbf{Z}: x^2 + y^2 = 2\}$. Then H is not a function, since $(1, 1) \in H$ and $(1, -1) \in H$, in violation of condition (ii).

To verify that a given relation f from A to B is a function from A to B, it must be shown that *every* element of A appears as a first coordinate of *exactly one* ordered pair in f. The fact that each $a \in A$ is used *at least* once as a first coordinate makes $\text{Dom}(f) = A$; the fact that a is used *only* once fulfills condition (ii).

It is condition (ii) of the definition that makes f into a rule of correspondence. The rule associated with f is that when $(x, y) \in f$, then y is the unique object that corresponds to x. Having $(1, 6)$ and $(1, 6)$ in a function is allowed (since writing an object twice in a set adds nothing to the set), but having $(1, 6)$ and $(1, 5)$ in f is not allowed, for this would give us two "answers" to the question "What corresponds to 1?"

It is worth noting that the definition of a function says a great deal about first coordinates but almost nothing about the second coordinates of ordered pairs. It may happen that some elements of the codomain are not used as second coordinates, or that some elements of the codomain are used as second coordinates more than once. The relation r_3 above, for example, has $\text{Rng}(r_3) = \{2, 5\} \neq B$, and both $(1, 5)$ and $(2, 5)$ are pairs in r_3. One-to-one and onto functions satisfy certain conditions on their second coordinates. They will be discussed in section 4.3.

Definition.   Let $f\colon A \to B$. If $(x, y) \in f$, then we write $f(x) = y$, and say that y is **the value of f at x.** Also, y is the **image** of x under f, and x is a **pre-image** of y under f. Elements of A are sometimes called **arguments** of the function f.

Each argument of a function has exactly one image, but elements of the codomain may have several pre-images or none at all. A function $f\colon A \to B$ may be expressed as

$$f = \{(x, f(x))\colon x \in A\}.$$

It is important to distinguish between the symbols f and $f(x)$. Technically, it is incorrect to speak of "the function $f(x)$." The symbol f denotes a set of ordered pairs, whereas $f(x)$ is simply an element of the range of f. Specifically, $f(x)$ is the element of $\text{Rng}(f)$ that corresponds to the element x of $\text{Dom}(f)$.

Example.   Let $F = \{(x, y) \in \mathbf{Z} \times \mathbf{Z}\colon y = x^2\}$. Then $F\colon \mathbf{Z} \to \mathbf{Z}$. The domain of F is $\mathbf{Z}$, the codomain of F is $\mathbf{Z}$, and $\text{Rng}(F) = \{0, 1, 4, 9, 16, 25, \ldots\}$. The image of 4 is 16, the value of F at -5 is 25, $F(12) = 144$, and both 3 and -3 are pre-images of 9. An element of the codomain that has no pre-image is 6. The function F is given by the rule $F(x) = x^2$.

Example.   The empty set $\varnothing$ is a function on $\varnothing$. Moreover, if $f\colon A \to B$ and any one of f, A, or $\text{Rng}(f)$ is empty, then all three are empty. (See exercise 6).

We next present several examples of functions that commonly occur in mathematics. You should remember them by name.

Example.   Let $K: \mathbf{R} \rightarrow \mathbf{R}$ be defined by $K(x) = 3$ for every $x \in \mathbf{R}$. Then $K = \{(x, 3): x \in \mathbf{R}\}$. This function has range $\{3\}$. The only element of the codomain $\mathbf{R}$ that has a pre-image is 3, and every element of $\mathbf{R}$ is a pre-image of 3. K is an example of a **constant function**—that is, it is a function whose *range consists of a single element.*

Example.   Assume that a universe U has been specified, and that $A \subseteq U$. Define $\chi_A: U \rightarrow \{0, 1\}$ by

$$\chi_A(x) = \begin{cases} 1 & \text{if } x \in A \\ 0 & \text{if } x \in U - A. \end{cases}$$

Then $\chi_A(x)$ is called the **characteristic function of A**. For example, if $A = [1, 4)$, with the universe being the real numbers, then $\chi_A(x) = 1$ iff $1 \leqslant x < 4$. Figure 4.1 is a graph of $\chi_{[1,4)}$.

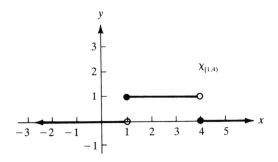

Figure 4.1

Example.   One generalization of the characteristic function is the **step function**. Suppose $\{C_1, C_2, \ldots, C_n\}$ is a partition of a set A and $\{b_1, b_2, \ldots, b_n\}$ is a subset of a set B. Define $f: A \rightarrow B$ by

for $x \in C_i$, let $f(x) = b_i$.

As an example, let $A = [1, 5]$ with $C_1 = [1, 2)$, $C_2 = \{2\}$, $C_3 = (2, 4)$, and $C_4 = [4, 5]$. Choose real numbers $b_1 = 3$, $b_2 = 1$, $b_3 = 4$, $b_4 = 2$. The graph of the corresponding step function is given in figure 4.2.

Example.   If R is an equivalence relation on the set X, then the function from X to X/R that sends each $a \in X$ to a/R is called the

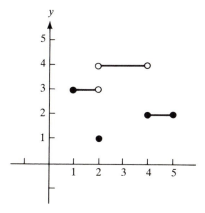

Figure 4.2

canonical map. As an example, let R be the relation of congruence modulo 5 on **Z**, and let f be the canonical map. The images of 9 and -3 are

$$f(9) \quad = 9/R = \{. \ . \ ., -6, -1, 4, 9, 14, \ . \ . \ .\}$$
$$f(-3) = -3/R = \{. \ . \ ., -8, -3, 2, 7, 12, \ . \ . \ .\}.$$

The equivalence classes under the last-name relation L on the set P of all people with last names (see page 103) are sets of people all having the same last name. Under the canonical map f from P to P/L, every person corresponds to his or her equivalence class. Thus $f(\text{Charlie Brown}) = f(\text{Buster Brown})$ is the set of all people with last name Brown. The canonical map is a natural function to consider, and it plays an essential role in the development of many mathematical structures.

Example.   Let A be any set and, for each $x \in A$, let $I_A(x) = x$. Then I_A is a function on A with range A, called the **identity function on A**. In chapter 3 the relation I_A was called the identity relation on A. If $A \subseteq B$, then the function $i: A \to B$ given by $i(x) = x$ for every $x \in A$ is the **inclusion map** from A to B. It is clear that $i = \{(x, x): x \in A\} = I_A$, but i is thought of as a function from A to B while I_A is thought of as a function from A to A.

Example.   The **greatest integer function, GI,** is a function with domain **R** and codomain **Z**. For $x \in$ **R**,

$$GI(x) = [\![x]\!] = \text{the largest integer } n \text{ so that } n \leqslant x.$$

Specifically, $[\![5.4]\!] = 5$, $[\![\sqrt{2}]\!] = 1$, $[\![10]\!] = 10$, $[\![-\frac{1}{3}]\!] = -1$, $[\![-\pi]\!] = -4$, and so on.

For functions whose domain and range are subsets of **R**, the domain is often left unspecified. It is assumed then that the domain is the largest possible subset of the reals. For example, since $\sqrt{x + 1}$ is a real number iff $x \geq -1$, the rule $G(x) = \sqrt{x + 1}$ has domain $[-1, \infty)$. Likewise, the domain of the function H given by $H(x) = 1/(\sin x)$ is **R** $- \{k\pi : k \in \mathbf{Z}\}$.

For a function f whose domain is a collection of ordered pairs we shall write $f(x, y)$ instead of the correct but cumbersome $f((x, y))$. For the multiplication function $f: \mathbf{Z} \times \mathbf{Z} \to \mathbf{Z}$ that associates the product xy with the pair of integers (x, y) we write $f(x, y) = xy$. Thus $f(3, 2) = 6$, $f(-10, 4) = -40$, etc.

Example. For a relation S from A to B we may consider two **projection** functions, π_1 and π_2, where

$$\pi_1: S \to A$$
$$\text{and } \pi_2: S \to B.$$

We define $\pi_1(a, b) = a$ for all $(a, b) \in S$. The effect of π_1 is to pick out the first coordinate of a given ordered pair. In a similar fashion, $\pi_2(a, b) = b$. Figure 4.3 shows the images of $(6, 4)$ under π_1 and π_2 on the relation **R** $\times$ **R**.

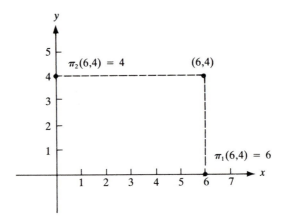

Figure 4.3

The projection function π_1 on a vertical line in **R** $\times$ **R** will send every point on the line to the x-intercept. All other lines are projected by π_1 to the horizontal axis (*i.e.*, have the horizontal axis as range).

Because functions are sets, we already have a definition of function equality: $f = g$ means $f \subseteq g$ and $g \subseteq f$. By this approach, the functions f and g, where $f(x) = (x^2 - 1)/(x + 1)$ and $g(x) = x - 1$, are not equal, because $(-1, -2) \in g$ and $(-1, -2) \notin f$, so $g \not\subseteq f$. A more natural and useful way to express the idea that two functions are equal is to assert that they should have the same domain (so that they act on the same objects) and that for each object in the common domain the functions should agree.

Theorem 4.1   Two functions f and g are equal iff

(i) $\text{Dom}(f) = \text{Dom}(g)$

and

(ii)   for all $x \in \text{Dom}(f)$, $f(x) = g(x)$.

Proof.   ⟨*We prove that conditions (i) and (ii) hold when $f = g$. The converse is left as exercise 10.*⟩ Assume $f = g$.

(i)   Suppose $x \in \text{Dom}(f)$. Then $(x, y) \in f$ for some y and, since $f = g$, $(x, y) \in g$. Therefore $x \in \text{Dom}(g)$. This shows $\text{Dom}(f) \subseteq \text{Dom}(g)$. Similarly, $\text{Dom}(g) \subseteq \text{Dom}(f)$; so $\text{Dom}(f) = \text{Dom}(g)$.

(ii)   Suppose $x \in \text{Dom}(f)$. Then for some y, $(x, y) \in f$. Since $f = g$, $(x, y) \in g$. Therefore $f(x) = y = g(x)$.   ∎

Exercises 4.1

1.   Which of the following relations are functions? For those relations that are functions indicate the domain and a possible codomain.
   ★   (a) $R_1 = \{(0, \triangle), (\triangle, \square), (\square, \cap), (\cap, \cup), (\cup, 0)\}$
      (b) $R_2 = \{(1, 2), (1, 3), (1, 4), (1, 5), (1, 6)\}$
      (c) $R_3 = \{(1, 2), (2, 1)\}$
      (d) $R_4 = \{(x, y) \in \mathbf{R} \times \mathbf{R}: x = \sin y\}$
      (e) $R_5 = \{(x, y) \in \mathbf{N} \times \mathbf{N}: x \leqslant y\}$
      (f) $R_6 = \{(x, y) \in \mathbf{Z} \times \mathbf{Z}: y^2 = x\}$
   ★   (g) $R_7 = \{(x, y) \in \mathbf{R} \times \mathbf{R}: y = x^2 + 2x + 1\}$
      (h) $R_8 = \{(x, y) \in \mathbf{N} \times \mathbf{N}: y^2 = x\}$
      (i) $R_9 = \{(1, 1), (2, 2), (3, 3), (4, 3), (5, 3)\}$
      (j) $R_{10} = \{(\varnothing, \{\varnothing\}), (\{\varnothing\}, \varnothing), (\varnothing, \varnothing), (\{\varnothing\}, \{\varnothing\})\}$

2.   Identify the domain, range, and a possible codomain for each of the following functions.
   ★   (a) $\{(x, y) \in \mathbf{R} \times \mathbf{R}: y = 1/(x + 1)\}$
      (b) $\{(x, y) \in \mathbf{R} \times \mathbf{R}: y = x^2 + 5\}$
      (c) $\{(x, y) \in \mathbf{N} \times \mathbf{N}: y = x + 5\}$
   ★   (d) $\{(x, y) \in \mathbf{R} \times \mathbf{R}: y = \tan x\}$
      (e) $\{(x, y) \in \mathbf{R} \times \mathbf{R}: y = 1/\cos x\}$

   (f) $\{(x, y) \in \mathbf{R} \times \mathbf{R}: y = \chi_{\mathbf{N}}(x)\}$
   (g) $\{(x, y) \in \mathbf{N} \times \mathbf{N}: y = 13\}$
   (h) $\{(x, y) \in \mathbf{R} \times \mathbf{R}: y = (e^x + e^{-x})/2\}$
   (i) $\{(x, y) \in \mathbf{R} \times \mathbf{R}: y = (x^2 - 4)/(x - 2)\}$
   (j) $\{(x, y) \in \mathbf{Z} \times \mathbf{Z}: y = (x^2 - 4)/(x - 2)\}$

3.   For the real function f given by $f(x) = x^2 - 1$,
   (a)   what is the image of 5 under f?
   (b)   what is a pre-image of 15?
★ (c)   find all pre-images of 24.
   (d)   what argument of f is associated with the value 20?
   (e)   what is the value of f at -1?
   (f)   what is a pre-image of -10?

4.   Assuming that the domain of each of the following functions is the largest possible subset of $\mathbf{R}$, find the domain and range of
★ (a) $f(x) = \dfrac{x^2 - 7x + 12}{x - 3}$ (b) $f(x) = 2x + 5$

   (c) $f(x) = \dfrac{1}{\sqrt{x + \pi}}$ (d) $f(x) = \sqrt{5 - x}$
   (e) $f(x) = \sqrt{5 - x} + \sqrt{x - 3}$ (f) $f(x) = \sqrt{x + 2} + \sqrt{-2 - x}$

5.   Show that the following relations are not functions on $\mathbf{R}$.
   (a) $\{(x, y) \in \mathbf{R} \times \mathbf{R}: x^2 = y^2\}$
   (b) $\{(x, y) \in \mathbf{R} \times \mathbf{R}: x^2 + y^2 = 1\}$
   (c) $\{(x, y) \in \mathbf{R} \times \mathbf{R}: x = \cos y\}$
   (d) $\{(x, y) \in \mathbf{R} \times \mathbf{R}: y = \sqrt{x}\}$

6.   (a)   Prove that $\varnothing: \varnothing \to \varnothing$.
   (b)   Prove that if $f: A \to B$ and any one of f, A, or $\text{Rng}(f)$ is empty, then all three are empty.

7.   Let the universe be $\mathbf{R}$ and $A = [1, 3)$. Sketch the graph of
   (a) χ_A (b) $\chi_{\tilde{A}}$ (c) $\chi_{\{1\}}$ (d) $\chi_{\mathbf{N}}$

8.   Let U be the universe and $A \subseteq U$ with $A \neq \varnothing$, $A \neq U$. Let χ_A be the characteristic function of A.
★ (a)   What is $\{x \in U: \chi_A(x) = 1\}$?   (b)   What is $\{x \in U: \chi_A(x) = 0\}$?
   (c)   What is $\{x \in U: \chi_A(x) = 2\}$?

9.   Explain why the functions $f(x) = (9 - x^2)/(x + 3)$ and $g(x) = 3 - x$ are not equal.

10.   Prove the converse of Theorem 4.1. That is, prove that if (i) $\text{Dom}(f) = \text{Dom}(g)$ and (ii) for all $x \in \text{Dom}(f)$, $f(x) = g(x)$, then $f = g$.

11.   For the canonical map $f: \mathbf{Z} \to \mathbf{Z}_6$, find
★ (a) $f(3)$ (b)   the image of 6
   (c)   a pre-image of $3/\equiv_6$ (d)   all pre-images of $1/\equiv_6$

12.   Let S be a relation from A to B. Let π_1 and π_2 be the projection functions. In terms of S, find
★ (a) $\text{Rng}(\pi_1)$ (b) $\text{Rng}(\pi_2)$

13. A **metric** on a set X is a function $d: X \times X \to \mathbf{R}$ such that for all $x, y, z \in X$,

   (i) $d(x, y) \geq 0$.
   (ii) $d(x, y) = 0$ iff $x = y$.
   (iii) $d(x, y) = d(y, x)$.
   (iv) $d(x, y) + d(y, z) \geq d(x, z)$.

   Prove each of the following is a metric for the indicated set.
   ★ (a) $X = \mathbf{N}, d(x, y) = |x - y|$
   (b) $X = \mathbf{R}, d(x, y) = \begin{cases} 0 & \text{if } x = y \\ 1 & \text{if } x \neq y \end{cases}$
   (c) $X = \mathbf{R} \times \mathbf{R}, d((x, y), (z, w)) = \sqrt{(x - z)^2 + (y - w)^2}$
   (d) $X = \mathbf{R} \times \mathbf{R}, d((x, y), (z, w)) = |x - z| + |y - w|$

14. Suppose $\#A = m$ and $\#B = n$. We have seen that $\#(A \times B) = mn$ and that there are 2^{mn} relations from A to B. Find the number of relations from A to B that are
   (a) functions from A to B.
   (b) functions with one element in the domain.
   ★ (c) functions with two elements in the domain.
   (d) functions whose domain is a subset of A.

15. **Proofs to Grade.**
   (a) **Claim.** The functions $f(x) = 1 + 1/x$ and $g(x) = (x + 1)/x$ are equal.
      "Proof." The domain of each function is assumed to be the largest possible subset of $\mathbf{R}$. Thus $\text{Dom}(f) = \text{Dom}(g) = \mathbf{R} - \{0\}$. For every $x \in \mathbf{R} - \{0\}$ we have $f(x) = 1 + (1/x) = (x/x) + (1/x) = (x + 1)/x = g(x)$. Therefore, by Theorem 4.1, $f = g$. ■
   (b) **Claim.** If $h: A \to B$ and $g: C \to D$, then $h \cup g: A \cup C \to B \cup D$.
      "Proof." Suppose $(x, y) \in h \cup g$ and $(x, z) \in h \cup g$. Then $(x, y) \in h$ or $(x, y) \in g$, and $(x, z) \in h$ or $(x, z) \in g$. If $(x, y) \in h$ and $(x, z) \in h$, then $x = z$. Otherwise, $(x, y) \in g$ and $(x, z) \in g$; so again $x = z$. Therefore, $h \cup g$ is a function. Also, $\text{Dom}(h \cup g) = \text{Dom}(h) \cup \text{Dom}(g) = A \cup C$, so $h \cup g: A \cup C \to B \cup D$. ■

4.2

Constructions of Functions

In this section we consider several methods for constructing new functions from given ones. The operations of composition and inversion of relations were discussed in chapter 3. Since every function is a relation, the operations of composition and inversion are performed in the same way as they were for relations. Thus if $F: A \to B$, then the inverse of the function F is the relation

$$F^{-1} = \{(x, y): (y, x) \in F\}.$$

It is necessary to say the *relation* F^{-1} because the inverse of a function is a relation, but might not be a function. Conditions under which F is a function will be given in the next section.

Example.   For the function $F = \{(x, y): y = 2x + 1\}$ (where x and y are understood to be real numbers), the inverse of F is

$$
\begin{aligned}
F^{-1} &= \{(x, y): (y, x) \in F\} \\
&= \{(x, y): x = 2y + 1\} \\
&= \{(x, y): y = (x - 1)/2\}
\end{aligned}
$$

which is also a function. However, for the function $G = \{(x, y): y = x^2\}$, we have

$$
\begin{aligned}
G^{-1} &= \{(x, y): (y, x) \in G\} \\
&= \{(x, y): x = y^2\} \\
&= \{(x, y): y = \pm\sqrt{x}\}
\end{aligned}
$$

which is not a function.

If $F: A \to B$ and $G: B \to C$, the composite of F and G is the relation

$$
G \circ F = \{(x, z) \in A \times C: \text{for some } y \in B, (x, y) \in F \text{ and } (y, z) \in G\}.
$$

Example.   A composite of the two functions F and G, above, is the relation

$$
\begin{aligned}
G \circ F &= \{(x, z): \text{for some } y \in \mathbf{R}, (x, y) \in F \text{ and } (y, z) \in G\} \\
&= \{(x, z): (\exists y \in \mathbf{R})(y = 2x + 1 \text{ and } z = y^2)\} \\
&= \{(x, z): z = (2x + 1)^2\}.
\end{aligned}
$$

Before going on with composition we will take advantage of the fact that each element of the domain of a function has a unique image. This will greatly simplify the notation for composition of functions.

For any functions H and K, $(x, z) \in K \circ H$ iff for some y, $(x, y) \in H$ and $(y, z) \in K$. This can be restated as $H(x) = y$ and $K(y) = z$. This means that $(x, z) \in K \circ H$ iff $z = K(H(x))$. The simplification, then, is that $(K \circ H)(x) = K(H(x))$. For example, the composite $G \circ F$ for the functions given above could be computed as follows:

$$
(G \circ F)(x) = G(F(x)) = G(2x + 1) = (2x + 1)^2
$$

Notice that the first function applied in composition is the function on the right, which is closer to the argument.

Example.   If $H(x) = \sin x$ and $K(x) = x^2 + 6x$, then

$$(H \circ K)(x) = H(K(x)) = H(x^2 + 6x) = \sin (x^2 + 6x)$$

and

$$(K \circ H)(x) = K(H(x)) = K(\sin x) = \sin^2 x + 6 \sin x$$

This example shows that $H \circ K$ and $K \circ H$ need not be equal. Thus composition of functions is not commutative.

Example.   In this example we consider functions on the sets $\mathbf{Z}_6$ and $\mathbf{Z}_3$ (see section 3.2). Denote by $\bar{0}, \bar{1}, \bar{2}, \bar{3}, \bar{4}, \bar{5}$ the six elements of $\mathbf{Z}_6$ and by $[0], [1], [2]$ the three elements of $\mathbf{Z}_3$. Let H be the function on $\mathbf{Z}_6$ to $\mathbf{Z}_3$ given by $H(\bar{0}) = H(\bar{3}) = [0]$, $H(\bar{1}) = H(\bar{4}) = [1]$, and $H(\bar{2}) = H(\bar{5}) = [2]$. Let K be the function $(([0], [0]), ([1], [2]), ([2], [1]))$ on $\mathbf{Z}_3$ to $\mathbf{Z}_3$. Then $K \circ H(\bar{4}) = K(H(\bar{4})) = K([1]) = [2]$ and $K \circ H(\bar{0}) = K([0]) = [0]$. Incidentally, notice that K^{-1} is a function. In fact, $K^{-1} = K$. The composite $K \circ K$ is $(([0], [0]), ([1], [1]), ([2], [2]))$, which is the identity function on $\mathbf{Z}_3$.

Theorem 4.2   If $F: A \to B$ and $G: B \to C$, then $G \circ F: A \to C$. Thus the composite of functions is a function whose domain is the domain of the first function applied.

Proof.   ⟨$G \circ F$ is a relation from A to C, and we know that $Dom(G \circ F) \subseteq A$ and $Rng(G \circ F) \subseteq C$. To show $G \circ F$ is a function from A to C, we must show (i) $A \subseteq Dom(G \circ F)$; and (ii) if $(x, y) \in G \circ F$ and $(x, z) \in G \circ F$, then $y = z$.⟩

(i)   Suppose $x \in A$. Since $A = Dom(F)$, there is $b \in B$ such that $F(x) = b$. But $B = Dom(G)$, so there is $c \in C$ such that $G(b) = c$. Then $c = G(b) = G(F(x)) = (G \circ F)(x)$, so $x \in Dom(G \circ F)$. Therefore $A \subseteq Dom(G \circ F)$.

(ii)   Assume $(x, y) \in G \circ F$ and $(x, z) \in G \circ F$. Then there is $u \in B$ such that $(x, u) \in F$ and $(u, y) \in G$; and there is $v \in B$ such that $(x, v) \in F$ and $(v, z) \in G$. Since F is a function, $(x, u) \in F$ and $(x, v) \in F$, $u = v$. Then because G is a function and $(u, y) \in G$ and $(v, z) = (u, z) \in G$, $y = z$. This shows $G \circ F$ is a function.   ∎

It has already been proved in chapter 3 that composition of relations is associative, and this result applies to functions as well. Similarly, forming the composite of a function with the appropriate identity function yields the function again. These results are restated here especially for functions in order to emphasize their importance and demonstrate the use of functional notation in their proofs.

Theorem 4.3 The composition of functions is associative. That is, if $f\colon A \to B$, $g\colon B \to C$, and $h\colon C \to D$, then $(h \circ g) \circ f = h \circ (g \circ f)$.

Proof. *⟨The idea behind this proof is the characterization of equal functions in Theorem 4.1. The proof also uses the result from Theorem 4.2 that the domain of a composite is the domain of the first function applied.⟩* The domain of each function is A, by Theorem 4.2. If $x \in A$, then $((h \circ g) \circ f)(x) = (h \circ g)(f(x)) = h(g(f(x))) = h((g \circ f)(x)) = (h \circ (g \circ f))(x)$. ∎

Theorem 4.4 Let $f\colon A \to B$. Then $f \circ I_A = f$ and $I_B \circ f = f$.

Proof. $\mathrm{Dom}(f \circ I_A) = \mathrm{Dom}(I_A) = A = \mathrm{Dom}(f)$. If $x \in A$, then $(f \circ I_A)(x) = f(I_A(x)) = f(x)$. Therefore $f \circ I_A = f$. The proof that $I_B \circ f = f$ is left as exercise 4. ∎

Theorem 4.5 Let $f\colon A \to B$ with $\mathrm{Rng}(f) = C$. If f^{-1} is a function, then $f^{-1} \circ f = I_A$ and $f \circ f^{-1} = I_C$.

Proof. Suppose $f\colon A \to B$ and f^{-1} is a function. Then $\mathrm{Dom}(f^{-1} \circ f) = \mathrm{Dom}(f)$ *⟨by Theorem 4.2⟩*. Thus $\mathrm{Dom}(f) = A = \mathrm{Dom}(I_A)$. Suppose $x \in A$. From the fact that $(x, f(x)) \in f$, we have $(f(x), x) \in f^{-1}$. Therefore $(f^{-1} \circ f)(x) = f^{-1}(f(x)) = x = I_A(x)$. This proves that $f^{-1} \circ f = I_A$. The proof of the second part of the theorem is left as exercise 5. ∎

It is sometimes desirable to create a new function from a given one by removing some of the original ordered pairs. You may recall having seen this done for the sine function with domain **R**. When this function is restricted to the domain $[-\pi/2, \pi/2]$, the result is usually referred to as the Sine function (with a capital S), abbreviated Sin, and is the principal branch of the sine function. Then $\mathrm{Sin}\,(\pi/3) = \sin\,(\pi/3) = \sqrt{3}/2$, but $\mathrm{Sin}\,(2\pi/3)$ is undefined.

Definition: Let $f\colon A \to B$, and let $D \subseteq A$. The **restriction of f to D,** denoted $f|_D$, is

$$\{(x, y)\colon (x, y) \in f \text{ and } x \in D\}.$$

We note that the restriction of any function to D will be a function with domain D. If f and g are mappings and g is a restriction of f, then we say f is an **extension** of g.

In this notation the function Sin is $\sin|_{[-\pi/2, \pi/2]}$. The graphs of sine and Sine are shown in figure 4.4.

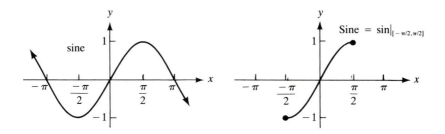

Figure 4.4

Example.   Let $A = \{1, 2, 3, 4\}$, $B = \{a, b, c, d\}$, and $g = \{(1, a), (2, a), (3, d), (4, c)\}$. Then $g|_{\{2\}} = \{(2, a)\}$, $g|_A = g$, and $g|_{\{1,4\}} = \{(1, a), (4, c)\}$.

Let $F: \mathbf{R} \to \mathbf{R}$ be given by $F(x) = 2x + 1$. Figure 4.5 shows the graphs of $F|_{[1,2]}$ and $F|_{\{-2, -1,0,1,2\}}$.

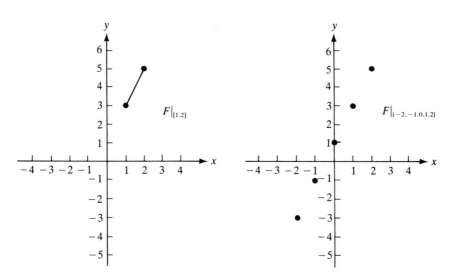

Figure 4.5

Let h and g be functions. Is $h \cap g$ always a function? Is $h \cup g$ always a function? Consider, for example, $h = \{(1, 2), (5, 7), (3, -9)\}$ and $g = \{(1, 8), (5, 7), (4, 8)\}$. Then $h \cap g = \{(5, 7)\}$, which is indeed a function. In general, if x is in the domain of both functions, and $g(x) = h(x) = y$, then $(x, y) \in h \cap g$. An object that is not in both domains or for which $g(x) \neq h(x)$ will not be in $\text{Dom}(h \cap g)$. It turns out that $h \cap g$ is a function (see exercise 8), but this function can just as easily be expressed by restricting the domain of either g or h.

The situation regarding $h \cup g$ is much more interesting and useful. First, for the functions given above, $h \cup g$ is not a function because $(1, 2)$ and $(1, 8)$ are both in $h \cup g$. If we are careful to be sure that two functions h and g have disjoint domains, however, we can make a new function that is an extension of both h and g by putting them together "piecewise." The proof of Theorem 4.6, which states this result, is left as exercise 9. See exercise 12 for a generalization stating that $h \cup g$ is a function when h and g agree on the intersection of their domains.

> **Theorem 4.6** Let h and g be functions such that $\text{Dom}(h) = A$, $\text{Dom}(g) = B$, and $A \cap B = \varnothing$. Then $h \cup g$ is a function with domain $A \cup B$.

For example, if $g: \{1, 2, 3\} \to \{a, b, c\}$ is the function $\{(1, b), (2, a), (3, c)\}$ and $h = \{(4, d)\}$, then $h \cup g: \{1, 2, 3, 4\} \to \{a, b, c, d\}$ is the function $\{(1, b), (2, a), (3, c), (4, d)\}$.

Let h and g be given by $h(x) = x^2$ and $g(x) = 6 - x$. Then $h|_{(-\infty, 2]}$ and $g|_{(2, \infty)}$ have disjoint domains, so their union f is a function that is an extension of each. It is not an extension of h or g (figure 4.6).

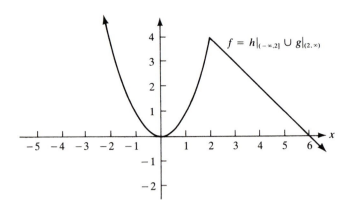

Figure 4.6

The function f can also be described as follows:

$$f(x) = \begin{cases} x^2 & \text{if } x \leq 2 \\ 6 - x & \text{if } x > 2 \end{cases}$$

Functions can also be constructed piecewise from three or more functions. For example, if

$$K(x) = \begin{cases} x + 1 & \text{if } x < -1 \\ \sin \pi x & \text{if } -1 \leqslant x \leqslant 0 \\ \dfrac{x + 3}{x - 3} & \text{if } 0 < x < 3 \\ 4 & \text{if } x \geqslant 3 \end{cases}$$

then K is a function with domain **R** (figure 4.7.) To check that the relation given is a function, it is necessary only to check that the conditions given on the right are mutually exclusive, so that the corresponding sets are pairwise disjoint. The vertical line test, which says that a graph represents a function so long as no vertical line touches the graph more than once, is useful so long as all the graph can be seen. It is not a rigorous proof that a given relation is a function.

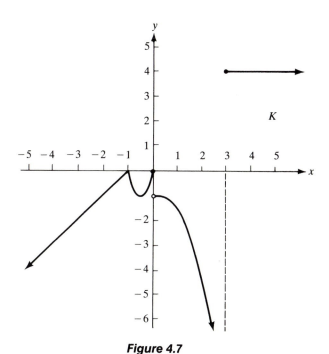

Figure 4.7

We have considered inversion, composition, restriction, extension, and union as means of constructing new functions from old ones. Other methods, especially for functions from the reals to the reals, may be found in exercises 14–17.

Exercises 4.2

1. Find $f \circ g$ and $g \circ f$ for each pair of functions f and g. Use understood domains for f and g.
    ★ (a) $f(x) = 2x + 5$, $g(x) = 6 - 7x$
       (b) $f(x) = x^2 + 2x$, $g(x) = 2x + 1$
    ★ (c) $f(x) = \sin x$, $g(x) = 2x^2 + 1$
       (d) $f(x) = \tan x$, $g(x) = \sin x$
       (e) $f(x) = \dfrac{x + 1}{x + 2}$, $g(x) = x^2 + 1$
       (f) $f(x) = 3x + 2$, $g(x) = |x|$

☆ 2. Find the domain and range of each composite in exercise 1.

3. For which of the following functions f is the relation f^{-1} a function? When f^{-1} is a function, write an explicit expression for $f^{-1}(x)$. Use the understood domain for each function.
    ★ (a) $f(x) = 5x + 2$ (b) $f(x) = 2x^2 + 1$ ★ (c) $f(x) = \dfrac{x + 1}{x + 2}$

       (d) $f(x) = \sin x$ ★ (e) $f(x) = e^{x+3}$ (f) $f(x) = \dfrac{1 - x}{-x}$

       (g) $f(x) = \dfrac{1}{1 - x}$ (h) $f(x) = -x + 3$ (i) $f(x) = \dfrac{-x}{3x - 4}$

4. Prove the remaining part of Theorem 4.4. That is, prove that if $f: A \to B$, then $I_B \circ f = f$.

5. Prove the remaining part of Theorem 4.5. That is, prove that if $f: A \to B$ with $\text{Rng}(f) = C$, and if f^{-1} is a function, then $f \circ f^{-1} = I_C$.

6. Let $f(x) = 4 - 3x$ with domain **R** and $A = \{1, 2, 3, 4\}$. Sketch the graphs of the functions $f|_A$, $f|_{[-1,3]}$, $f|_{(2,4]}$, and $f|_{\{6\}}$. What is the range of $f|_\mathbf{N}$?

7. Describe two extensions of f with domain **R** for the function
    ★ (a) $f = \{(x, y) \in \mathbf{N} \times \mathbf{N}: y = x^2\}$.   (b) $f = \{(x, y) \in \mathbf{N} \times \mathbf{N}: y = 3\}$.

8. Prove that, if f and g are functions, then $f \cap g$ is a function by showing that $f \cap g = g|_A$ where $A = \{x: g(x) = f(x)\}$.

9. Prove Theorem 4.6.

10. Let $f(x) = x^2 + 2$ and $g(x) = x + 5$. Describe the function $f|_{(-\infty, 0]} \cup g|_{(0, \infty)}$.

11. For each pair of functions h and g, determine whether $h \cup g$ is a function. In each case sketch a graph of $h \cup g$.
    ★ (a) $h: (-\infty, 0] \to \mathbf{R}, h(x) = 3x + 4$ (b) $h: [-1, \infty) \to \mathbf{R}, h(x) = x^2 + 1$
          $g: (0, \infty) \to \mathbf{R}, g(x) = \dfrac{1}{x}$ $g: (-\infty, -1] \to \mathbf{R}, g(x) = x + 3$

       (c) $h: (-\infty, 1] \to \mathbf{R}, h(x) = |x|$ (d) $h: (-\infty, 2] \to \mathbf{R}, h(x) = \cos x$
          $g: [0, \infty) \to \mathbf{R}, g(x) = 3 - |x - 3|$ $g: [2, \infty) \to \mathbf{R}, g(x) = x^2$

       (e) $h: (-\infty, 3) \to \mathbf{R}, h(x) = 3 - x$
          $g: (0, \infty) \to \mathbf{R}, g(x) = x + 1$

☆ 12. Let $h: A \to B$, $g: C \to D$ and suppose $E = A \cap C$. Prove $h \cup g$ is a function from $A \cup C$ to $B \cup D$ if and only if $h|_E = g|_E$.

13. Let $f: A \to B$ and $g: C \to D$. Define

$$f \times g = \{((a, c), (b, d)): (a, b) \in f \text{ and } (c, d) \in g\}.$$

Prove that $f \times g: A \times C \to B \times D$. Find an explicit expression for $f \times g (a, c)$.

14. Let $f_1: \mathbf{R} \to \mathbf{R}$ and $f_2: \mathbf{R} \to \mathbf{R}$. Define the **pointwise sum** $f_1 + f_2$ and **pointwise product** $f_1 \cdot f_2$ as follows:

$$f_1 + f_2 = \{(a, c + d): (a, c) \in f_1 \text{ and } (a, d) \in f_2\}$$
$$f_1 \cdot f_2 = \{(a, cd): (a, c) \in f_1 \text{ and } (a, d) \in f_2\}$$

☆ (a) Prove $f_1 + f_2$ and $f_1 \cdot f_2$ are functions with domain $\mathbf{R}$.
   (b) Show that $(f_1 + f_2)(x) = f_1(x) + f_2(x)$ and that $(f_1 \cdot f_2)(x) = f_1(x) \cdot f_2(x)$.

15. Let $f: \mathbf{R} \to \mathbf{R}$ and $c \in \mathbf{R}$. Define the **scalar product** cf by

$$cf = \{(a, cd): (a, d) \in f\}.$$

Prove $cf: \mathbf{R} \to \mathbf{R}$ and find an explicit expression for $(cf)(x)$.

16. State and prove by induction a generalization of exercise 14 for a finite collection of n functions $f_1, f_2, \ldots, f_n$.

17. Let $f_i: \mathbf{R} \to \mathbf{R}$ for $i = 1, 2, \ldots, n$. A **linear combination** of $f_1, f_2, \ldots, f_n$ is a function of the form

$$c_1 f_1 + c_2 f_2 + \cdots c_n f_n$$

where $c_i \in \mathbf{R}$, $i = 1, 2, \ldots, n$. Prove that a step function $g: \mathbf{R} \to \mathbf{R}$ is a linear combination of characteristic functions.

18. **Proofs to Grade.**
   ★ (a) **Claim.** Let $f: A \to B$. If f^{-1} is a function, then $f^{-1} \circ f = I_A$.
      "Proof." Suppose $(x, y) \in f^{-1} \circ f$. Then there is z such that $(x, z) \in f$ and $(z, y) \in f^{-1}$. But this means that $(z, x) \in f^{-1}$ and $(z, y) \in f^{-1}$. Since f^{-1} is a function, $x = y$. Hence $(x, y) \in f^{-1} \circ f$ implies $(x, y) \in I_A$; that is, $f^{-1} \circ f \subseteq I_A$. Now suppose $(x, y) \in I_A$. Since $A = \text{Dom}(f)$, there is a $w \in B$ such that $(x, w) \in f$. Hence $(w, x) \in f^{-1}$. But $(x, y) \in I_A$ implies $x = y$ and $(w, y) \in f^{-1}$. But from $(x, w) \in f$ and $(w, y) \in f^{-1}$, we have $(x, y) \in f^{-1} \circ f$. This shows $I_A \subseteq f^{-1} \circ f$. Therefore $I_A = f^{-1} \circ f$. ∎
      (b) **Claim.** If f and f^{-1} are functions on A, and $f \circ f = f$, then $f = I_A$.
      "Proof." Since $f = f \circ f$, $f^{-1} \circ f = f^{-1} \circ (f \circ f)$. By associativity, we have $f^{-1} \circ f = (f^{-1} \circ f) \circ f$. This gives $I_A = I_A \circ f$. Since $I_A \circ f = f$, we have $I_A = f$. ∎
      (c) **Claim.** If f, g, and f^{-1} are functions on A, then $g = f^{-1} \circ (g \circ f)$.
      "Proof." Using associativity and Theorems 4.4 and 4.5, $f^{-1} \circ (g \circ f) = f^{-1} \circ (f \circ g) = (f^{-1} \circ f) \circ g = I_A \circ g = g$. ∎

4.3

Onto Functions; One-to-One Functions

For every function $f: A \rightarrow B$, $\text{Rng}(f) \subseteq B$. It must not be assumed that $\text{Rng}(f) = B$. In the case when the codomain and range are equal, we say the function maps **onto the codomain.**

Definition.   A function $f: A \rightarrow B$ is **onto** B iff $\text{Rng}(f) = B$. We write $f: A \xrightarrow{\text{onto}} B$. A function that maps onto its codomain is also called a **surjection.**

Whether a function f maps onto its codomain or not, it is still correct to say f maps *to* the codomain. It is not good style to say that a function "is onto" without some indication of the codomain. If the codomain is clear from the context, then simply saying "is onto" is common practice.

Up to this point there has been little need to emphasize the codomain in the definition of a function. After all, the distinction between

$f: \mathbf{N} \rightarrow \mathbf{N}$, where $f(n) = 2n$

and

$g: \mathbf{N} \rightarrow \mathbf{Z}$, where $g(n) = 2n$

seems remote. In this example $\text{Rng}(f) = \text{Rng}(g) = E$, the set of even natural numbers, and any set that contains E is a perfectly acceptable codomain. The function $h: \mathbf{N} \rightarrow E$ with $h(n) = 2n$ has the added feature that every element of the codomain (E) is an image; that is, h is onto E.

As an immediate consequence of the definition, *every function maps onto its range.* That is, if we have a function given as a set of ordered pairs and we wish to be sure the function "is onto," we can do so by choosing the range as the codomain.

To prove that a given function $f: A \rightarrow B$ is onto B, we choose an arbitrary $y \in B$. We then show $y \in \text{Rng}(f)$ by showing there is $x \in A$ such that $f(x) = y$. This shows $B \subseteq \text{Rng}(f)$, which is sufficient to satisfy the definition, since $\text{Rng}(f) \subseteq B$ is always true.

Example.   We will show that $F: \mathbf{R} \rightarrow \mathbf{R}$, defined by $F(x) = x + 2$, is onto $\mathbf{R}$ by showing $\mathbf{R} \subseteq \text{Rng}(F)$. To do this, we show that for every $w \in \mathbf{R}$, there is $x \in \mathbf{R}$ such that $F(x) = w$. ⟨*This x must be chosen such that $F(x) = x + 2 = w$.*⟩ Our x must be

$w - 2$, for with this choice $F(x) = F(w - 2) = (w - 2) + 2 = w$. Hence F is onto **R**.

Example. Let $G: \mathbf{R} \to \mathbf{R}$ be defined by $G(x) = x^2 + 1$. Then G is not onto **R**. To show this, we find an element y in the codomain **R** that has no pre-image in the domain **R**. Let y be -2. Since $x^2 + 1 \geqslant 1$ for every real number x, there is no $x \in \mathbf{R}$ such that $G(x) = -2$. Hence G is not onto **R**.

Example. Let $f: \mathbf{R} \to \mathbf{R}$ be the polynomial function $f(x) = x^3 + 3x^2 - 24x$ (figure 4.8). To prove f is onto **R**, let $w \in \mathbf{R}$. We must show the existence of $a \in \mathbf{R}$ such that $f(a) = w$. The equation

$$f(x) - w = 0$$

is a third-degree polynomial equation in one variable x. Since its degree is odd, and since complex, nonreal roots occur in conjugate pairs, the polynomial $f(x) - w$ has an odd number of real roots. Let a be a real root to $f(x) - w = 0$. Then $f(a) - w = 0$ and hence $f(a) = w$. Thus f is onto **R**.

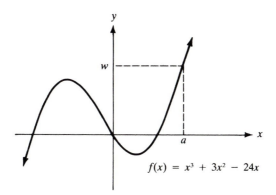

Figure 4.8

Example. We will show that $f: \mathbf{Z} \times \mathbf{Z} \to \mathbf{Z}$ given by $f(x, y) = xy$ is onto **Z**. Let $z \in \mathbf{Z}$. We choose $(z, 1) \in \mathbf{Z} \times \mathbf{Z}$. Then

$$f(z, 1) = z \cdot 1 = z.$$

Thus f is onto **Z**.

Example. Let $F: \mathbf{N} \times \mathbf{N} \to \mathbf{N}$ be defined by $F(m, n) = 2^{m-1}(2n - 1)$. To show that F is onto **N**, let $s \in \mathbf{N}$. We must show that there is $(m, n) \in \mathbf{N} \times \mathbf{N}$ such that $F(m, n) = s$. If s is even, then s may be written as $2^k t$, where t is odd. Since t is odd, $t = 2n - 1$ for some $n \in \mathbf{N}$. Choosing $m = k + 1$, we have $F(m, n) = 2^{m-1}(2n - 1) = 2^k t = s$. If s is odd, then $s = 2n - 1$ for some $n \in \mathbf{N}$. For this n and $m = 1$, we find $F(m, n) = 2^0(2n - 1) = s$. Therefore F is onto $\mathbf{N}$.

The next two theorems relate composition and the property of being onto.

Theorem 4.7   If $f: A \xrightarrow{\text{onto}} B$, and $g: B \xrightarrow{\text{onto}} C$, then $g \circ f: A \xrightarrow{\text{onto}} C$. That is, the composite of onto functions is an onto function.

Proof.   Exercise 4.   ∎

Theorem 4.8   If $f: A \rightarrow B$, $g: B \rightarrow C$, and $g \circ f: A \xrightarrow{\text{onto}} C$, then g is onto C.

Proof. $\langle$*We must show* $C \subseteq Rng(g)$.$\rangle$ Suppose $c \in C$. Since $g \circ f$ maps onto C, there is $a \in A$ such that $(g \circ f)(a) = c$. Let $b = f(a)$, which is in B. Then $(g \circ f)(a) = g(f(a)) = g(b) = c$. Thus there is $b \in B$ such that $g(b) = c$, and g maps onto C.   ∎

For a relation f to be a function from A to B, every element of A must appear exactly once as a first coordinate. No restrictions were made for second coordinates, except that they be elements of B. When every element of B appears *at least once* as a second coordinate, we have called the function *onto* B. Functions for which every element of B appears *at most once* as a second coordinate are called *one-to-one.*

Definition.   A function $f: A \rightarrow B$ is said to be **one-to-one,** written as $f: A \xrightarrow{1-1} B$, iff $(x, y) \in f$ and $(z, y) \in f$ imply $x = z$. In functional notation, that f is one-to-one means $f(x) = f(z)$ implies $x = z$. A one-to-one function is also called an **injection.**

To prove that a given function $f: A \rightarrow B$ is one-to-one, we assume x and z are elements of A such that $f(x) = f(z)$. We then show that $x = z$. The alternative, using the contrapositive, is to assume $x \neq z$ and to show that $f(x) \neq f(z)$. To show that f is not one-to-one, it suffices to exhibit two different elements of A with the same image.

Example.   The function $F: \mathbf{R} \to \mathbf{R}$ defined by $F(x) = 2x + 1$ is one-to-one. To show this, assume $F(x) = F(z)$. Then $2x + 1 = 2z + 1$. Therefore $2x = 2z$, so $x = z$.

Example.   Let $G(x) = 1/(x^2 + 1)$. We attempt to show G is one-to-one by assuming that $G(x) = G(y)$. Then $1/(x^2 + 1) = 1/(y^2 + 1)$. Therefore $x^2 + 1 = y^2 + 1$, so $x^2 = y^2$. It does not follow from this that $x = y$. In fact, this failed "proof" suggests a way to find distinct real numbers with equal images. Indeed, $G(3) = G(-3) = \frac{1}{10}$. Therefore G is not one-to-one.

Example.   Let $f: [0, \infty) \to [0, \infty)$ be the function $f(x) = x^2$. To show that f is one-to-one, suppose $f(x) = f(y)$. Then $x^2 = y^2$, so $x = \pm y$. Since both $x, y \in \text{Dom}(f)$, $x \geq 0$ and $y \geq 0$. Thus from $x^2 = y^2$ we can conclude $x = y$. Therefore f is one-to-one. This example shows that the domain may be quite important in determining whether a function is one-to-one.

Example.   Define $F: \mathbf{N} \times \mathbf{N} \to \mathbf{N}$ by $F(m, n) = 2^{m-1}(2n - 1)$. We will show F is one-to-one. Assume that $F(m, n) = F(r, s)$. We first prove that $m = r$. We may assume that $m \geq r$. ⟨*If $m \leq r$, we could relabel the arguments.*⟩ From $F(m, n) = F(r, s)$, we have $2^{m-1}(2n - 1) = 2^{r-1}(2s - 1)$, which implies $(2^{m-1}/2^{r-1})(2n - 1) = 2s - 1$. Therefore $2^{(m-1)-(r-1)}(2n - 1) = 2s - 1$; that is, $2^{m-r}(2n - 1) = 2s - 1$. Since the right side of the equality is odd, the left side is odd. Thus $2^{m-r} = 1$. Therefore $m - r = 0$, and we conclude that $m = r$.
   Dividing both sides of the equation $2^{m-1}(2n - 1) = 2^{r-1}(2s - 1)$ by 2^{m-1} ⟨$2^{m-1} = 2^{r-1}$⟩, we have $2n - 1 = 2s - 1$, which implies $2n = 2s$, or $n = s$. Thus $m = r$ and $n = s$, which gives $(m, n) = (r, s)$. Hence the function F is one-to-one.

It was observed in the previous section that the inverse of a function is not always a function. The situation is clarified when we understand the connection between inverses and one-to-one functions.

Theorem 4.9   Let $F: A \to B$. F^{-1} is a function from $\text{Rng}(F)$ to A iff F is one-to-one. Furthermore, if F^{-1} is a function, then F^{-1} is one-to-one.

Proof.   Assume that $F: A \xrightarrow{1-1} B$. To show that F^{-1} is a function, assume $(x, y) \in F^{-1}$ and $(x, z) \in F^{-1}$. Then $(y, x) \in F$ and $(z, x) \in F$. Since F is one-to-one, $y = z$. Therefore if F is one-to-one, F^{-1} is a function. By Theorem 3.2(b), the domain of F^{-1} is $\text{Rng}(F)$.

Assume now that F^{-1} is a function. To show that F is one-to-one, assume that $(x, y) \in F$ and $(z, y) \in F$. Then $(y, x) \in F^{-1}$ and $(y, z) \in F^{-1}$. Since F^{-1} is a function, $x = z$. Therefore if F^{-1} is a function, then F is one-to-one.

The proof that if F and F^{-1} are functions, then F^{-1} is one-to-one, is left as exercise 5. ∎

We must be careful not to conclude that if $F: A \xrightarrow{1-1} B$, then $F^{-1}: B \xrightarrow{1-1} A$, since F may not be onto B. Recall that the domain and range of a relation and its inverse are interchanged. Therefore if $F: A \xrightarrow{1-1} B$, then $F^{-1}: \text{Rng}(F) \xrightarrow{1-1} A$.

Corollary 4.10   If $F: A \xrightarrow[\text{onto}]{1-1} B$, then $F^{-1}: B \xrightarrow[\text{onto}]{1-1} A$.

Theorem 4.11   If $f: A \xrightarrow{1-1} B$ and $g: B \xrightarrow{1-1} C$, then $g \circ f: A \xrightarrow{1-1} C$. That is, the composite of one-to-one functions is a one-to-one function.

Proof.   Assume that $(g \circ f)(x) = (g \circ f)(z)$; that is, $g(f(x)) = g(f(z))$. Then $f(x) = f(z)$ since g is one-to-one. Then $x = z$ since f is one-to-one. Therefore $g \circ f$ is one-to-one. ∎

There exist functions that are one-to-one but not onto, onto but not one-to-one, neither, and both (see the exercises). A function that is both one-to-one and maps onto its codomain is called a **one-to-one correspondence** or a **bijection**.

Example.   Let $A = \{a, b, c\}$ and $B = \{p, q, r\}$. The function $f = \{(a, p), (b, r), (c, q)\}$ is a bijection from A onto B.

Example.   In previous examples we have seen $F: \mathbf{N} \times \mathbf{N} \to \mathbf{N}$ given by $F(m, n) = 2^{m-1}(2n - 1)$ is both one-to-one and onto. Thus F is a one-to-one correspondence between $\mathbf{N} \times \mathbf{N}$ and $\mathbf{N}$.

Combining Theorems 4.7, 4.10, and 4.11, we have the following theorem.

Theorem 4.12   If $f: A \to B$ and $g: B \to C$, and each is a one-to-one correspondence, then

(a) $g \circ f: A \to C$ is a one-to-one correspondence.
(b) $f^{-1}: B \to A$ is a one-to-one correspondence.

Analogous to Theorem 4.8 for onto functions, we have the following theorem for one-to-one functions.

Theorem 4.13   If $f: A \to B$, $g: B \to C$, and $g \circ f: A \xrightarrow{1-1} C$, then $f: A \xrightarrow{1-1} B$.

Proof.   Exercise 6.   ∎

We conclude this section with a result that relates the concepts of one-to-one, onto, composition, and inversion, and that also gives a practical method for determining whether two given functions are inverses.

Theorem 4.14   Let $F: A \xrightarrow[\text{onto}]{1-1} B$ and $G: B \xrightarrow[\text{onto}]{1-1} A$. Then $G = F^{-1}$ iff $G \circ F = I_A$ (or $F \circ G = I_B$).

Proof.   If $G = F^{-1}$, then $G \circ F = I_A$ and $F \circ G = I_B$, by Theorem 4.5. Assume now that $G \circ F = I_A$. Then

$$G = G \circ I_B = G \circ (F \circ F^{-1}) = (G \circ F) \circ F^{-1}$$
$$= I_A \circ F^{-1} = F^{-1}.$$

That $G = F^{-1}$ follows similarly from $F \circ G = I_B$.   ∎

Let $F(x) = 2x + 1$, and let $G(x) = (x - 1)/2$. Then $F: \mathbf{R} \xrightarrow[\text{onto}]{1-1} \mathbf{R}$ and $G: \mathbf{R} \xrightarrow[\text{onto}]{1-1} \mathbf{R}$. We calculate the two composites

$$(G \circ F)(x) = G(F(x)) = G(2x + 1) = [(2x + 1) - 1]/2$$
$$= 2x/2 = x.$$

$$(F \circ G)(x) = F(G(x)) = F[(x - 1)/2] = 2[(x - 1)/2] + 1$$
$$= (x - 1) + 1 = x.$$

Therefore $G \circ F = I_\mathbf{R}$ and $F \circ G = I_\mathbf{R}$. Either computation implies that $G = F^{-1}$.

Constructions of functions by restrictions and unions can also be related to the one-to-one and onto properties of functions. These results will be used in the study of cardinality in chapter 5.

Theorem 4.15

(a)   A restriction of a one-to-one function is one-to-one.

(b)   If $h: A \xrightarrow{\text{onto}} C$, $g: B \xrightarrow{\text{onto}} D$, and $A \cap B = \emptyset$, then $h \cup g: A \cup B \xrightarrow{\text{onto}} C \cup D$.

(c)   If $h: A \xrightarrow{1-1} C$, $g: B \xrightarrow{1-1} D$, $A \cap B = \emptyset$, and $C \cap D = \emptyset$, then $h \cup g: A \cup B \xrightarrow{1-1} C \cup D$.

Proof.   Parts (a) and (b) are left as exercise 7.

(c)   Suppose $h: A \xrightarrow{1-1} C$, $g: B \xrightarrow{1-1} D$, $A \cap B = \varnothing$, and $C \cap D = \varnothing$. Then by Theorem 4.6, $h \cup g$ is a function with domain $A \cup B$.

   Suppose $x, y \in A \cup B$. Assume $(h \cup g)(x) = (h \cup g)(y)$.

   (i)   If $x, y \in A$, then $h(x) = (h \cup g)(x) = (h \cup g)(y) = h(y)$. Since h is one-to-one, $x = y$.
   (ii)   If $x, y \in B$, then $g(x) = g(y)$, and g is one-to-one; so $x = y$.
   (iii) Suppose $x \in A$ and $y \in B$. Then $h(x) = g(y)$ and $h(x) \in C$ and $g(y) \in D$. But $C \cap D = \varnothing$. This case is impossible.
   (iv)   Similarly, $x \in B$ and $y \in A$ is impossible.

   In every possible case, $x = y$. Therefore $h \cup g$ is one-to-one.   ■

Exercises 4.3

1.   Which of the following functions map onto their indicated codomains? Prove each of your answers.
   ★ (a) $f: \mathbf{R} \to \mathbf{R}$, $f(x) = \frac{1}{2}x + 6$ (b) $f: \mathbf{Z} \to \mathbf{Z}$, $f(x) = -x + 1000$
   ★ (c) $f: \mathbf{N} \to \mathbf{N} \times \mathbf{N}$, $f(x) = (x, x)$ (d) $f: \mathbf{R} \to \mathbf{R}$, $f(x) = x^3$
     (e) $f: \mathbf{R} \to \mathbf{R}$, $f(x) = \sqrt{x^2 + 5}$ (f) $f: \mathbf{R} \to \mathbf{R}$, $f(x) = 2^x$
     (g) $f: \mathbf{R} \to \mathbf{R}$, $f(x) = \sin x$ (h) $f: \mathbf{R} \times \mathbf{R} \to \mathbf{R}$, $f(x, y) = x - y$
     (i) $f: \mathbf{R} \to [-1, 1]$, $f(x) = \cos x$ (j) $f: \mathbf{R} \to [1, \infty)$, $f(x) = x^2 + 1$

☆ 2.   Which of the functions in exercise 1 are one-to-one? Prove each of your answers.

3.   Let $A = \{1, 2, 3, 4\}$. Describe a codomain B and a function $f: A \to B$ such that f is
   ★ (a)   onto B but not one-to-one.
     (b)   one-to-one but not onto B.
     (c)   both one-to-one and onto B.
     (d)   neither one-to-one nor onto B.

4.   Prove Theorem 4.7.

5.   Prove the remaining part of Theorem 4.9. That is, prove that if F and F^{-1} are functions, then F^{-1} is one-to-one.

6.   Prove Theorem 4.13.

7.   Prove parts (a) and (b) of Theorem 4.15.

8.   Give an example of functions $f: A \to B$ and $g: B \to C$ such that
★   (a) f is onto B, but $g \circ f$ is not onto C.
    (b) g is onto C, but $g \circ f$ is not onto C.
    (c) $g \circ f$ is onto C, but f is not onto B.
    (d) f is one-to-one, but $g \circ f$ is not one-to-one.
★   (e) g is one-to-one, but $g \circ f$ is not one-to-one.
    (f) $g \circ f$ is one-to-one, but g is not one-to-one.

9.   Prove that
    (a) $f(x) = \begin{cases} 2 - x & \text{if } x \leq 1 \\ 1/x & \text{if } x > 1 \end{cases}$ is one-to-one but not onto **R**.

    (b) $f(x) = \begin{cases} x + 4 & \text{if } x \leq -2 \\ -x & \text{if } -2 < x < 2 \\ x - 4 & \text{if } x \geq 2 \end{cases}$ is onto **R** but not one-to-one.

10.   Let A and B be sets and $S \subseteq A \times B$. Let π_1 be the projection function on S to A and π_2 be the projection function on S to B. Give an example to show that
★   (a) π_1 need not be one-to-one.   (b) π_1 need not be onto A.
    (c) π_2 need not be one-to-one.   (d) π_2 need not be onto B.

11.   Suppose $S: A \to B$ is a function. Then $S \subseteq A \times B$. Let π_1 and π_2 be as in exercise 10.
★   (a)   Is $\pi_1: S \to A$ onto A?   (b)   Is $\pi_1: S \to A$ one-to-one?
★   (c)   Is $\pi_2: S \to B$ one-to-one?   (d)   Is $\pi_2: S \to B$ onto B?

    Prove your answers.

12.   Suppose $\#A = m$ and $\#B = n$. Then there are 2^{mn} relations from A to B and n^m functions from A to B (see exercise 14 in section 4.1).
    (a)   Assuming that $m \leq n$, find the number of one-to-one functions from A to B.
★   (b)   If $m > n$, how many one-to-one functions are there from A to B?
★   (c)   Assuming that $m \geq n$, find the number of functions from A onto B.
    (d)   If $m < n$, how many functions are there from A onto B?
    (e)   How many bijections are there from A to A? Check that your answer is consistent with your answers for parts (a) and (c).

13.   **Proofs to Grade.**
    (a)   **Claim.**   If $f: A \xrightarrow{1-1} B$ and $g: B \xrightarrow{1-1} C$, then $g \circ f: A \xrightarrow{1-1} C$ (Theorem 4.11.)
    "Proof."   We must show that if (x, y) and (z, y) are elements of $g \circ f$, then $x = z$. If $(x, y) \in g \circ f$, then there is $u \in B$ such that $(x, u) \in f$ and $(u, y) \in g$. If $(z, y) \in g \circ f$, then there is $v \in B$ such that $(z, v) \in f$ and $(v, y) \in g$. However, $(u, y) \in g$ and $(v, y) \in g$ imply $u = v$ since g is one-to-one. Then $(x, u) \in f$ and $(z, v) \in f$ and $u = v$; therefore, $x = z$, since f is one-to-one. Hence, (x, y) and (z, y) in $g \circ f$ imply $x = z$. Therefore, $g \circ f$ is one-to-one.   ■
    (b)   **Claim.**   The function $f: \mathbf{R} \to \mathbf{R}$ given by $f(x) = 2x + 7$ is one-to-one.
    "Proof."   Suppose x_1 and x_2 are real numbers with $f(x_1) \neq f(x_2)$. Then $2x_1 + 7 \neq 2x_2 + 7$ and thus $2x_1 \neq 2x_2$. Hence $x_1 \neq x_2$, which shows that f is one-to-one.   ■
★   (c)   **Claim.**   The function $f: \mathbf{R} \to \mathbf{R}$ given by $f(x) = 2x + 7$ is onto **R**.
    "Proof."   Suppose f is not onto **R**. Then there exists $b \in \mathbf{R}$ with

$b \notin \text{Rng}(f)$. Thus $b \neq 2x + 7$ for all real numbers x. But $a = \frac{1}{2}(b - 7)$ is a real number and $f(a) = b$. This is a contradiction. Thus f is onto **R**. ∎

★ (d) **Claim.** The function $f: \mathbf{R} \rightarrow (-\pi/2, \pi/2)$ given by $f(x) = \arctan(x)$ maps onto $(-\pi/2, \pi/2)$.

"*Proof.*" Let $x \in \mathbf{R}$. Then $f(x) = \arctan(x)$ is a real number such that $-\pi/2 < f(x) < \pi/2$. Thus f maps onto $(-\pi/2, \pi/2)$. ∎

4.4

Induced Set Functions

Up to this point a function f from set A to set B has always been considered "pointwise." That is, we have considered the mapping of individual elements in A to their images in B or else considered pre-images of individual elements in B. The next step is to ask about collections of points in A or in B and what corresponds to them in the other set. Every function f from A to B *induces* a function that maps subsets of A to subsets of B. Also, regardless of whether f^{-1} is a function, there is a well-behaved induced function in the other direction mapping subsets of B to subsets of A.

Definition. Let $f: A \rightarrow B$. If $X \subseteq A$, then the **image of X** or **image set of X** is

$$f(X) = \{y \in B: y = f(x) \text{ for some } x \in X\}.$$

If $Y \subseteq B$, then the **inverse image of Y** is

$$f^{-1}(Y) = \{x \in A: f(x) \in Y\}.$$

There need not be any confusion about the fact that the induced functions $f: \mathscr{P}(A) \rightarrow \mathscr{P}(B)$ and $f^{-1}: \mathscr{P}(B) \rightarrow \mathscr{P}(A)$ have the same names as the function f from A to B and the relation f^{-1} (which may not be a function). The induced functions f and f^{-1} have as domain elements subsets of A and B, respectively. The image set $f(X)$ is just the set of all images of elements of X, and the inverse image $f^{-1}(Y)$ is the set of all pre-images of elements of Y.

Example. Let $A = \{0, 1, 2, 3, -1, -2, -3\}$, $B = \{0, 1, 2, 4, 6, 9\}$, and $f: A \rightarrow B$ be given by $f(x) = x^2$. Figure 4.9 shows that $f(\{-1, 3\}) = \{1, 9\}$ and $f^{-1}(\{4, 6\}) = \{2, -2\}$. Also, $f(A) = \{0, 1, 4, 9\}$, $f^{-1}(B) = A$, $f^{-1}(\{6\}) = \varnothing$, and $f(\{3, -3\}) = \{9\}$. Note that f^{-1} is not a function from B to A, so it would not make sense to consider $f^{-1}(1)$. However, $f^{-1}(\{1\})$ is meaningful and equal to $\{1, -1\}$.

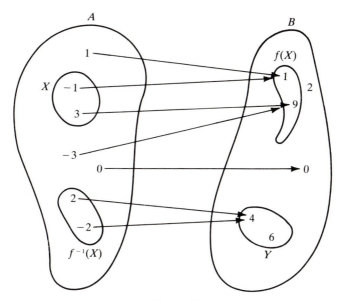

Figure 4.9

Example.   Let $F: \mathbf{R} \to \mathbf{R}$ be given by $F(x) = 2x + 1$. Let $X = \{1, 2, 3\}$. Then X is a subset of the domain and $F(X) = \{3, 5, 7\}$. Also, X is a subset of the codomain and $F^{-1}(X) = \{0, \frac{1}{2}, 1\}$. We also have $F^{-1}(\{2\}) = \{\frac{1}{2}\}$, $F(\mathbf{N}) = \{3, 5, 7, 9, \ldots\}$, and $F^{-1}([3, 5]) = [1, 2]$. Figure 4.10 shows that for $D = [1, 2]$, $F(D) = [3, 5]$.

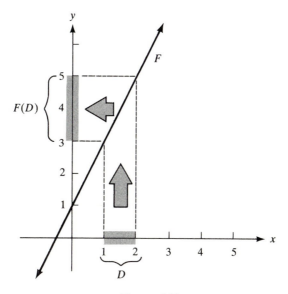

Figure 4.10

Example.   Let $f: \mathbf{R} \to \mathbf{R}$ be given by $f(x) = x^2$. Then $f^{-1}([-4, -3]) = \varnothing$ and $f([1, 2]) = [1, 4]$. However, $f^{-1}([1, 4]) \neq [1, 2]$. Figure 4.11 shows that $f^{-1}([1, 4]) = [-2, -1] \cup [1, 2]$.

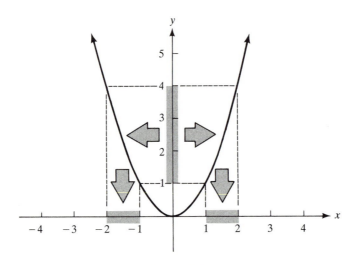

Figure 4.11

Proofs involving induced set functions are likely to be more troublesome than others we have seen thus far. Before tackling such proofs study carefully the definitions to see that each of these facts follows immediately from the definitions.

If $f: A \to B$, $D \subseteq A$, $E \subseteq B$, and $a \in A$, then:

$a \in D \Rightarrow f(a) \in f(D)$
$a \in f^{-1}(E) \Rightarrow f(a) \in E$
$f(a) \in E \Rightarrow a \in f^{-1}(E)$
$f(a) \in f(D) \Rightarrow a \in D$, provided that f is one-to-one.

You should verify that the last implication is false when f is not one-to-one. *Familiarity with these ideas is crucial to understanding induced set functions.*

By the algebraic properties of the induced set functions we mean their behavior (properties) as they interact with the operations of union, intersection, and set difference. The next theorems collect results about these properties.

Theorem 4.16   Let $f: A \to B$, and let $\{D_\alpha: \alpha \in \Delta\}$ and $\{E_\beta: \beta \in \Gamma\}$ be families of subsets of A and B, respectively. Then

(a) $f(\bigcap_{\alpha \in \Delta} D_\alpha) \subseteq \bigcap_{\alpha \in \Delta} f(D_\alpha)$.

(b) $f(\bigcup_{\alpha \in \Delta} D_\alpha) = \bigcup_{\alpha \in \Delta} f(D_\alpha)$.

(c) $f^{-1}(\underset{\beta\in\Gamma}{\cap} E_\beta) = \underset{\beta\in\Gamma}{\cap} f^{-1}(E_\beta)$.

(d) $f^{-1}(\underset{\beta\in\Gamma}{\cup} E_\beta) = \underset{\beta\in\Gamma}{\cup} f^{-1}(E_\beta)$.

Proof.

(a)   Suppose $b \in f(\underset{\alpha\in\Delta}{\cap} D_\alpha)$. Then $b = f(a)$ for some
$a \in \underset{\alpha\in\Delta}{\cap} D_\alpha$. Thus $a \in D_\alpha$ for every $\alpha \in \Delta$. Since
$b = f(a)$, $b \in f(D_\alpha)$ for every $\alpha \in \Delta$. Therefore
$b \in \underset{\alpha\in\Delta}{\cap} f(D_\alpha)$. This proves that $f(\underset{\alpha\in\Delta}{\cap} D_\alpha) \subseteq \underset{\alpha\in\Delta}{\cap} f(D_\alpha)$.

Parts (b), (c), and (d) are left as exercise 9.   ■

Theorem 4.17   Let $f: A \to B$ and $E \subseteq B$. Then $f^{-1}(B - E)$
$= A - f^{-1}(E)$.

Proof.   Suppose $a \in f^{-1}(B - E)$. Then $f(a) \in B - E$. That is,
$f(a) \in B$ and $f(a) \notin E$. Therefore $a \in A$ and $a \notin f^{-1}(E)$. ⟨*This
is the contrapositive of the second statement recommended above
for study.*⟩ Thus $a \in A - f^{-1}(E)$. Therefore $f^{-1}(B - E) \subseteq$
$A - f^{-1}(E)$. The opposite inclusion is left as exercise 10.   ■

Theorem 4.18   Let $f: A \to B$, $D \subseteq A$, and $E \subseteq B$. Then

(a) $f(f^{-1}(E)) \subseteq E$.
(b) $E = f(f^{-1}(E))$ iff $E \subseteq \text{Rng}(f)$.
(c) $D \subseteq f^{-1}(f(D))$.
(d) $f^{-1}(f(D)) = D$ iff $f(A - D) \subseteq B - f(D)$.

Proof.

(a)   Suppose $b \in f(f^{-1}(E))$. Then there is $a \in f^{-1}(E)$ such
that $f(a) = b$. Since $a \in f^{-1}(E)$, $f(a) \in E$. But $f(a) = b$,
so $b \in E$. Therefore $f(f^{-1}(E)) \subseteq E$.
(b)   First, suppose $E = f(f^{-1}(E))$. Suppose $b \in E$. Then
$b \in f(f^{-1}(E))$. Thus there is $a \in f^{-1}(E)$ such that $b =$
$f(a)$, so $b \in \text{Rng}(f)$. Therefore $E \subseteq \text{Rng}(f)$.
      Now assume $E \subseteq \text{Rng}(f)$. We know by part (a)
that $f(f^{-1}(E)) \subseteq E$, so to prove equality we need only
$E \subseteq f(f^{-1}(E))$. Suppose $b \in E$. Then $b \in \text{Rng}(f)$, so
$b = f(a)$ for some $a \in A$. Since $b = f(a) \in E$, $a \in f^{-1}(E)$.
Thus $b = f(a)$ and $a \in f^{-1}(E)$, so $b \in f(f^{-1}(E))$. Therefore $E \subseteq f(f^{-1}(E))$.

Parts (c) and (d) constitute exercise 11.   ■

Exercises 4.4

1. Let $A = \{1, 2, 3\}$, $B = \{4, 5, 6\}$, and $h = \{(1, 4), (2, 4), (3, 5)\}$.
   ★ (a) List the eight ordered pairs in the induced function on $\mathcal{P}(A)$ to $\mathcal{P}(B)$.
     (b) List the eight ordered pairs in the induced function on $\mathcal{P}(B)$ to $\mathcal{P}(A)$.

2. Let $f(x) = x^2 + 1$. Find
   ★ (a) $f([1, 3])$ (b) $f([-1, 0] \cup [2, 4])$ ★ (c) $f^{-1}([-1, 1])$
     (d) $f^{-1}([-2, 3])$ (e) $f^{-1}([5, 10])$

3. Let $f(x) = 1 - 2x$. Find
     (a) $f(A)$ where $A = \{-1, 0, 1, 2, 3\}$ (b) $f(\mathbf{N})$
     (c) $f^{-1}(\mathbf{R})$ ✓(d) $f^{-1}([2, 5])$
     ✓(e) $f((1, 4])$

4. Let $f: \mathbf{N} \times \mathbf{N} \to \mathbf{N}$ be given by $f(m, n) = 2^m(2n + 1)$. Find
   ★ (a) $f^{-1}(A)$ where $A = \{1, 2, 3, 4, 5, 6\}$.
     (b) $f^{-1}(B)$ where $B = \{4, 6, 8, 10, 12, 14\}$.
     (c) $f(C)$ where $C = \{(1, 1), (3, 3), (3, 1), (1, 3)\}$.

5. Let $f: A \to B$ where $A = \{1, 2, 3, 4, 5, 6\}$, $B = \{p, q, r, s, t, z\}$, and $f = \{(1, p), (2, p), (3, s), (4, t), (5, z), (6, t)\}$. Find
     (a) $f^{-1}(\{p, q, s\})$ ★ (b) $f(\{1, 3, 4, 6\})$ (c) $f(\{3, 5\})$
   ★ (d) $f^{-1}(\{p, r, s, z\})$ (e) $f(f^{-1}(\{p, r, s, z\}))$ (f) $f^{-1}(f(\{1, 4, 5\}))$

6. Let $f: \mathbf{R} - \{0\} \to \mathbf{R}$ be given by $f(x) = x + 1/x$. Find
     (a) $f((0, 2))$ (b) $f((-1, 1])$ ★ (c) $f^{-1}((3, 4])$
     (d) $f^{-1}([0, 1))$ (e) $f(f^{-1}((-4, 10)))$

7. Let $f: \mathbf{R} \to \mathbf{R}$ be given by $f(x) = 10x - x^2$. Find
   ★ (a) $f([1, 6))$.   (b) $f^{-1}((0, 21])$.   ★ (c) $f^{-1}((3, 4])$.
     (d) $f^{-1}([24, 50])$.   (e) $f([4, 7])$.

8. Let $f: \mathbf{N} \times \mathbf{N} \to \mathbf{N}$ be given by $f(m, n) = 2^m 3^n$. Find
     (a) $f(A \times B)$ where $A = \{1, 2, 3\}$, $B = \{3, 4\}$.
     (b) $f^{-1}(\{5, 6, 7, 8, 9, 10\})$.

9. Prove parts (b), (c), and (d) of Theorem 4.16.

10. Prove the remaining inclusion of Theorem 4.17. That is, prove that if $f: A \to B$ and $E \subseteq B$, then $A - f^{-1}(E) \subseteq f^{-1}(B - E)$.

11. Prove parts (c) and (d) of Theorem 4.18.

12. Let $f: A \to B$ and let $X, Y \subseteq A$ and $U, V \subseteq B$. Prove
     (a) $f(X) \subseteq U$ iff $X \subseteq f^{-1}(U)$.   ★ (b) $f(X) - f(Y) \subseteq f(X - Y)$.
     (c) $f^{-1}(U) - f^{-1}(V) = f^{-1}(U - V)$.

☆ 13. Let $f: A \to B$. Prove that, if f is one-to-one, then $f(X) \cap f(Y) = f(X \cap Y)$ for all $X, Y \subseteq A$. Is the converse true? Explain.

14. Let $f: A \to B$. Prove that, if $X \subseteq A$ and f is one-to-one, then $f(A - X) = f(A) - f(X)$.

15. Let $f: A \to B$. Prove that if $X \subseteq A$, $Y \subseteq B$, and f is one-to-one and onto, then $f(X) = Y$ iff $f^{-1}(Y) = X$.

16. Let $f: A \to B$. Consider the function on $\mathscr{P}(A)$ to $\mathscr{P}(B)$ induced by f.
    ★ (a) What condition on f will make the induced function one-to-one?
        (b) What condition on f will make the induced function onto $\mathscr{P}(B)$?

17. Let $f: A \to B$ and $K \subseteq B$. Prove that $f(f^{-1}(K)) = K \cap \text{Rng}(f)$.

18. Let $f: A \to B$. Let R be the relation on A defined by $x \, R \, y$ iff $f(x) = f(y)$.
    (a) Show that R is an equivalence relation.
    (b) Describe the partition of A associated with R.

19. **Proofs to Grade.**
    ★ (a) **Claim.** If $f: A \to B$ and $X \subseteq A$, then $f^{-1}(f(X)) \subseteq X$.
        "Proof."   If $x \in f^{-1}(f(X))$, then by definition of f^{-1}, $f(x) \in f(X)$. Therefore $x \in X$. Thus $f^{-1}(f(X)) \subseteq X$. ■
        (b) **Claim.** If $f: A \to B$ and $X \subseteq A$, then $X \subseteq f^{-1}(f(X))$.
        "Proof."   Suppose $z \in X$. Then $f(z) \in f(X)$. Therefore $z \in f^{-1}(f(X))$, which proves the set inclusion. ■
        (c) **Claim.** If $f: A \to B$ and $\{D_\alpha: \alpha \in \Delta\}$ is a family of subsets of A, then $\bigcap_{\alpha \in \Delta} f(D_\alpha) \subseteq f(\bigcap_{\alpha \in \Delta} D_\alpha)$.
        "Proof."   Suppose $y \in \bigcap_{\alpha \in \Delta} f(D_\alpha)$. Then $y \in f(D_\alpha)$ for all α. Thus there exists $x \in D_\alpha$ such that $f(x) = y$, for all α. Then $x \in \bigcap_{\alpha \in \Delta} D_\alpha$ and $f(x) = y$, so $y \in f(\bigcap_{\alpha \in \Delta} D_\alpha)$. Therefore $\bigcap_{\alpha \in \Delta} f(D_\alpha) \subseteq f(\bigcap_{\alpha \in \Delta} D_\alpha)$. ■

5

Cardinality

How many elements are in the set

$$A = \{\pi, 28, \sqrt{2}, \tfrac{1}{2}, -3, \Delta, \alpha, 0\}?$$

After a short pause you said "eight." Right? Consider for a moment how you arrived at that answer. You probably looked at π and thought "1," then looked at 28 and thought "2," and so on up through 0, which is "8." What you have done is set up a one-to-one correspondence between the set A and the "known" set of eight elements $\{1, 2, 3, 4, 5, 6, 7, 8\}$. Counting the number of elements in sets is essentially a matter of one-to-one correspondences. This process will be extended when we "count" the number of elements in infinite sets in this chapter. Here is another counting problem.

A certain shepherd has more than 400 sheep in his flock, but he cannot count beyond 10. Each day he takes his sheep out to graze, and each night he brings them back into the fold. How can he be sure all the sheep have returned? The answer is that he can count them with a one-to-one correspondence. He needs two containers and a pile of pebbles, one pebble for each sheep. When the sheep return in the evening, he transfers pebbles from one container to the other, one at a time for each returning sheep. Whenever there are pebbles left over, he knows there are lost sheep. The solution to the shepherd's problem illustrates the point that even though we have not counted the sheep, we know that the set of missing sheep and the set of left-over pebbles have the same number of elements—because there is a one-to-one correspondence between them.

5.1

Equivalent Sets; Finite Sets

To determine whether two sets have the same number of elements, we see if it is possible to match the elements of the sets in a one-to-one fashion. This idea may be conveniently described in terms of a one-to-one correspondence (a bijection) from one set to another.

Definition. Two sets A and B are **equivalent** iff there exists a one-to-one function from A onto B. A and B are also said to be **in one-to-one correspondence,** and we write $A \approx B$.

If A and B are not equivalent, we write $A \not\approx B$.

Example. The sets $A = \{5, 8, \phi\}$ and $B = \{r, p, m\}$ are equivalent. The function $f: A \to B$ given by $f(5) = r$, $f(8) = p$, and $f(\phi) = m$ is one of six such functions that verify this.

Example. The sets $C = \{x, y\}$ and $D = \{q, r, s\}$ are not equivalent. There are nine different functions from C to D. An examination of all nine will show that none of them is a bijection. Thus $C \not\approx D$.

Example. The set E of even integers is equivalent to D, the set of odd integers. To prove this, we employ the function $f: E \to D$ given by $f(x) = x + 1$. The function is one-to-one, because $f(x) = f(y)$ implies $x + 1 = y + 1$, which yields $x = y$. Also, f is onto D because if z is any odd integer, then $w = z - 1$ is even and $f(w) = w + 1 = (z - 1) + 1 = z$.

Example. Let $\mathcal{F}$ be the set of all functions from $\mathbf{N}$ to $\{0, 1\}$. This set is sometimes denoted $\{0, 1\}^{\mathbf{N}}$. We will show $\mathcal{F} \approx \mathcal{P}(\mathbf{N})$, the power set of $\mathbf{N}$.

Proof. $\langle$*This proof will note that $\mathcal{F}$ is precisely the set of all characteristic functions with domain $\mathbf{N}$, which is in a one-to-one correspondence with the subsets of $\mathbf{N}$.*$\rangle$ To show $\mathcal{F} \approx \mathcal{P}(\mathbf{N})$, we define $H: \mathcal{F} \to \mathcal{P}(\mathbf{N})$ as follows:

$$\text{for } g \in \mathcal{F}, H(g) = \{x \in \mathbf{N}: g(x) = 1\}.$$

$\langle$*Note that under the function H, every function in $\mathcal{F}$ has an image in $\mathcal{P}(\mathbf{N})$.*$\rangle$

To show H is one-to-one, let $g_1, g_2 \in \mathscr{F}$. Suppose $g_1 \neq g_2$. Then there exists $n \in \mathbf{N}$ such that $g_1(n) \neq g_2(n)$. Both g_1 and g_2 have codomain $\{0, 1\}$, so we may assume $g_1(n) = 1$ and $g_2(n) = 0$. ⟨*The case $g_1(n) = 0$ and $g_2(n) = 1$ is similar.*⟩ But then $n \in \{x \in \mathbf{N}\colon g_1(x) = 1\}$ and $n \notin \{x \in \mathbf{N}\colon g_2(x) = 1\}$. Thus $H(g_1) \neq H(g_2)$.

To show that H is onto $\mathscr{P}(\mathbf{N})$, let $A \in \mathscr{P}(\mathbf{N})$. Then $A \subseteq \mathbf{N}$. We note that $\chi_A\colon \mathbf{N} \to \{0, 1\}$ and thus $\chi_A \in \mathscr{F}$. Furthermore,

$$H(\chi_A) = \{x \in \mathbf{N}\colon \chi_A(x) = 1\} = A$$

by definition of χ_A. Thus H is onto.

Because H is a bijection, $\mathscr{F} \approx \mathscr{P}(\mathbf{N})$. ∎

Theorem 5.1   The relation $\approx$ is reflexive, symmetric, and transitive. Thus $\approx$ is an equivalence relation on the class of all sets.

Proof.   Exercise 1. ∎

The next theorem will be particularly useful for showing equivalences of sets.

Theorem 5.2   Suppose A, B, C, and D are sets with $A \approx C$ and $B \approx D$. If A and B are disjoint and C and D are disjoint, then $A \cup B \approx C \cup D$.

Proof.   Since $A \approx C$, there is a one-to-one correspondence $h\colon A \to C$. Similarly, let $g\colon B \to D$ be a one-to-one correspondence from B to D. Then by Theorem 4.15, $h \cup g\colon A \cup B \to C \cup D$ is one-to-one and onto $C \cup D$. Therefore $A \cup B \approx C \cup D$. ∎

We shall use the symbol $\mathbf{N}_k$ to denote the set $\{1, 2, 3, \ldots, k\}$. Think of $\mathbf{N}_k$ as the standard set with k elements against which the sizes of other sets may be compared.

Definition.   A set S is **finite** iff $S = \varnothing$ or S is equivalent to $\mathbf{N}_k$ for some natural number k. In the case $S = \varnothing$, we say $\varnothing$ **has cardinal number 0** and write $\overline{\overline{\varnothing}} = 0$. If $S \approx \mathbf{N}_k$, then S **has cardinal number k,** and we write $\overline{\overline{S}} = k$. A set S is **infinite** iff it is not finite.

The set $X = \{98.6, c, \pi\}$ is finite and has cardinal number 3. Exhibiting any one-to-one correspondence from X to $\mathbf{N}_3$ will prove this. One such correspondence f is $f(98.6) = 1$, $f(c) = 2$, $f(\pi) = 3$.

The set $\{8, 7, 3, 7, 2\}$ is finite and has cardinal number 4, since it is equal to $\{8, 7, 3, 2\}$, which is equivalent to $\mathbf{N}_4$.

We use the symbol $\overline{\overline{A}}$ to represent the cardinality of A so as to distinguish what we prove here from the more informal results in section 2.5, where we used the symbol $\#A$. We shall show that for finite sets the cardinality $\overline{\overline{A}}$ of A corresponds to our intuitive notion of the number of elements in A.

The empty set is finite by definition. Also, each set $\mathbf{N}_k$ is finite and has cardinal number k because the identity function $I_{\mathbf{N}_k}$ is a one-to-one function from $\mathbf{N}_k$ onto $\mathbf{N}_k$. In addition, any set equivalent to a finite set must also be finite. Suppose A is a finite set and $A \approx B$. If $A = \emptyset$, then $B = \emptyset$ (see exercise 2). Otherwise, $A \approx \mathbf{N}_k$ for some k and thus $B \approx \mathbf{N}_k$ by transitivity of $\approx$. In either case B is finite.

The remaining theorems of this section give other properties of finite sets. One goal is to make sure that the notion of cardinality for a finite set A corresponds to our notion of the number of elements in A in the following important way: *The cardinality of a finite set should be unique.* That is, if A has cardinality m (that is, $A \approx \mathbf{N}_m$) and A has cardinality n ($A \approx \mathbf{N}_n$), then $m = n$. You are asked to show this in exercise 13, using a property of finite sets called the pigeonhole principle, discussed later in this section. Our immediate goal is the key theorem that every subset of a finite set is finite. This theorem uses two lemmas.

Lemma 5.3   If S is finite with cardinality k and x is any object not in S, then $S \cup \{x\}$ is finite and has cardinality $k + 1$.

Proof.   If $S = \emptyset$, then $S \cup \{x\} = \{x\}$, which is equivalent to $\mathbf{N}_1$ and thus finite. In this case S has cardinality 0 and $S \cup \{x\}$ has cardinality $0 + 1$. If $S \neq \emptyset$, then $S \approx \mathbf{N}_k$ for some k. Also, $\{x\} \approx \{k + 1\}$. Therefore by Theorem 5.2, $S \cup \{x\} \approx \mathbf{N}_k \cup \{k + 1\} = \mathbf{N}_{k+1}$. This proves $S \cup \{x\}$ is finite and has cardinality $k + 1$.   ∎

Lemma 5.4   Every subset of $\mathbf{N}_k$ is finite.

Proof.   Let A be a subset of $\mathbf{N}_k$. ⟨*We prove A is finite by induction on the number k.*⟩

Let $k = 1$. Then either $A = \emptyset$ or $A = \mathbf{N}_1$, both of which are finite.

Assume all subsets of $\mathbf{N}_k$ are finite and let $A \subseteq \mathbf{N}_{k+1}$. Then $A - \{k + 1\}$ is a subset of $\mathbf{N}_k$ and is finite by the hypothesis of induction. If $A = A - \{k + 1\}$, then A is finite. Otherwise, $A = (A - \{k + 1\}) \cup \{k + 1\}$, which is finite by Lemma 5.3.   ∎

Theorem 5.5   Every subset of a finite set is finite.

Proof.   Assume S is a finite set and $T \subseteq S$. If $T = \emptyset$, then T is finite. Thus we may assume $T \neq \emptyset$ and hence $S \neq \emptyset$. Since $S \approx$

$\mathbf{N}_k$ for some $k \in \mathbf{N}$, let f be a one-to-one correspondence from S onto $\mathbf{N}_k$. Then $f|_T$ is a one-to-one correspondence from T onto the set $f(T)$ (figure 5.1). Therefore $T \approx f(T)$. But $f(T)$ is a subset of $\mathbf{N}_k$ and is finite. Thus T is finite. ∎

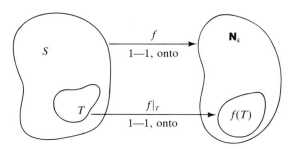

Figure 5.1

At this point you may feel that Lemmas 5.3 and 5.4 and Theorem 5.5 are a lot of hard work to prove the very obvious result that subsets of finite sets are finite. You may be right. The value of these results lies in the reasoning and in the use of functions to establish facts about cardinalities. This work will be helpful when we deal with infinite sets because there our intuition often fails us.

The next result of this section is that the union of a finite number of finite sets is finite. To this end the next theorem is a special case: the union of two disjoint finite sets is finite. Its proof is a rigorous development of the sum rule (Theorem 2.15), which states that if A has m elements, B has n elements, and $A \cap B = \varnothing$, then $A \cup B$ has $m + n$ elements.

Theorem 5.6 If A and B are finite disjoint sets, then $A \cup B$ is finite and $\overline{\overline{A \cup B}} = \overline{\overline{A}} + \overline{\overline{B}}$.

Proof. Suppose A and B are finite sets and $A \cap B = \varnothing$. If $A = \varnothing$, then $A \cup B = B$; if $B = \varnothing$, then $A \cup B = A$. In either case $A \cup B$ is finite, and since $\overline{\overline{\varnothing}} = 0$, $\overline{\overline{A \cup B}} = \overline{\overline{A}} + \overline{\overline{B}}$. Now suppose that $A \neq \varnothing$ and $B \neq \varnothing$. Let $A \approx \mathbf{N}_m$ and $B \approx \mathbf{N}_n$, and suppose that $f: A \to \mathbf{N}_m$ and $g: B \to \mathbf{N}_n$ are one-to-one correspondences. Let $H = \{m + 1, m + 2, \ldots, m + n\}$. Then $h: \mathbf{N}_n \to H$ given by $h(x) = m + x$ is a one-to-one correspondence, and thus $\mathbf{N}_n \approx H$. Therefore $B \approx H$ by transitivity. Finally, by Theorem 5.2, $A \cup B \approx \mathbf{N}_m \cup H = \mathbf{N}_{m+n}$, which proves that $A \cup B$ is finite and that $\overline{\overline{A \cup B}} = m + n$. ∎

Corollary 5.7 (a) If A and B are finite sets, then $A \cup B$ is finite. (b) If A_1, A_2, . . ., A_n are finite sets, then $\bigcup\limits_{i=1}^{n} A_i$ is finite.

Proof.   We prove part (a) and leave part (b) as an exercise in mathematical induction (exercise 5). Assume that both A and B are finite. Since $B - A \subseteq B$, $B - A$ is finite. Thus by Theorem 5.6, $A \cup B = A \cup (B - A)$ is a finite set.   ∎

Lemma 5.3 shows that adding one element to a finite set increases its cardinality by one. It is also true that removing one element from a finite set reduces the cardinality by one. The proof of Lemma 5.8 is left as exercise 12.

Lemma 5.8   Let $r \in \mathbf{N}$ with $r > 1$. For all $x \in \mathbf{N}_r$,
$\mathbf{N}_r - \{x\} \approx \mathbf{N}_{r-1}$.

The final property of finite sets we consider is popularly known as the **pigeonhole principle.** In its informal version it says "If a flock of n pigeons comes to roost in a house with r pigeonholes and $n > r$, then at least one hole contains more than one pigeon." If we think of the set of pigeons as $\mathbf{N}_n$ and the set of holes as $\mathbf{N}_r$, the formal version is Theorem 5.9.

Theorem 5.9   Let $n, r \in \mathbf{N}$. If $f: \mathbf{N}_n \to \mathbf{N}_r$ and $n > r$, then f is not one-to-one.

Proof.   The proof proceeds by induction on the number n. Since $n > r$, we begin with $n = 2$. If $n = 2$, then $r = 1$. In this case f is the constant function $f(x) = 1$, which is not one-to-one.
　　　Suppose the pigeonhole principle holds for some n; that is, suppose for $r < n$, if $f: \mathbf{N}_n \to \mathbf{N}_r$, then f is not one-to-one. Let $r < n + 1$. The case $r = 1$ is treated just as in the case $n = 2$ above, so we may assume $r > 1$. Suppose there is a one-to-one function $f: \mathbf{N}_{n+1} \to \mathbf{N}_r$. Then $f|_{\mathbf{N}_n}$ is a one-to-one function. The range of this function may not contain $f(n + 1)$, but by Lemma 5.8 there is a one-to-one correspondence $g: \mathbf{N}_r - \{f(n + 1)\} \to \mathbf{N}_{r-1}$. Therefore the composite $g \circ f|_{\mathbf{N}_n}: \mathbf{N}_n \to \mathbf{N}_{r-1}$ is one-to-one. This contradicts the induction hypothesis.
　　　By the PMI, there is no one-to-one function $f: \mathbf{N}_n \to \mathbf{N}_r$ when $n > r$.   ∎

The pigeonhole principle can be used to prove several important results about finite sets (see, for example, exercise 13). We use it here to give a characterization of finite sets.

Corollary 5.10   A finite set is not equivalent to any of its proper subsets.

Proof.   We will show that $\mathbf{N}_k$ is not equivalent to any proper subset and leave the general case as exercise 14. The case $k = 1$ is trivial, so let $k > 1$. Suppose A is a proper subset of $\mathbf{N}_k$ and $f\colon \mathbf{N}_k \to A$ is one-to-one and onto A.

Case 1.   Suppose $k \notin A$. Then $A \subseteq \mathbf{N}_{k-1}$ and the inclusion function $i\colon A \to \mathbf{N}_{k-1}$ is one-to-one. But then $i \circ f\colon \mathbf{N}_k \to \mathbf{N}_{k-1}$ is one-to-one, which contradicts the pigeonhole principle.

Case 2.   Suppose $k \in A$. Since f is onto A, there exists $x \in \mathbf{N}_k$ such that $f(x) = k$. Choose an element $y \in \mathbf{N}_k - A$ and let $A' = (A - \{k\}) \cup \{y\}$. Then $A \approx A'$ (because the function $I_{A-\{k\}} \cup \{(y, y)\}$ is a one-to-one correspondence). Thus $A' \approx \mathbf{N}_k$, $A' \subsetneqq \mathbf{N}_k$, and $k \notin A'$. This is the situation of Case 1 with $\mathbf{N}_k$ and A' and again yields a contradiction.   ∎

Exercises 5.1

1.   Prove Theorem 5.1. That is, show that the relation $\approx$ is reflexive, symmetric, and transitive on the class of all sets.

2.   Show that if $A \approx \varnothing$, then $A = \varnothing$. [See also exercise 6(b), section 4.1.]

☆ 3.   Show that $A \approx A \times \{x\}$.

4.   Which of the following sets are finite?
  ★ (a)   the set of all grains of sand on the earth.
  (b)   the set of all positive integer powers of 2.
  ★ (c)   the set of four-letter words in English.
  (d)   the set of rational numbers.
  ★ (e)   the set of rationals in $(0, 1)$ with denominator 2^k for some $k \in \mathbf{N}$.
  (f) $\{x \in \mathbf{R}\colon x^2 + 1 = 0\}$.
  (g)   the set of all turkeys eaten in the year 1620.
  (h) $\{1, 3, 5\} \times \{2, 4, 6, 8\}$.
  (i) $\{x \in \mathbf{N}\colon x \text{ is a prime}\}$.

5.   Complete the proof of Corollary 5.7.

6.   Let A and B be sets. Prove that
  ★ (a)   if A is finite then $A \cap B$ is finite.
  (b)   if A is infinite, and $A \subseteq B$, then B is infinite.

7.   ★ (a)   Prove that for all $k, m \in \mathbf{N}$, $\mathbf{N}_k \times \mathbf{N}_m$ is finite.
  (b)   Suppose A and B are finite. Prove $A \times B$ is finite.

8.   Define B^A to be the set of all functions from A to B. Show that if A and B are finite, then B^A is finite.

9.   If possible, give an example of each:
  (a)   an infinite subset of a finite set.
  (b)   a collection $\{A_i\colon i \in \mathbf{N}\}$ of finite sets whose union is finite.

★ (c)   a finite collection of finite sets whose union is infinite.

(d)   finite sets A and B such that $\overline{\overline{A \cup B}} \neq \overline{\overline{A}} + \overline{\overline{B}}$.

☆ 10. Using the methods of this section, prove that if A and B are finite sets, then $\overline{\overline{A \cup B}} = \overline{\overline{A}} + \overline{\overline{B}} - \overline{\overline{A \cap B}}$. This fact is a restatement of Theorem 2.17.

11. Prove that if A is finite and B is infinite, then $B - A$ is infinite.

12. Prove Lemma 5.8.

☆ 13. Show that if a finite set S has cardinal number m and cardinal number n, then $m = n$.

14. Complete the proof of Corollary 5.10 by showing that if A is finite and B is a proper subset of A, then $B \neq A$.

15. Prove by induction on n that if $r < n$ and $f: \mathbf{N}_r \rightarrow \mathbf{N}_n$, then f is not onto $\mathbf{N}_n$.

16. Let A and B be finite sets with $A \approx B$. Suppose $f: A \rightarrow B$.
   ☆ (a)   If f is one-to-one, show that f is onto B.
   ☆ (b)   If f is onto B, prove that f is one-to-one.

☆ 17. Prove that if the domain of a function is finite, then the range is finite.

18. **Proofs to Grade.**
   (a)   **Claim.**   If A and B are finite, then $A \cup B$ is finite.
      "Proof."   If A and B are finite, then there exist $m, n \in \mathbf{N}$ such that $A \approx \mathbf{N}_m$, $B \approx \mathbf{N}_n$. Let $f: A \xrightarrow[\text{onto}]{1-1} \mathbf{N}_m$ and $h: B \xrightarrow[\text{onto}]{1-1} \mathbf{N}_n$. Then $f \cup h: A \cup B \xrightarrow[\text{onto}]{1-1} \mathbf{N}_{m+n}$, which shows that $A \cup B \approx \mathbf{N}_{m+n}$. Thus $A \cup B$ is finite.   ∎

   ★ (b)   **Claim.**   If S is a finite, nonempty set, then $S \cup \{x\}$ is finite.
      "Proof."   Suppose S is finite and nonempty. Then $S \approx \mathbf{N}_k$ for some integer k.
      Case 1. $x \in S$. Then $S \cup \{x\} = S$, so $S \cup \{x\}$ has k elements and is finite.
      Case 2. $x \notin S$. Then $S \cup \{x\} \approx \mathbf{N}_k \cup \{x\} \approx \mathbf{N}_k \cup \mathbf{N}_1 \approx \mathbf{N}_{k+1}$. Thus $S \cup \{x\}$ is finite.   ∎

   (c)   **Claim.**   If $A \times B$ is finite, then A is finite.
      "Proof."   Choose any $b^* \in B$. Then $A \approx A \times \{b^*\}$. But $A \times \{b^*\} = \{(a, b^*): a \in A\} \subseteq A \times B$. Since $A \times B$ is finite, $A \times \{b^*\}$ is finite. Since A is equivalent to a finite set, A is finite.   ∎

5.2

Infinite Sets

In the previous section an infinite set was defined as a nonempty set that cannot be put into a one-to-one correspondence with any $\mathbf{N}_k$. According to the next theorem, the set $\mathbf{N}$ is one such set. Since infinite means not finite, the proof, as might be expected, is by contradiction.

Theorem 5.11 The set of natural numbers is infinite.

Proof. Suppose $\mathbf{N}$ is finite. Clearly, $\mathbf{N} \neq \varnothing$. Therefore for some natural number k, there exists a one-to-one function f from $\mathbf{N}_k$ onto $\mathbf{N}$. ⟨*We will contradict that f is onto* $\mathbf{N}$.⟩
   Let $n = f(1) + f(2) + \cdots + f(k) + 1$. Since each $f(i) > 0$, n is a natural number larger than each $f(i)$. Thus $n \neq f(i)$ for any $i \in \mathbf{N}_k$. Hence $n \not\in \text{Rng}(f)$. Therefore f is not onto $\mathbf{N}$, a contradiction. ∎

There are many other infinite sets, some—but not all—of which are equivalent to $\mathbf{N}$. It is *not* true that all infinite sets are equivalent.

Definition. A set is **denumerable** iff it is equivalent to $\mathbf{N}$. A denumerable set S has **cardinal number** $\aleph_0$, and we write $\overline{\overline{S}} = \aleph_0$. If a set is finite or denumerable, it is **countable**; otherwise, the set is **uncountable**.

The symbol $\aleph$, aleph, is the first symbol in the Hebrew alphabet. The subscript $_0$ (read "naught") indicates $\aleph_0$ is the first infinite cardinal number. Other infinite cardinal numbers are associated with uncountable sets.
   These definitions along with the ideas of finite and infinite are related in figure 5.2. Note that denumerable sets are those that are countable and infinite.

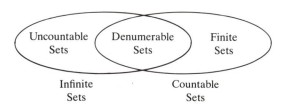

Figure 5.2

The set E^+ of positive even integers is an example of a denumerable set. The function $f: \mathbf{N} \rightarrow E^+$ defined by setting $f(x) = 2x$ is clearly one-to-one and onto E^+. Of course, there are many other one-to-one correspondences; the one given here is the simplest.
   What we have shown here is that E^+ has the same number of elements as $\mathbf{N}$; that is, E^+ has $\aleph_0$ elements. Thus although E^+ is a proper subset of $\mathbf{N}$, it could be misleading to say that $\mathbf{N}$ has more elements than E^+. In section 5.4 we will show that *every infinite set is equivalent to one of its proper subsets.*

Thus we have a distinction between finite and infinite sets; by Corollary 5.10, *no finite set is equivalent to any of its proper subsets.*

The next theorem will show that **N** and **Z** have the same number of elements. This, coupled with our knowledge that $\approx$ is transitive, will show that E^+ and **Z** are equivalent. Thus E^+, **N**, and **Z** all have cardinal number $\aleph_0$.

Theorem 5.12   The set of integers is denumerable.

Proof.   We define $f: \mathbf{N} \to \mathbf{Z}$ by

$$f(x) = \begin{cases} \dfrac{x}{2} & \text{if } x \text{ is even} \\[2ex] \dfrac{1-x}{2} & \text{if } x \text{ is odd.} \end{cases}$$

Thus $f(1) = 0$, $f(2) = 1$, $f(3) = -1$, $f(4) = 2$, $f(5) = -2$, and so on. Pictorially, f is represented as a one-to-one matching.

$$\mathbf{N} = \{1, 2, \quad 3, 4, \quad 5, 6, \quad 7, \ldots\}$$
$$\downarrow \downarrow \quad \downarrow \downarrow \quad \downarrow \downarrow \quad \downarrow$$
$$\mathbf{Z} = \{0, 1, \ -1, 2, \ -2, 3, \ -3, \ldots\}$$

We claim f is one-to-one. Suppose $f(x) = f(y)$. If x and y are both even, then $x/2 = y/2$, and thus $x = y$. If x and y are both odd, then $(1 - x)/2 = (1 - y)/2$, so $1 - x = 1 - y$ and $x = y$. If one of x and y is even and the other is odd, then $f(x)$ and $f(y)$ have opposite signs, so $f(x) \neq f(y)$. Thus whenever $f(x) = f(y)$, $x = y$.

It remains to show that the function maps onto **Z**. If $w \in \mathbf{Z}$ and $w > 0$, then $2w$ is even and $f(2w) = (2w)/2 = w$. If $w \in \mathbf{Z}$ and $w \leq 0$, then $1 - 2w$ is an odd natural number and $f(1 - 2w) = [1 - (1 - 2w)]/2 = (2w)/2 = w$. In either case if $w \in \mathbf{Z}$, then $w \in \text{Rng}(f)$.   ∎

Example.   The set P of reciprocals of positive integer powers of 2 is denumerable. Writing the set P as

$$P = \left\{ \frac{1}{2^k} : k \in \mathbf{N} \right\}$$

exhibits the one-to-one correspondence

$f: \mathbf{N} \to P$ given by $f(k) = \dfrac{1}{2^k}$.

Example.   The set $\mathbf{N} \times \mathbf{N}$ is denumerable. To see this, recall the function $f: \mathbf{N} \times \mathbf{N} \to \mathbf{N}$ on page 146 given by $f(m, n) = 2^{m-1}(2n - 1)$. Since this function is a bijection, $\mathbf{N} \times \mathbf{N} \approx \mathbf{N}$.

We have seen examples of infinite sets that are denumerable, but no example, as yet, of a set that is uncountable (infinite but not denumerable). Before considering the next theorem, which states that the set of real numbers in the interval $(0, 1)$ is such an example, we need to review decimal expressions for real numbers. In its decimal form, any real number in $(0, 1)$ may be written as $0.a_1a_2a_3a_4 \ldots$, where each a_i is an integer, $0 \leq a_i \leq 9$. In this form, $\frac{7}{12} = 0.583333 \ldots$, which is abbreviated to $0.58\overline{3}$ to indicate that the 3 is repeated. A block of digits may also be repeated, as in $\frac{23}{28} = 0.82\overline{142857}$. The number $x = 0.a_1a_2a_3 \ldots$ is said to be in **normalized form** iff there is no k such that for all $n > k$, $a_n = 9$. For example, $0.82\overline{142857}$ and $\frac{2}{5} = 0.4\overline{0}$ are in normalized form, but $0.4\overline{9}$ is not. *Every real number can be expressed uniquely in normalized form.* Both $0.4\overline{9}$ and $0.5\overline{0}$ represent the same real number $\frac{1}{2}$, but only $0.5\overline{0}$ is normalized. The importance of normalizing decimals is that *two decimal numbers in normalized form are equal iff they have identical digits in each decimal position.*

Theorem 5.13   The interval $(0, 1)$ is uncountable.

Proof.   We must show that $(0, 1)$ is neither finite nor denumerable. The interval $(0, 1)$ is not finite since it contains the infinite subset $\{\frac{1}{2}, \frac{1}{3}, \frac{1}{4}, \ldots\}$. ⟨*See Theorem 5.5.*⟩

Suppose there is a function $f: \mathbf{N} \rightarrow (0, 1)$ that is one-to-one. We will show that f does not map onto $(0, 1)$. Writing the images of the elements of $\mathbf{N}$ in normalized form, we have

$$f(1) = 0.a_{11}a_{12}a_{13}a_{14}a_{15} \ldots$$
$$f(2) = 0.a_{21}a_{22}a_{23}a_{24}a_{25} \ldots$$
$$f(3) = 0.a_{31}a_{32}a_{33}a_{34}a_{35} \ldots$$
$$f(4) = 0.a_{41}a_{42}a_{43}a_{44}a_{45} \ldots$$

.
.
.

$$f(n) = 0.a_{n1}a_{n2}a_{n3}a_{n4}a_{n5} \ldots$$

.
.
.

Now let b be the number $b = 0.b_1b_2b_3b_4b_5 \ldots$, where

$$b_i = \begin{cases} 5 & \text{if } a_{ii} \neq 5 \\ 3 & \text{if } a_{ii} = 5. \end{cases} \quad \langle \textit{The choice of 3 and 5 is arbitrary.} \rangle$$

Then b is not the image of any $n \in \mathbf{N}$, because it differs from $f(n)$ in the nth decimal place. We conclude there is no one-to-one and onto function from $\mathbf{N}$ to $(0, 1)$ and hence $(0, 1)$ is not denumerable.   ∎

The interval $(0, 1)$ is our first example of an uncountable set. The cardinal number of $(0, 1)$ is defined to be **c** (which stands for **continuum**) and is the only infinite cardinal other than $\aleph_0$ we will mention by name. There is nothing special about the numbers 0 and 1 in this example. Every open interval has cardinality **c**.

Theorem 5.14   Let $a, b \in \mathbf{R}$ with $a < b$. Then the open interval (a, b) is uncountable and has cardinality **c**.

Proof.   The linear function $f: (0, 1) \to (a, b)$ given by $f(x) = (b - a)x + a$ (figure 5.3) is a bijection. Thus $(a, b) \approx (0, 1)$.   ∎

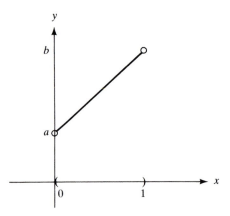

Figure 5.3

Theorem 5.15   The set **R** is uncountable and has cardinality **c**.

Proof.   Define $f: (0, 1) \to \mathbf{R}$ by $f(x) = \tan(\pi x - \pi/2)$. See figure 5.4. The function f is a contraction and translation of one branch of the tangent function and is one-to-one and onto **R**. Thus $(0, 1) \approx \mathbf{R}$.   ∎

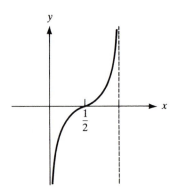

Figure 5.4

We turn now to the set of rational numbers. Since $\mathbf{N} \subseteq \mathbf{Q} \subseteq \mathbf{R}$, you should suspect that the cardinality of $\mathbf{Q}$ is $\aleph_0$, or $\mathbf{c}$, or some infinite cardinal in between (the ordering of cardinal numbers is discussed in the next section). You might also suspect that, since there are infinitely many rationals between any two rationals, $\mathbf{Q}$ is not denumerable, but this is not the case. Georg Cantor (1845–1918) first showed that $\mathbf{Q}^+$ (the positive rationals) is indeed denumerable through a clever rearrangement of $\mathbf{Q}^+$.

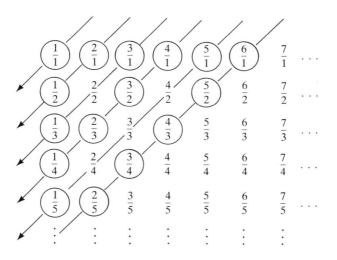

Figure 5.5

Every element in $\mathbf{Q}^+$ may be expressed as p/q for some $p, q \in \mathbf{N}$. Thus the elements of this set can be presented as in figure 5.5, where the nth row contains all the positive fractions with denominator n.

To show that $\mathbf{Q}^+$ is denumerable, we list its elements in the order indicated by the diagonal arrows. First are all fractions in which the sum of the numerator and denominator is 2 (only $\frac{1}{1}$), then those whose sum is 3 ($\frac{2}{1}$, $\frac{1}{2}$), then those whose sum is 4, and so on. We omit from the list $\frac{2}{2}$ ($= \frac{1}{1}$), $\frac{2}{4}$ ($= \frac{2}{1}$), $\frac{2}{3}$ ($= \frac{2}{2} = \frac{1}{1}$), $\frac{2}{4}$ ($= \frac{1}{2}$), and all other fractions not in lowest terms. The remaining numbers are circled in the array. The result is the one-to-one correspondence

$$
\begin{array}{ccccccccccccccc}
\mathbf{N} & = & \{1, & 2, & 3, & 4, & 5, & 6, & 7, & 8, & 9, & 10, & 11, & 12, & 13, & 14, & \ldots\} \\
& & \downarrow & \downarrow & \downarrow & \downarrow & \downarrow & \downarrow & \downarrow & \downarrow & \downarrow & \downarrow & \downarrow & \downarrow & \downarrow & \downarrow \\
\mathbf{Q}^+ & = & \{\frac{1}{1}, & \frac{2}{1}, & \frac{1}{2}, & \frac{3}{1}, & \frac{1}{3}, & \frac{4}{1}, & \frac{3}{2}, & \frac{2}{3}, & \frac{1}{4}, & \frac{5}{1}, & \frac{1}{5}, & \frac{6}{1}, & \frac{5}{2}, & \frac{4}{3}, & \ldots\}
\end{array}
$$

which suggests a proof of the next theorem. See also the example following Theorem 5.32.

Theorem 5.16 The set $\mathbf{Q}^+$ of positive rationals is denumerable.

Adding one or any finite number of elements to a finite set yields a finite set. Our next theorems provide analogs of these results for denumerable sets. The important distinction to be made is that adding finitely many or denumerably many elements does not change the cardinality of a denumerable set.

Theorem 5.17 If A is denumerable, then $A \cup \{x\}$ is denumerable.

Proof. Let $f: \mathbf{N} \xrightarrow[\text{onto}]{1-1} A$. If $x \in A$, $A \cup \{x\} = A$, which is denumerable. If $x \notin A$, define $g: \mathbf{N} \to A \cup \{x\}$ by

$$g(n) = \begin{cases} x & \text{if } n = 1 \\ f(n-1) & \text{if } n > 1 \end{cases}$$

It is now straightforward to verify that g is a one-to-one correspondence between $\mathbf{N}$ and $A \cup \{x\}$, which proves $A \cup \{x\}$ is denumerable. ∎

Theorem 5.17 may be loosely restated as $\aleph_0 + 1 = \aleph_0$. Its proof is illustrated by the story of the Infinite Hotel.†

The Infinite Hotel has $\aleph_0$ rooms and is full to capacity with one person in each room. You approach the desk clerk and ask for a room. When the clerk explains that each room is already occupied, you say, "There is room for me! For each n let the person in room n move to room $n + 1$. Then I will move into room 1, and everyone will have a room as before." There are $\aleph_0 + 1$ people and they fit exactly into the $\aleph_0$ rooms.

Rooms can also be found for any finite number of additional people (Theorem 5.18) or any denumerable number of additional people (Theorem 5.19).

Theorem 5.18 If A is denumerable and B is finite, then $A \cup B$ is denumerable.

Proof. Exercise 2. ∎

Theorem 5.19 If A and B are disjoint denumerable sets, then $A \cup B$ is denumerable.

---

†The Infinite Hotel is one of the topics discussed in *Aha! Gotcha: Paradoxes to Puzzle and Delight*, by Martin Gardner, Freeman, 1981.

Proof. Let $f\colon \mathbf{N} \xrightarrow[\text{onto}]{1-1} A$ and $g\colon \mathbf{N} \xrightarrow[\text{onto}]{1-1} B$. Define $h\colon \mathbf{N} \to A \cup B$ via

$$
h(n) = \begin{cases} f\!\left(\dfrac{n+1}{2}\right) & \text{if } n \text{ is odd} \\[2ex] g(n/2) & \text{if } n \text{ is even.} \end{cases}
$$

It is left as exercise 3 to show that h is a one-to-one correspondence from $\mathbf{N}$ onto $A \cup B$. ∎

Theorems 5.17 and 5.19 provide a simple proof that the set of all rationals is denumerable.

Theorem 5.20   The set $\mathbf{Q}$ of all rationals is denumerable.

Proof.   By Theorem 5.16, $\mathbf{Q}^+$ is denumerable. By Theorem 5.17, $\mathbf{Q}^+ \cup \{0\}$ is denumerable. The function associating each rational with its negative is a one-to-one correspondence from $\mathbf{Q}^+$ to $\mathbf{Q}^-$ (the negative rationals). Thus $\mathbf{Q}^-$ is denumerable. Therefore by Theorem 5.19, $\mathbf{Q} = (\mathbf{Q}^+ \cup \{0\}) \cup \mathbf{Q}^-$ is denumerable. ∎

Exercises 5.2

1. Prove that the following sets are denumerable.
    ★ (a) D^+, the odd positive integers
    (b) T^+, the positive integer multiples of 3
    (c) T, the integer multiples of 3
    (d) $\{n\colon n \in \mathbf{N} \text{ and } n \geqslant 7\}$
    ☆ (e) $\{x\colon x \in \mathbf{Z} \text{ and } x < -12\}$
    (f) $\mathbf{N} - \{5, 6\}$

2. Prove Theorem 5.18.

3. Prove that the function h of Theorem 5.19 is one-to-one.

4. Prove that
    ☆ (a) $(0, 1) \approx (1, \infty)$.
    (b) $(1, \infty) \approx (a, \infty)$ for any $a \in \mathbf{R}$.
    (c) $(0, 1) \approx (a, \infty)$ for any $a \in \mathbf{R}$.
    (d) $(-\infty, a) \approx (b, \infty)$ for any $a, b \in \mathbf{R}$.

5. What is the 28th term in the sequence of rationals given in the discussion of Theorem 5.16?

6. Prove that a set is infinite iff it contains an infinite subset.

7. Let A be a denumerable set and let x and y be objects not in A. Prove that $A \cup \{x, y\}$ is denumerable by
    (a) constructing a function to show equivalence with $\mathbf{N}$.
    (b) applying Theorem 5.17.

8. Give an example of a bijection f from $\mathbf{N}$ to $\mathbf{N} \cup \{\sqrt{2}\}$ such that
    (a) $f(1) = \sqrt{2}$.
    (b) $f(1) = 2$.
    (c) $f(1) > 500$.
    (d) for every odd $x \in \mathbf{N}$, $f(x)$ is even.

9. Which sets have cardinal number $\aleph_0$? $\mathbf{c}$?
    ★ (a) $\mathbf{R} - \mathbf{Q}$ (b) $(5, \infty)$
    ★ (c) $\{1/n: n \in \mathbf{N}\}$ (d) $\{2^x: x \in \mathbf{N}\}$
    ★ (e) $\{(p, q) \in \mathbf{R} \times \mathbf{R}: p + q = 1\}$
      (f) $\{(p, q) \in \mathbf{R} \times \mathbf{R}: q = \sqrt{1 - p^2}\}$

10. (a) Show that for all $k \in \mathbf{N}$, $\mathbf{N} - \mathbf{N}_k$ is denumerable.
    ★ (b) Show that if $S \subseteq \mathbf{N}$ and S is finite, then $\mathbf{N} - S$ is denumerable.
    (c) Show that if $B \subseteq A$, A is denumerable, and B is finite, then $A - B$ is denumerable.

☆ 11. Let S be the set of all sequences of 0's and 1's. For example, 1010101 . . . and 010110111 . . . are in S. Prove that S is uncountable.

12. Give another proof of Theorem 5.15 by showing that $f(x) = (x - 1/2)/[x(x - 1)]$ is a one-to-one correspondence from $(0, 1)$ onto $\mathbf{R}$.

☆ 13. Let $\{A_i: i = 1, 2, \ldots, n\}$ be a finite pairwise disjoint collection of denumerable sets. Prove that $\bigcup_{i=1}^{n} A_i$ is denumerable.

14. Give an example of two distinct infinite sets A and B such that
    (a) $A - B$ is empty.   (b) $A - B$ is finite and nonempty.
    (c) $A - B$ is denumerable.   (d) $A - B$ is uncountable.

15. **Proofs to Grade.**
    ★ (a) **Claim.** The sets E^+ of even natural numbers and D^+ of odd natural numbers are equivalent.
      "Proof." E^+ is an infinite subset of $\mathbf{N}$. Thus E^+ is denumerable. Likewise D^+ is an infinite subset of $\mathbf{N}$ and is denumerable. Therefore $E^+ \approx D^+$. ∎

    (b) **Claim.** If A is infinite and $x \notin A$, then $A \cup \{x\}$ is infinite.
      "Proof." Let A be infinite. Then $A \approx \mathbf{N}$. Let $f: \mathbf{N} \to A$ be a one-to-one correspondence. Then $g: \mathbf{N} \to A \cup \{x\}$, defined by

    $$g(t) = \begin{cases} x & \text{if } t = 1 \\ f(t - 1) & \text{if } t > 1 \end{cases}$$

    is one-to-one and onto $A \cup \{x\}$. Thus $\mathbf{N} \approx A \cup \{x\}$, so $A \cup \{x\}$ is infinite. ∎

    ★ (c) **Claim.** If $A \cup B$ is infinite, then A and B are infinite.

"Proof."   This is a proof by contrapositive so assume the denial of the consequent. Thus assume A and B are finite. Then by Theorem 5.6, $A \cup B$ is finite, which is a denial of the antecedent. Therefore the result is proved. ∎

★ (d) **Claim.**   If a set A is infinite, then A is equivalent to a proper subset of A.
"Proof."   Let $A = \{x_1, x_2, \ldots\}$. Choose $B = \{x_2, x_3, \ldots\}$. Then B is a proper subset of A. The function $f: A \to B$ defined by $f(x_k) = x_{k+1}$ is clearly one-to-one and onto B. Thus $A \approx B$. ∎

(e) **Claim.**   (Theorem 5.16) The set $\mathbf{Q}^+$ of positive rationals is denumerable.
"Proof."   Consider the positive rationals in the array on page 169. Consider the order formed by listing all the rationals in the first row, then the second row, and so forth. Omitting fractions that are not in lowest terms, we have an ordering of $\mathbf{Q}^+$ in which every positive rational appears. Therefore $\mathbf{Q}^+$ is denumerable. ∎

(f) **Claim.**   If A and B are infinite, then $A \approx B$.
"Proof."   Suppose A and B are infinite sets. Let $A = \{a_1, a_2, a_3, \ldots\}$ and $B = \{b_1, b_2, b_3, \ldots\}$. Define $f: A \to B$ as in the picture

$$\{a_1, a_2, a_3, a_4, \ldots\}$$
$$\downarrow \quad \downarrow \quad \downarrow \quad \downarrow$$
$$\{b_1, b_2, b_3, b_4, \ldots\}$$

Then since we never run out of elements in either set, f is one-to-one and onto B, so $A \approx B$. ∎

5.3

The Ordering of Cardinal Numbers

The theory of infinite sets was developed by Georg Cantor over a twenty-year span, culminating primarily in papers that appeared during 1895 and 1897. He described a cardinal number of a set M as "the general concept which, with the aid of our intelligence, results from M when we abstract from the nature of its various elements and from the order of their being given." This definition was criticized as being less precise and more mystical than a definition in mathematics ought to be. Other definitions were given, and eventually the concept of cardinal number was made precise. One way this may be done is to determine a fixed set from each equivalence class of sets under the relation $\approx$, and then to call this set the cardinal number of each set in the class. Under such a procedure we would think of the number 0 as being the empty set and the number 1 as being the set whose only element is the number 0. That is, $1 = \{0\}$; $2 = \{0, 1\}$; $3 = \{0, 1, 2\}$, and so on.

We will not be concerned with a precise definition of a cardinal number. For our purposes the essential point is that the **cardinal number** $\overline{\overline{A}}$ of a set A is an object associated with all sets equivalent to A and with no other set. The double bars on $\overline{\overline{A}}$ are suggestive of the double abstraction referred to by Cantor. For example, $\overline{\overline{\{5\}}} = \overline{\overline{\{1\}}} = 1$ and $\overline{\overline{\{p, q, r, s\}}} = \overline{\overline{\{1, 2, 3, 4\}}} = 4$.

Cardinal numbers may be ordered (compared) in the following manner:

Definitions.

(a) $\overline{\overline{A}} = \overline{\overline{B}}$ if and only if $A \approx B$; otherwise $\overline{\overline{A}} \neq \overline{\overline{B}}$.

(b) $\overline{\overline{A}} \leqslant \overline{\overline{B}}$ if and only if there exists $f\colon A \xrightarrow{1-1} B$.

(c) $\overline{\overline{A}} < \overline{\overline{B}}$ if and only if $\overline{\overline{A}} \leqslant \overline{\overline{B}}$ and $\overline{\overline{A}} \neq \overline{\overline{B}}$.

$\overline{\overline{A}} < \overline{\overline{B}}$ is read "the cardinality of A is strictly less than the cardinality of B" while $\leqslant$ is read "less than or equal to." In addition, we use $\overline{\overline{A}} \not< \overline{\overline{B}}$ and $\overline{\overline{A}} \not\leqslant \overline{\overline{B}}$ to denote the denials of $\overline{\overline{A}} < \overline{\overline{B}}$ and $\overline{\overline{A}} \leqslant \overline{\overline{B}}$, respectively.

Usually proof of $\overline{\overline{A}} \leqslant \overline{\overline{B}}$ will involve constructing a one-to-one function from A to B, while a proof of $\overline{\overline{A}} < \overline{\overline{B}}$ will have a proof of $\overline{\overline{A}} \leqslant \overline{\overline{B}}$ together with a proof, generally by contradiction, that $\overline{\overline{A}} \neq \overline{\overline{B}}$. Once we have developed some properties of cardinal inequalities those facts can be used to prove statements of the form $\overline{\overline{A}} \leqslant \overline{\overline{B}}$ without resorting to the construction of functions.

Since 1, 2, 3, . . . are cardinal numbers, the natural numbers may be viewed as a subset of the collection of all cardinal numbers. In this sense the properties of $\leqslant$ and $<$ that we will develop for cardinal numbers may be viewed as extensions of those same properties of $\leqslant$ and $<$ that hold for **N**. Those not proved are left as exercise 6.

Theorem 5.21   For sets A, B, and C,

(a)   (Reflexivity) $\overline{\overline{A}} \leqslant \overline{\overline{A}}$.

(b)   (Transitivity of $=$) If $\overline{\overline{A}} = \overline{\overline{B}}$ and $\overline{\overline{B}} = \overline{\overline{C}}$, then $\overline{\overline{A}} = \overline{\overline{C}}$.

(c)   (Transitivity of $\leqslant$) If $\overline{\overline{A}} \leqslant \overline{\overline{B}}$ and $\overline{\overline{B}} \leqslant \overline{\overline{C}}$, then $\overline{\overline{A}} \leqslant \overline{\overline{C}}$.

(d) $\overline{\overline{A}} \leqslant \overline{\overline{B}}$ iff $\overline{\overline{A}} < \overline{\overline{B}}$ or $\overline{\overline{A}} = \overline{\overline{B}}$.

(e)   If $A \subseteq B$, then $\overline{\overline{A}} \leqslant \overline{\overline{B}}$.

(f) $\overline{\overline{A}} \leqslant \overline{\overline{B}}$ iff there is a subset W of B such that $\overline{\overline{W}} = \overline{\overline{A}}$.

Proof.

(c)   Suppose $\overline{\overline{A}} \leqslant \overline{\overline{B}}$ and $\overline{\overline{B}} \leqslant \overline{\overline{C}}$. Then there exist functions $f\colon A \xrightarrow{1-1} B$ and $g\colon B \xrightarrow{1-1} C$. Since the composite $g \circ f\colon A \to C$ is one-to-one, we conclude $\overline{\overline{A}} \leqslant \overline{\overline{C}}$.

(e)   Let $A \subseteq B$. We note that the inclusion map $i: A \to B$, given by $i(a) = a$, is one-to-one, whence $\overline{\overline{A}} \leqslant \overline{\overline{B}}$.   ∎

Although Cantor developed many aspects of the theory of infinite sets, his name remains attached particularly to the next theorem, which states that the cardinality of a set A is strictly less than the cardinality of its power set. We already know the result to be true if A is a finite set with n elements, since $\mathcal{P}(A)$ has 2^n elements by Theorem 2.4. The proof given here holds for all sets A.

Theorem 5.22   (Cantor's Theorem)   For every set A, $\overline{\overline{A}} < \overline{\overline{\mathcal{P}(A)}}$.

Proof.   To show $\overline{\overline{A}} < \overline{\overline{\mathcal{P}(A)}}$, we must show that (i) $\overline{\overline{A}} \leqslant \overline{\overline{\mathcal{P}(A)}}$, and (ii) $\overline{\overline{A}} \neq \overline{\overline{\mathcal{P}(A)}}$. Part (i) follows from the fact that $F: A \to \mathcal{P}(A)$ defined by $F(x) = \{x\}$ is one-to-one.

To prove (ii), suppose $\overline{\overline{A}} = \overline{\overline{\mathcal{P}(A)}}$; that is, assume $A \approx \mathcal{P}(A)$. Then there exists $g: A \xrightarrow[\text{onto}]{1-1} \mathcal{P}(A)$. Let $B = \{y \in A : y \notin g(y)\}$. Since $B \subseteq A$, $B \in \mathcal{P}(A)$, and since g is onto $\mathcal{P}(A)$, $B = g(z)$ for some $z \in A$. Now either $z \in B$ or $z \notin B$. If $z \in B$, then $z \notin g(z) = B$, a contradiction. Similarly, $z \notin B$ implies $z \in g(z)$, which implies $z \in B$, again a contradiction. We conclude $A \neq \mathcal{P}(A)$ and hence $\overline{\overline{A}} < \overline{\overline{\mathcal{P}(A)}}$.   ∎

Cantor's Theorem has some interesting consequences. First, there are an infinite number of infinite cardinal numbers. We know one, $\aleph_0$, which corresponds to $\overline{\overline{\mathbf{N}}}$. By Cantor's Theorem, $\aleph_0 < \overline{\overline{\mathcal{P}(\mathbf{N})}}$. Since $\mathcal{P}(\mathbf{N})$ is a set, its power set $\mathcal{P}(\mathcal{P}(\mathbf{N}))$ has a strictly greater cardinality than that of $\mathcal{P}(\mathbf{N})$. In this fashion we may generate a denumerable set of cardinal numbers:

$$\aleph_0 < \overline{\overline{\mathcal{P}(\mathbf{N})}} < \overline{\overline{\mathcal{P}(\mathcal{P}(\mathbf{N}))}} < \overline{\overline{\mathcal{P}(\mathcal{P}(\mathcal{P}(\mathbf{N})))}} < \cdots$$

It is also an immediate consequence of Cantor's Theorem that there can be no largest cardinal number (see exercise 7).

It appears to be obvious that if B has at least as many elements as A $(\overline{\overline{A}} \leqslant \overline{\overline{B}})$, and A has at least as many elements as B $(\overline{\overline{B}} \leqslant \overline{\overline{A}})$, then A and B are equivalent $(\overline{\overline{A}} = \overline{\overline{B}})$. The proof, however, is not obvious. The situation may be represented as in figure 5.6. From $\overline{\overline{A}} \leqslant \overline{\overline{B}}$ and $\overline{\overline{B}} \leqslant \overline{\overline{A}}$, there are functions $F: A \xrightarrow{1-1} B$ and $G: B \xrightarrow{1-1} A$. The problem is to construct $H: A \to B$, which is both one-to-one and onto B. Cantor (1895) solved this problem, but his proof used the controversial Axiom of Choice (section 5.4). Proofs not depending on the Axiom of Choice were given independently by Ernest Schröder in 1896 and two years later by Felix Bernstein.

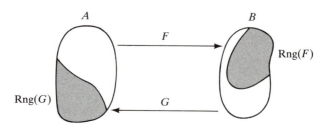

Figure 5.6

Theorem 5.23 (Cantor-Schröder-Bernstein *Theorem*)   If $\overline{\overline{A}} \leqslant \overline{\overline{B}}$ and $\overline{\overline{B}} \leqslant \overline{\overline{A}}$, then $\overline{\overline{A}} = \overline{\overline{B}}$.

Proof.   We may assume that A and B are disjoint, for otherwise we could replace A and B with the equivalent sets $A \times \{0\}$ and $B \times \{1\}$, respectively. Let $F: A \xrightarrow{1-1} B$, with $D = \text{Rng}(F)$, and let $G: B \xrightarrow{1-1} A$, with $C = \text{Rng}(G)$. Define a **string** to be a function $f: \mathbf{N} \to A \cup B$ such that

$f(1) \in B - D$,
$f(n) \in B$ implies $f(n + 1) = G(f(n))$, and
$f(n) \in A$ implies $f(n + 1) = F(f(n))$.

We think of a string as a sequence of elements of $A \cup B$ with first term in $B - D$, and such that the terms are alternately in B and in A. Each element of $B - D$ is the first term of some string.
     Let $W = \{x \in A: x$ is a term of some string$\}$ and define $H: A \to B$ by

$$H(x) = \begin{cases} F(x) & \text{if } x \in A - W \\ G^{-1}(x) & \text{if } x \in W. \end{cases}$$

     We will show that $H: A \xrightarrow[\text{onto}]{1-1} B$. Because F and G^{-1} are one-to-one, to show H is one-to-one we need consider only the possibility that $F(a) = G^{-1}(b)$ for some $a \in A - W$ and $b \in W$. Suppose $b = G(F(a))$ and b is in a string f. Then $b = f(2n)$ for some $n \geqslant 1$ and $F(a) = f(2n - 1)$. If $n = 1$, then $F(a) = f(1)$. But $f(1) \notin \text{Rng}(F)$. Thus $n \geqslant 2$, so $a = f(2n - 2)$. Since we are assuming that $a \in A - W$, this is a contradiction. Therefore H is one-to-one.
     We next show that H is onto B. Let $b \in B$. We must show that for some $x \in A$, $H(x) = b$.

Case 1.   If $G(b) \in W$, then $H(G(b)) = G^{-1}(G(b)) = b$. Hence we may take $x = G(b)$.

Case 2.   If $G(b) \notin W$, then there exists $x \in A$ such that $F(x) = b$. ⟨*For if there were no such x, then $b \in B - D$, and so some string would begin at b, and $G(b)$ would be a term of that string, hence in W.*⟩ Furthermore, $x \in A - W$, for if x were a term of some string, then also $F(x)$ and $G(F(x)) = G(b)$ would be terms of the same string, which would imply $G(b) \in W$. Since $x \in A - W$, $H(x) = F(x) = b$. ∎

The Cantor-Schröder-Bernstein Theorem may be used to prove equivalence between sets in cases where it would be difficult to exhibit a one-to-one correspondence.

Example.   We will show that $(0, 1) \approx [0, 1]$. First, note that $(0, 1) \subseteq [0, 1]$, so by Theorem 5.21(e), $\overline{\overline{(0, 1)}} \leqslant \overline{\overline{[0, 1]}}$. Likewise, since $[0, 1] \subseteq (-1, 2)$, we have $\overline{\overline{[0, 1]}} \subseteq \overline{\overline{(-1, 2)}}$. But we know $(0, 1) \approx (-1, 2)$ and thus $\overline{\overline{(0, 1)}} = \overline{\overline{(-1, 2)}}$. Therefore we may write $\overline{\overline{[0, 1]}} \leqslant \overline{\overline{(0, 1)}}$. We conclude $\overline{\overline{(0, 1)}} = \overline{\overline{[0, 1]}}$ by the Cantor-Schröder-Bernstein Theorem and thus $(0, 1) \approx [0, 1]$.

The Cantor-Schröder-Bernstein Theorem is another result in the extension of the familiar ordering properties of **N** to properties for all cardinal numbers. It, in turn, leads to others. In the following, parts (a) and (c) are proved; (b) and (d) are given as exercise 11.

Corollary 5.24   For sets A, B, and C,

(a)   if $\overline{\overline{A}} \leqslant \overline{\overline{B}}$, then $\overline{\overline{B}} \not< \overline{\overline{A}}$.
(b)   if $\overline{\overline{A}} \leqslant \overline{\overline{B}}$ and $\overline{\overline{B}} < \overline{\overline{C}}$, then $\overline{\overline{A}} < \overline{\overline{C}}$.
(c)   if $\overline{\overline{A}} < \overline{\overline{B}}$ and $\overline{\overline{B}} \leqslant \overline{\overline{C}}$, then $\overline{\overline{A}} < \overline{\overline{C}}$.
(d)   if $\overline{\overline{A}} < \overline{\overline{B}}$ and $\overline{\overline{B}} < \overline{\overline{C}}$, then $\overline{\overline{A}} < \overline{\overline{C}}$.

Proof.

(a)   Suppose $\overline{\overline{B}} < \overline{\overline{A}}$. Then $\overline{\overline{A}} \neq \overline{\overline{B}}$ and $\overline{\overline{B}} \leqslant \overline{\overline{A}}$. Combining this with the hypothesis that $\overline{\overline{A}} \leqslant \overline{\overline{B}}$, we conclude by the Cantor-Schröder-Bernstein Theorem that $\overline{\overline{A}} = \overline{\overline{B}}$, which is a contradiction. Therefore $\overline{\overline{B}} \not< \overline{\overline{A}}$.

(c)   Suppose $\overline{\overline{A}} < \overline{\overline{B}}$ and $\overline{\overline{B}} \leqslant \overline{\overline{C}}$. Then $\overline{\overline{A}} \leqslant \overline{\overline{B}}$; so by Theorem 5.21(c), $\overline{\overline{A}} \leqslant \overline{\overline{C}}$. Suppose $\overline{\overline{A}} \not< \overline{\overline{C}}$. Then $\overline{\overline{A}} = \overline{\overline{C}}$, which implies $\overline{\overline{C}} \leqslant \overline{\overline{A}}$. But $\overline{\overline{B}} \leqslant \overline{\overline{C}}$ and $\overline{\overline{C}} \leqslant \overline{\overline{A}}$ imply $\overline{\overline{B}} \leqslant \overline{\overline{A}}$. Combining this with $\overline{\overline{A}} \leqslant \overline{\overline{B}}$, we conclude by the Cantor-Schröder-Bernstein Theorem that $\overline{\overline{A}} = \overline{\overline{B}}$. Since this contradicts $\overline{\overline{A}} < \overline{\overline{B}}$, we have $\overline{\overline{A}} < \overline{\overline{C}}$. ∎

It is tempting to extend our results even further to include the converse of Corollary 5.24(a): "If $\overline{\overline{B}} \not< \overline{\overline{A}}$, then $\overline{\overline{A}} \leqslant \overline{\overline{B}}$." (As far as we know now, for two given sets A and B, both $\overline{\overline{A}} \leqslant \overline{\overline{B}}$ and $\overline{\overline{B}} \leqslant \overline{\overline{A}}$ may be false.) The Cantor-Schröder-Bernstein Theorem turned out to be more difficult to prove than one would have guessed from its simple statement, but the situation regarding the converse of Corollary 5.24(a) is even more remarkable. This is discussed in section 5.4, where "If $\overline{\overline{B}} \not< \overline{\overline{A}}$, then $\overline{\overline{A}} \leqslant \overline{\overline{B}}$" is rephrased as "Either $\overline{\overline{A}} < \overline{\overline{B}}$ or $\overline{\overline{A}} = \overline{\overline{B}}$ or $\overline{\overline{B}} < \overline{\overline{A}}$."

Exercises 5.3

★   1.   Prove that if $n \in \mathbf{N}$, then $n < \aleph_0$.

2.   Prove $\aleph_0 < \mathbf{c}$.

3.   Prove that if $\overline{\overline{A}} \leqslant \overline{\overline{B}}$ and $\overline{\overline{B}} = \overline{\overline{C}}$, then $\overline{\overline{A}} \leqslant \overline{\overline{C}}$.

4.   Prove that if $\overline{\overline{A}} \leqslant \overline{\overline{B}}$ and $\overline{\overline{A}} = \overline{\overline{C}}$, then $\overline{\overline{C}} \leqslant \overline{\overline{B}}$.

5.   Arrange the following cardinal numbers in order:
★   (a) $\overline{\overline{(0, 1)}}, \overline{\overline{[0, 1]}}, \overline{\overline{\{0, 1\}}}, \overline{\overline{\{0\}}}, \overline{\overline{\mathcal{P}(\mathbf{R})}}, \overline{\overline{\mathbf{Q}}}, \overline{\overline{\varnothing}}, \overline{\overline{\mathbf{R} - \mathbf{Q}}}, \overline{\overline{\mathcal{P}(\mathcal{P}(\mathbf{R}))}}, \overline{\overline{\mathbf{R}}}$
   (b) $\overline{\overline{\mathbf{Q} \cup \{\pi\}}}, \overline{\overline{\mathbf{R} - \{\pi\}}}, \overline{\overline{\mathcal{P}(\{0, 1\})}}, \overline{\overline{[0, 2]}}, \overline{\overline{(0, \infty)}}, \overline{\overline{\mathbf{Z}}}, \overline{\overline{\mathbf{R} - \mathbf{Z}}}, \overline{\overline{\mathcal{P}(\mathbf{R})}}$

6.   Prove the remaining parts of Theorem 5.21.

☆   7.   Prove that there is no largest cardinal number.

8.   Apply the proof of the Cantor-Schröder-Bernstein Theorem to this situation:
   $A = \{2, 3, 4, 5, \ldots\}$, $B = \{\frac{1}{2}, \frac{1}{3}, \frac{1}{4}, \ldots\}$, $F: A \to B$ where $F(x) = \dfrac{1}{x + 6}$,
   and $G: B \to A$ where $G(x) = \dfrac{1}{x} + 5$. Note that $\frac{1}{3}$ and $\frac{1}{4}$ are in $B - \text{Rng}(F)$. Let f be the string that begins at $\frac{1}{3}$, and let g be the string that begins at $\frac{1}{4}$.
   (a)   Find $f(1), f(2), f(3), f(4)$.
   (b) Find $g(1), g(2), g(3), g(4)$.
   (c)   Define H as in the proof of the Cantor-Schröder-Bernstein Theorem and find $H(2), H(8), H(13)$, and $H(20)$.

9.   Suppose there exist functions $f: A \xrightarrow{1-1} B$, $g: B \xrightarrow{1-1} C$, and $h: C \xrightarrow{1-1} A$. Prove $A \approx B \approx C$. Do not assume that the functions map onto their codomains.

10.   If possible, give an example of
   (a)   functions f and g such that $f: \mathbf{Q} \xrightarrow{1-1} \mathbf{N}$, $g: \mathbf{N} \xrightarrow{1-1} \mathbf{Q}$, but neither f nor g is an onto map.
★   (b)   a function $f: \mathbf{R} \xrightarrow{1-1} \mathbf{N}$.
   (c)   a function $f: \mathcal{P}(\mathbf{N}) \xrightarrow{1-1} \mathbf{N}$.

11.   Prove parts (b) and (d) of Corollary 5.24.

12.   Use a cardinality argument to prove that there is no universal set U of all sets.

13. Use the Cantor-Schröder-Bernstein Theorem to prove the following.
     (a)   The set of all integers whose digits are 6, 7, or 8 is denumerable.
     (b) $\mathbf{R} \times \mathbf{R} \approx \mathbf{R}$. Note that this means there are just as many points on the real line as there are in the Cartesian plane.

14. **Proofs to Grade.**
     (a)   **Claim.**   If $\overline{\overline{A}} \leq \overline{\overline{B}}$ and $\overline{\overline{A}} = \overline{\overline{C}}$, then $\overline{\overline{C}} \leq \overline{\overline{B}}$.
           "Proof."   Assume $\overline{\overline{A}} \leq \overline{\overline{B}}$ and $\overline{\overline{A}} = \overline{\overline{C}}$. Then there exists a function f such that $f\colon A \xrightarrow{\ 1-1\ } B$. Since $\overline{\overline{A}} = \overline{\overline{C}}$, $f\colon C \xrightarrow{\ 1-1\ } B$. Therefore $\overline{\overline{C}} \leq \overline{\overline{B}}$.   ∎
   ★ (b)   **Claim.**   If $B \subseteq C$ and $\overline{\overline{B}} = \overline{\overline{C}}$, then $B = C$.
           "Proof."   Suppose $B \neq C$. Then B is a proper subset of C. Thus $C - B \neq \varnothing$. This implies $\overline{\overline{C - B}} > 0$. But $C = B \cup (C - B)$ and, since B and $C - B$ are disjoint, $\overline{\overline{C}} = \overline{\overline{B}} + \overline{\overline{(C - B)}}$. By hypothesis, $\overline{\overline{B}} = \overline{\overline{C}}$. Thus $\overline{\overline{(C - B)}} = 0$, a contradiction.   ∎
     (c)   **Claim.**   If $\overline{\overline{A}} \leq \overline{\overline{B}}$ and $\overline{\overline{B}} = \overline{\overline{C}}$, then $\overline{\overline{A}} \leq \overline{\overline{C}}$.
           "Proof."   Assume $\overline{\overline{A}} \leq \overline{\overline{B}}$. Then, since $\overline{\overline{B}} = \overline{\overline{C}}$, we have $\overline{\overline{A}} \leq \overline{\overline{C}}$ by substitution.   ∎
   ★ (d)   **Claim.**   If $A \neq \varnothing$ and $\overline{\overline{A}} \leq \overline{\overline{B}}$, then there exists a function $f\colon B \xrightarrow{\ \text{onto}\ } A$.
           "Proof."   From $\overline{\overline{A}} \leq \overline{\overline{B}}$, we know there exists $g\colon A \xrightarrow{\ 1-1\ } B$. Fix a particular $a^* \in A$. Define $f\colon B \to A$ as follows:
           (1) For $b \in B$, if $b \in \text{Rng}(g)$, then $g^{-1}(\{b\})$ consists of a single element of A, since g is one-to-one. Define $f(b)$ to be equal to that element of A.
           (2) If $b \notin \text{Rng}(g)$, define $f(b) = a^*$.
           The function f is onto A, for if $a \in A$, let $g(a) = b$. Then $b \in B$ and $g^{-1}(\{b\}) = a$, so $f(b) = a$.   ∎

5.4

Comparability of Cardinals and the Axiom of Choice

The goal of section 5.3 was to establish results for the relations $\leq$ and $<$ on the class of all cardinal numbers while thinking of these results as extensions of properties of $\mathbf{N}$. One of the most useful ordering properties of $\mathbf{N}$ is the **trichotomy property**: if m and n are any two natural numbers, then $m > n$, $m = n$, or $m < n$. The analog for cardinal numbers is stated in the Comparability Theorem.

Theorem 5.25   *(The Comparability Theorem)*   If A and B are any two sets, then $\overline{\overline{A}} < \overline{\overline{B}}$, $\overline{\overline{A}} = \overline{\overline{B}}$, or $\overline{\overline{B}} < \overline{\overline{A}}$.

Surprisingly, all our knowledge of sets and functions is insufficient to prove the Comparability Theorem. On the other hand, it is also insufficient to prove that the Comparability Theorem is false. Theorem 5.25 is an undecidable sentence.

The solution to our problem is that we may either assume Theorem 5.25 to be true (or assume true some other statement that implies comparability) or else assume the truth of some statement from which we can show comparability is false. It has become standard practice by most mathematicians to assume the Comparability Theorem is true by assuming the truth of the following statement:

> **The Axiom of Choice.**   If $\mathscr{A}$ is any collection of nonempty sets, then there exists a function F (called a **choice function**) from $\mathscr{A}$ to $\bigcup_{A \in \mathscr{A}} A$ such that for every $A \in \mathscr{A}$, $F(A) \in A$.

The Axiom of Choice at first appears to have little significance: From a collection of nonempty sets we can choose an element from each set. If the collection is finite, then this axiom is not needed to prove the existence of a choice function. It is only for infinite collections of sets that the result is not obvious and for which the Axiom of Choice is independent of other axioms of set theory.

Many examples and uses of the Axiom of Choice require more advanced knowledge of mathematics. The example we present is not mathematical in content yet has become part of mathematical folklore.

A shoe store has in the stockroom an infinite number of pairs of shoes and an infinite number of pairs of socks. A customer asks to see one shoe of each pair and one sock of each pair. The clerk does not need the Axiom of Choice to select a shoe from each pair of shoes. His choice rule might be "From each pair of shoes, choose the left shoe." However, since the socks of any pair of socks are indistinguishable, and since there are an infinite number of pairs of socks, the clerk must employ the Axiom of Choice to show the customer one sock from each pair.

With the Axiom of Choice, we could, but will not, prove Theorem 5.25. For a proof, see R. L. Wilder, *Introduction to the Foundation of Mathematics,* 2nd Ed. (Krieger, 1980).

Many other important theorems, in many areas of mathematics, cannot be proved without the use of the Axiom of Choice.† In fact, several crucial results are equivalent to it. Some of the consequences of the axiom are not so natural as the Comparability Theorem however, and some of them are extremely difficult to believe. (One of these is that the real numbers can be rearranged in such a way that every nonempty subset of **R** has a first element—in other words, that the reals can be well ordered.) The Axiom of Choice has been objected to because of such consequences, and also because of a lack of precision in the statement of the axiom, which does not provide any hint of a rule for constructing the choice function F. Because of these

---

†See H. Rubin and J. E. Rubin, *Equivalents of the Axiom of Choice* (North Holland Pub. Co., New Amsterdam, 1963).

objections it is common practice to call attention to the fact that the Axiom of Choice has been used in a proof, so that anyone who is interested can attempt to find an alternate proof that does not use the axiom.

We present two theorems whose proofs require the Axiom of Choice. The first is that if there is a function f from A onto B, then A must have at least as many elements as B. The proof chooses for each $b \in B$, an $a \in A$ such that $f(a) = b$. The second theorem is that every infinite set has a denumerable subset. The axiom is used to define the denumerable subset inductively.

Theorem 5.26   If there exists a function from A onto B, then $\overline{\overline{B}} \leqslant \overline{\overline{A}}$.

Proof.   If $B = \varnothing$, then $B \subseteq A$, so $\overline{\overline{B}} \leqslant \overline{\overline{A}}$. Let $f: A \to B$ be onto B where $B \neq \varnothing$. To show $\overline{\overline{B}} \leqslant \overline{\overline{A}}$, we must construct a function $h: B \to A$ which is one-to-one. Since f is onto B, for each $b \in B$, $b \in Rng(f)$ or, equivalently, $f^{-1}(\{b\}) \neq \varnothing$. Thus $\mathcal{A} = \{f^{-1}(\{b\}): b \in B\}$ is a nonempty collection of nonempty sets. By the Axiom of Choice, there exists a function g from $\mathcal{A}$ to $\bigcup_{b \in B} f^{-1}(\{b\}) = A$ such that $g(f^{-1}(\{b\})) \in f^{-1}(\{b\})$ for all $b \in B$.

Define $h: B \to A$ by $h(b) = g(f^{-1}(\{b\}))$. It remains to prove h is one-to-one. Suppose $h(b_1) = h(b_2)$. Then $g(f^{-1}(\{b_1\})) = g(f^{-1}(\{b_2\}))$. By definition of g, this element is in both $f^{-1}(\{b_1\})$ and $f^{-1}(\{b_2\})$. Thus $f^{-1}(\{b_1\}) \cap f^{-1}(\{b_2\}) \neq \varnothing$. For any $x \in f^{-1}(\{b_1\}) \cap f^{-1}(\{b_2\})$, $f(x) = b_1$, and $f(x) = b_2$. Therefore, $b_1 = b_2$. This proves that h is one-to-one and that $\overline{\overline{B}} \leqslant \overline{\overline{A}}$.   ∎

Theorem 5.27   Every infinite set A has a denumerable subset.

Proof.   We shall inductively define a denumerable subset of A. First, since A is infinite, $A \neq \varnothing$. Choose $a_1 \in A$. Then $A - \{a_1\}$ is infinite, hence nonempty. Choose $a_2 \in A - \{a_1\}$. Note that $a_2 \neq a_1$ and $a_2 \in A$. Continuing in this fashion, suppose $a_1, \ldots, a_k$ have been defined. Then $A - \{a_1, \ldots, a_k\} \neq \varnothing$, so select any a_{k+1} from this set. By the Axiom of Choice, a_n is defined for all $n \in \mathbf{N}$. The a_n have been constructed so that each $a_n \in A$ and $a_i \neq a_j$ for $i \neq j$. Thus $B = \{a_n: n \in \mathbf{N}\}$ is a subset of A, and the function f given by $f(n) = a_n$ is a one-to-one correspondence from $\mathbf{N}$ to B. Thus B is denumerable.   ∎

Theorem 5.27 can be used to prove that every infinite set is equivalent to one of its proper subsets. (See exercise 8.) This characterizes infinite sets because, as we saw in section 5.1, no finite set is equivalent to any of its proper subsets.

Exercises 5.4

1. Indicate whether the Axiom of Choice must be employed to select one element from each set in the following collections.
   (a) $\mathcal{A}$ is an infinite collection of sets, each set containing one odd and one even integer.
   (b) $\mathcal{A}$ is a finite collection of sets with each set uncountable.
   (c) $\mathcal{A}$ is an infinite collection of sets, each set containing exactly one integer.
   (d) $\mathcal{A}$ is a denumerable collection of uncountable sets.

2. Prove this partial converse of Theorem 5.25 without using the Axiom of Choice. Let A and B be sets with $B \neq \varnothing$. If $\overline{\overline{B}} \leq \overline{\overline{A}}$ then there exists $g: A \to B$ that is onto B.

3. Let A and B be any two nonempty sets. Prove that there exists $f: A \to B$ that has at least one of these properties:
   (a) f is one-to-one.   (b) f is onto B.

4. Prove that if $f: A \to B$, then $\overline{\overline{\text{Rng}(f)}} \leq \overline{\overline{A}}$.

★ 5. Suppose A is a denumerable set and B is an infinite subset of A. Prove $A \approx B$.

6. Suppose $\overline{\overline{B}} < \overline{\overline{C}}$ and $\overline{\overline{B}} \nleq \overline{\overline{A}}$. Prove $\overline{\overline{A}} < \overline{\overline{C}}$.

7. Let $\{A_i: i \in \mathbf{N}\}$ be a collection of pairwise disjoint nonempty sets. Prove that $\bigcup_{i \in \mathbf{N}} A_i$ includes a denumerable subset.

☆ 8. Let A be an infinite set. Prove that A is equivalent to a proper subset of A.

9. **Proofs to Grade.**
   ★ (a) **Claim.** There exists an infinite set of irrational numbers, no two of which differ by a rational number.
      "Proof." For $x, y \in \mathbf{R}$, define the relation S by $x \, S \, y$ iff $x - y \in \mathbf{Q}$. It is easy to show that S is an equivalence relation. For the family of equivalence classes $\{x/S: x \in \mathbf{R}\}$, choose one element from each equivalence class except the equivalence class $\mathbf{Q}$. The set of all such chosen elements is an infinite set of irrational numbers, no two of which differ by a rational. (You may accept, without proof, the claim that this set is infinite.) ■
   (b) **Claim.** Every infinite set A has a denumerable subset.
      "Proof." Suppose no subset of A is denumerable. Then all subsets of A must be finite. In particular $A \subseteq A$. Thus A is finite, contradicting the assumption. ■
   ★ (c) **Claim.** Every infinite set A has a denumerable subset B.
      "Proof." If A is denumerable, let $B = A$, and we are done. Otherwise, A is uncountable. Choose $x_1 \in A$. If $A - \{x_1\}$ is denumerable, let $B = A - \{x_1\}$. Otherwise, choose $x_2 \in A - \{x_1\}$. If $A - \{x_1, x_2\}$ is denumerable, let $B = A - \{x_1, x_2\}$. Continuing in this manner, using the Axiom of Choice, we obtain a subset $C = \{x_1, x_2, \ldots\}$ such that $B = A - C$ is denumerable. ■
   (d) **Claim.** Every infinite set has two disjoint denumerable subsets.
      "Proof." Let A be an infinite set. By Theorem 5.27, A has a denumerable subset B. Then $A - B$ is infinite, because A is infinite, and is disjoint from B. By Theorem 5.27, $A - B$ has a denumerable subset C. Then B and C are disjoint denumerable subsets of A. ■

(e) **Claim.**   Every infinite set has two disjoint denumerable subsets.
   "Proof."   Let A be an infinite set. By Theorem 5.27, A has a denumerable
   subset B. Since B is denumerable, there is a function $f: \mathbf{N} \xrightarrow[\text{onto}]{1-1} B$. Let
   $C = \{f(2n): n \in \mathbf{N}\}$ and $D = \{f(2n - 1): n \in \mathbf{N}\}$. Then $C = \{f(2), f(4),$
   $f(6), \ldots\}$ and $D = \{f(1), f(3), f(5), \ldots\}$ are disjoint denumerable subsets
   of A.   ∎

5.5

Countable Sets

Countable sets have been defined as those sets that are finite or denumerable. Such sets are countable in the sense that they can be "counted" by using
subsets of $\mathbf{N}$: A finite set is counted by using exactly the elements of some
$\mathbf{N}_k$, while a denumerable set may be counted by using exactly the elements
of $\mathbf{N}$. This section will survey some of the important results on countability.
We note that many of the results in this section depend on the Axiom of
Choice, because their proofs use theorems such as Theorem 5.27 and 5.26
that are proved using the Choice Axiom.

Theorem 5.28   Every subset of a countable set is countable.

Proof.   Let A be a countable set and let $B \subseteq A$. If B is finite, then
B is countable by definition. If B is infinite, since $B \subseteq A$, A is infinite. Thus A is denumerable. By Theorem 5.27, B has a denu-
merable subset C. Thus $C \subseteq B \subseteq A$, which implies $\aleph_0 = \overline{\overline{C}}$ and
$\overline{\overline{C}} \leqslant \overline{\overline{B}} \leqslant \overline{\overline{A}} = \aleph_0$. Therefore $\overline{\overline{A}} = \overline{\overline{B}} = \aleph_0$. Thus B is denumerable
and hence countable.   ∎

We have seen that the set $\mathbf{Q}$ of all rational numbers is countable. Thus
such subsets as $\{1/n: n \in \mathbf{N}\}$, $\mathbf{Q} \cap (0, 1)$, $\{\frac{5}{6}, \frac{6}{5}, \frac{3}{7}\}$, and $\mathbf{Z}$ are countable sets.

Corollary 5.29.   Let A be a set.

(a) A is countable iff A is equivalent to a subset of $\mathbf{N}$.
(b) A is countable iff $\overline{\overline{A}} \leqslant \aleph_0$.

Proof.   Exercise 2.   ∎

Theorem 5.30   If A and B are countable sets, then $A \cup B$ is
countable.

Proof.   This theorem has been proved in the cases where A and
B are finite (Corollary 5.7), where one set is denumerable and the

other is finite (Theorem 5.18), and where A and B are denumerable and disjoint (Theorem 5.19). The only remaining case is where A and B are denumerable and not disjoint. Write $A \cup B$ as $A \cup (B - A)$, a union of disjoint sets. Since $B - A \subseteq B$, $B - A$ is either finite or denumerable by Theorem 5.28. If $B - A$ is finite, then $A \cup B$ is denumerable by Theorem 5.18 and if $B - A$ is denumerable, then $A \cup B$ is denumerable by Theorem 5.19. ∎

Theorem 5.30 may be extended (by induction) to any finite union of countable sets and finally to a countable union of countable sets.

Theorem 5.31   Let $\mathcal{A}$ be a finite collection of countable sets. Then $\underset{A \in \mathcal{A}}{\cup} A$ is countable.

Proof.   Exercise 3. ∎

Theorem 5.32   Let $\mathcal{A}$ be a denumerable collection of denumerable sets. Then $\underset{A \in \mathcal{A}}{\cup} A$ is denumerable.

Proof.   Since $\mathcal{A}$ is denumerable, we may write $\mathcal{A}$ as $\{A_i : i \in \mathbf{N}\}$. We first consider the case where $\mathcal{A}$ is a pairwise disjoint collection. Since each A_i is denumerable, for each i there exists a function $f_i : \mathbf{N} \xrightarrow[\text{onto}]{1-1} A_i$. Let $f : \mathbf{N} \to \underset{A \in \mathcal{A}}{\cup} A$ be defined by $f(n) = f_1(n)$. Since f_1 is one-to-one, f is one-to-one and, therefore, $\overline{\overline{\mathbf{N}}} \leqslant \overline{\overline{\underset{A \in \mathcal{A}}{\cup} A}}$.

For each $a \in \underset{A \in \mathcal{A}}{\cup} A$, there is a unique n such that $a \in A_n$ and a unique $k \in \mathbf{N}$ such that $f_n(k) = a$. Thus we may define $g : \underset{A \in \mathcal{A}}{\cup} A \to \mathbf{N}$ by setting $g(a) = g(f_n(k)) = 2^n 3^k$. Since the prime factorization of natural numbers is unique, g is one-to-one. Hence, $\overline{\overline{\underset{A \in \mathcal{A}}{\cup} A}} \leqslant \overline{\overline{\mathbf{N}}}$. By the Cantor-Schröder-Bernstein Theorem, $\underset{A \in \mathcal{A}}{\cup} A$ is denumerable.

Finally, in the case that the A_i's are not pairwise disjoint, we may consider the denumerable collection of disjoint sets $\{A_i \times \{i\} : i \in \mathbf{N}\}$. By the first part of this proof, $\underset{i \in \mathbf{N}}{\cup} (A_i \times \{i\})$ is denumerable. Define $h : \underset{i \in \mathbf{N}}{\cup} (A_i \times \{i\}) \to \underset{i \in \mathbf{N}}{\cup} A_i$ by letting the image of (a, i) be a. We leave the proof that h is onto $\underset{i \in \mathbf{N}}{\cup} A_i$ to exercise 4. Then by Theorem 5.26,

$$\overline{\overline{\mathbf{N}}} = \overline{\overline{A_1}} \leqslant \overline{\overline{\underset{i \in \mathbf{N}}{\cup} A_i}} \leqslant \overline{\overline{\underset{i \in \mathbf{N}}{\cup} (A_i \times \{i\})}} = \overline{\overline{\mathbf{N}}},$$

so that $\underset{i \in \mathbf{N}}{\cup} A_i$ is also denumerable. ∎

Example.   For each $n \in \mathbf{N}$, let $A_n = \{a/n : a \in \mathbf{N}\}$. Then $\bigcup_{n \in \mathbf{N}} A_n = \mathbf{Q}^+$. Since each A_n is denumerable, Theorem 5.32 provides an alternate proof that $\mathbf{Q}^+$ is denumerable (Theorem 5.16).

Many of the results of this chapter can be amalgamated into our final theorem on cardinality. In terms of the Infinite Hotel it tells us that if there are countably many rooms, each accommodating countably many people, then there is room for all in the Infinite Hotel.

Theorem 5.33   The union of a countable collection of countable sets is countable.

Proof.   The only cases that remain unproved are the denumerable union of a collection of finite sets and the denumerable union of a collection of countable sets (exercise 5). ∎

Exercises 5.5

1. Give an example, if possible, of
   ★ (a)   a denumerable collection of finite sets whose union is denumerable.
   (b)   a denumerable collection of finite sets whose union is finite.
   (c)   a denumerable collection of pairwise disjoint finite sets whose union is finite.

2. Prove Corollary 5.29.

3. Prove Theorem 5.31.

4. Complete the proof of Theorem 5.32 by showing that the function h maps onto $\bigcup_{i \in \mathbf{N}} A_i$.

5. Complete the proof of Theorem 5.33.

6. ☆ (a)   Use Theorem 5.32 to prove that $\mathbf{N} \times \mathbf{N}$ is denumerable.
   (b)   Prove that if A and B are denumerable, then $A \times B$ is denumerable.
   (c)   Prove that if A and B are countable, then $A \times B$ is countable.

☆ 7. Use exercise 6 and the Cantor-Schröder-Bernstein Theorem to give an alternate proof that $\mathbf{Q}^+$ is denumerable.

8. Prove that
   (a)   the set of all singleton subsets of $\mathbf{N}$ is denumerable.
   ☆ (b)   the set of all two-element subsets of $\mathbf{N}$ is denumerable.
   (c)   the set of all n-element subsets of $\mathbf{N}$ is denumerable.
   (d)   the set of all finite subsets of $\mathbf{N}$ is denumerable.

☆ 9. Let S' be the set of all sequences of zeros and ones where all but a finite number of terms are 0. (See exercise 11 of section 5.2.) Prove S' is denumerable.

10. Prove that if B is uncountable and $\overline{\overline{B}} \leqslant \overline{\overline{A}}$, then A is uncountable.

11. **Proofs to Grade.**

★ (a) **Claim.**   If A is denumerable, then $A - \{x\}$ is denumerable.

   "Proof."   Assume A is denumerable.

   Case 1.   If $x \notin A$, then $A - \{x\} = A$, which is denumerable by hypothesis.

   Case 2.   Assume $x \in A$. Since A is denumerable, there exists $f: \mathbf{N} \xrightarrow[\text{onto}]{1-1} A$. Define g by setting $g(n) = f(n + 1)$. Then $g: \mathbf{N} \xrightarrow[\text{onto}]{1-1} A - \{x\}$, so $\mathbf{N} \approx A - \{x\}$. Therefore $A - \{x\}$ is denumerable. ■

   (b) **Claim.**   If A and B are denumerable, then $A \times B$ is denumerable.

   "Proof."   Assume A and B are denumerable, but that $A \times B$ is not denumerable. Then $A \times B$ is finite. Since A and B are denumerable, they are not empty; therefore choose $a \in A$ and $b \in B$. By exercise 3 of section 5.1, $A \approx A \times \{b\}$, and by an obvious modification of that exercise, $B \approx \{a\} \times B$. Since $A \times B$ is finite, the subsets $A \times \{b\}$ and $\{a\} \times B$ are finite. Therefore A and B are finite. This contradicts the statement that A and B are denumerable. We conclude that $A \times B$ is denumerable. ■

6

Concepts of Algebra

Much of modern mathematics is algebraic in nature. We have in mind a broad meaning of algebra as a system of computation and the study of properties of such a system. In this chapter we make precise the idea of an algebraic structure; study an especially important structure, the group; and investigate the notions of substructure and quotient structure.

6.1

Algebraic Structures

Let A be a nonempty set. A **binary operation on** A is a function from $A \times A$ to A. We will usually denote an operation by one of the symbols $+$, $\cdot$, $\circ$, or $*$. If the operation is $\circ$ and the image $\circ (x, y)$ of the ordered pair (x, y) is z, we usually write $x \circ y = z$. This notation is familiar from the operations of addition and multiplication on the set of real numbers, where it is more natural to write $4 + 7 = 11$ than $+(4, 7) = 11$. Often we omit completely the operation symbol and write $xy = z$, as is done with multiplication.

The images xy, $x \circ y$, and $x * y$ are usually called products, regardless of whether the operations involved have anything to do with multiplication. Similarly, $s + t$ is referred to as the sum of s and t, even when the function $+$ has nothing to do with addition.

Besides the usual arithmetic operations on sets of numbers, you are already familiar with several binary operations. Composition is an operation on the set of all functions that map a set A onto itself. The operation $\cup$, union, is a binary operation on the power set of A. Intersection, set difference, and symmetric difference are other operations on $\mathcal{P}(A)$.

There are operations more complicated than binary operations. Ternary operations, for example, map $A \times A \times A$ to A. An **algebraic structure** or **system** is a nonempty set A together with a collection of (at least one) operations on A and a collection (possibly empty) of relations on A. For example, the system of real numbers with addition and multiplication and the relation "less than" is an algebraic structure. We could as well consider the system of reals with just addition, or the system of reals with multiplication and relation "less than."

In this chapter when we say "operation," we mean a binary operation, and when we say "algebraic structure," we mean a structure with one binary operation.

Definition.   The set A is said to **be closed under the binary operation** $*$ iff for all $x, y \in A$, $x * y \in A$. The statements "A is closed under $*$," "$*$ is an operation on A," and "$(A, *)$ is an algebraic system" are all equivalent.

Definition.   If A is a finite set, the **order** of $(A, *)$ is $\overline{\overline{A}}$, which is the number of elements in A. When A is infinite, we simply say $(A, *)$ has infinite order.

A convenient way to display information about a binary operation, at least for a system of small finite order, is by means of its **operation table,** or **Cayley table.** An operation table for a system $(A, *)$ of order n is an $n \times n$ array of products such that $x * y$ appears in row x and column y. Table 6.1 represents a system $(A, *)$ with $A = \{1, 2, 3\}$ in which, for example, $2 * 1 = 3$. As an example of computation in this system, notice that $(3 * 2) * (1 * 3) = 3 * 1 = 2$.

Table 6.1

$*$	1	2	3
1	3	2	1
2	3	1	3
3	2	3	3

Definitions. Let $(A, *)$ be an algebraic system. Then

(i) $*$ is **commutative** on A iff for all x, $y \in A$, $x * y = y * x$.
(ii) $*$ is **associative** on A iff for all x, y, $z \in A$, $(x * y) * z = x * (y * z)$.
(iii) an element e of A is an **identity element** for $*$ iff for all $x \in A$, $x * e = e * x = x$.
(iv) if A has an identity element e, and a and b are in A, then b is an **inverse** of a iff $a * b = b * a = e$. In this case a would also be an inverse of b.

You are familiar with the fact that the system $(\mathbf{Z}, \cdot)$, with the usual multiplication of integers, is commutative and associative. In this system 1 is the identity element and only the elements 1 and -1 have inverses. For the system consisting of the real numbers with addition, the operation is commutative and associative, 0 is the identity, and every element has an inverse (its negative).

Our study of functions provides the information needed to consider what can be regarded as the most important kind of algebraic system. The set of all functions mapping a set one-to-one onto itself is closed under the operation of function composition. This operation is associative, but generally not commutative. The identity function is the identity element, and every element f has an inverse f^{-1} (see Theorem 4.9). This system will be studied in greater detail in section 6.3.

The operation $*$ of table 6.1 is not commutative because, for example, $1 * 3 \neq 3 * 1$. It is not associative because $(1 * 1) * 2 \neq 1 * (1 * 2)$. There is no identity element, so the question of inverses does not even arise.

Three different operations on $A = \{1, 2, 3\}$ are shown in tables 6.2, 6.3, and 6.4.

Table 6.2

$\circ$	1	2	3
1	1	2	3
2	1	2	3
3	1	2	3

Table 6.3

$\cdot$	1	2	3
1	3	1	2
2	1	2	3
3	2	3	1

Table 6.4

$+$	1	2	3
1	3	3	1
2	1	1	2
3	1	2	3

Tables 6.2 and 6.4 are not commutative. The fact that the operation of table 6.3 is commutative on A can be seen by noting that the table is symmetric about its **main diagonal,** from the upper left to lower right.

It is not easy to tell by looking at a table whether an operation is associative. For a system of order n, verification of associativity may require checking n^3 products of three elements, each grouped two ways. The operations in tables 6.2 and 6.3 are associative, but $+$ is not associative on A. (Why?)

The associative property is a great convenience in computing products. First, it means that so long as factors appear in the same order, we need no parentheses. For both $x(yz)$ and $(xy)z$ we can write xyz. This can be extended inductively to products of four or more factors:$(xy)(zw) = (x(yz)w) = x(y(zw))$, and so forth. Second, for an associative operation we can define powers. Without associativity, $(xx)x$ might be different from $x(xx)$, but with associativity they are equal, and both can be denoted by x^3.

The element 2 is an identity for $\cdot$ of table 6.3, and in this system the inverses of 1, 2, and 3 are, respectively, 3, 2, and 1. Table 6.2 has no identity element. In table 6.4, where 3 is the identity element, only 1 and 3 have inverses.

One of the most interesting concepts in algebra involves mappings between systems. Most interesting are the **operation preserving mappings.** As we shall see, such mappings actually preserve the structure of an algebraic system.

Definition.   Let $(A, \circ)$ and $(B, *)$ be algebraic systems. The mapping $f: A \rightarrow B$ is called **operation preserving** (or OP) iff for all $x, y \in A$, $f(x \circ y) = f(x) * f(y)$.

A simplified statement of how an OP map works is "the image of a product is the product of the images." That is, the result is the same whether operating or mapping is done first. That "the image of the product is the product of the images" must be carefully understood. To compute $f(x \circ y)$, one must use the operation of $(A, \circ)$, while in $f(x) * f(y)$, the operation is the operation of $(B, *)$. The final equation is an equality of elements of B, because $f(x \circ y)$ and $f(x) * f(y)$ are elements of B.

Examples of OP maps abound and mathematicians delight in finding them. The familiar equation

$$\log (x \cdot y) = \log x + \log y$$

tells us that the logarithm function from $(\mathbf{R}^+, \cdot)$ to $(\mathbf{R}, +)$ is operation preserving. As another example, let $(P, +)$ be the set of polynomial functions on $\mathbf{R}$ with the operation of function addition. Then the differentiation map $D: (P, +) \rightarrow (P, +)$ defined by $D(F) = F'$, the first derivative of F, is operation preserving. This mapping preserves addition because we know from calculus that $D(F + G) = (F + G)' = F' + G' = D(F) + D(G)$. In other words, the derivative of the sum is the sum of the derivatives. This same mapping from $(P, \cdot)$ onto itself, where $\cdot$ is the operation of function multiplication, is not operation preserving, because $D(F \cdot G)$ is not usually equal to $D(F) \cdot D(G)$.

These examples involve functions with properties you already know. To prove that a function f from $(A, \circ)$ to $(B, *)$ is OP, we begin with two arbitrary elements x and y of A and either evaluate the two terms $f(x \circ y)$ and $f(x) * f(y)$ or rewrite one until it has the form of the other.

Let the operation $\circ$ on $\mathbf{R} \times \mathbf{R}$ be defined by setting $(a, b) \circ (c, d) = (a + c, b + d)$ and let $\cdot$ be the usual multiplication on $\mathbf{R}$. Let f be the function from $(\mathbf{R} \times \mathbf{R}, \circ)$ to $(\mathbf{R}, \cdot)$ given by $f(a, b) = 2^a \cdot 3^b$. To prove f is operation preserving, let (x, y) and (u, v) be in $\mathbf{R} \times \mathbf{R}$. Then

$$f((x, y) \circ (u, v)) = f(x + u, y + v) = 2^{x+u} \cdot 3^{y+v}.$$

Also

$$f(x, y) \cdot f(u, v) = (2^x \cdot 3^y) \cdot (2^u \cdot 3^v) = 2^{x+u} \cdot 3^{y+v}.$$

Therefore f is operation preserving.

The next series of theorems is a part of the explanation of what we mean by saying that an OP map preserves the structure of an algebra. We will see that under an OP map the image set (range) of an algebraic system is an algebraic system and that, if the original operation is commutative or has an identity element, then so does the image set. More information about OP maps appears in sections 6.2, 6.4, and 6.7.

Theorem 6.1   Let f be an OP map from $(A, \circ)$ to $(B, *)$. Then $(\mathrm{Rng}(f), *)$ is an algebraic structure.

Proof.   ⟨*What we must show is that $*$ is closed on $\mathrm{Rng}(f)$; that is, if $u, v \in \mathrm{Rng}(f)$, then $u * v \in \mathrm{Rng}(f)$.*⟩

First, note that because A is nonempty, $\mathrm{Rng}(f)$ is nonempty. Assume $u, v \in \mathrm{Rng}(f)$. Then there exist elements x and y of A such that $f(x) = u$ and $f(y) = v$. Then $u * v = f(x) * f(y) = f(x \circ y)$, so $u * v$ is the image of $x \circ y$, which is in A. Therefore $u * v \in \mathrm{Rng}(f)$.   ∎

Now that we know that the range of an OP map is an algebraic system, we can simplify things by ignoring the part of the codomain that is not in the range. This amounts to assuming that f maps onto B. It does not mean that every OP map is onto every possible codomain.

Theorem 6.2   Let f be an OP map from $(A, \circ)$ onto $(B, *)$. If $\circ$ is commutative on A, then $*$ is commutative on B.

Proof.   Assume that f is OP and $\circ$ is commutative on A. Let u and v be elements of B. ⟨*We want to show $u * v = v * u$.*⟩

From the fact that f maps onto B we know that there are x and y in A such that $u = f(x)$ and $v = f(y)$. Then $u * v = f(x) * f(y) = f(x \circ y) = f(y \circ x) = f(y) * f(x) = v * u$. ⟨*This equation uses the fact that $\circ$ is commutative and (twice) that f is OP.*⟩ Therefore $u * v = v * u$.   ∎

The properties of associativity, existence of identities, and existence of inverses are all preserved by an OP mapping.

Theorem 6.3 Let f be an OP map from $(A, \circ)$ onto $(B, *)$.

(a) If $\circ$ is associative on A, then $*$ is associative on B.
(b) If e is an identity for A, then $f(e)$ is an identity for B.
(c) If x^{-1} is an inverse for x in A, then $f(x^{-1})$ is an inverse for $f(x)$ in B.

Proof. Exercise 7. ∎

Exercises 6.1

1. Which of the following are algebraic structures? (The symbols $+$, $-$, $\cup$, $\cap$, and so on, have their usual meanings.)
    ★ (a) $(\mathbf{Z}, -)$ (b) $(\mathbf{Z}, \div)$ (c) $(\mathbf{R}, -)$
    (d) $(\mathbf{R}, \div)$ ★ (e) $(\mathbf{N}, -)$ (f) $(\mathbf{Q}, \div)$
    (g) $(\mathbf{Q} - \{0\}, \div)$ (h) $(\mathcal{P}(A), \cap)$ (i) $(\mathcal{P}(A), \cup)$
    (j) $(\mathcal{P}(A) - \{\varnothing\}, -)$

☆ 2. Which of the operations in exercise 1 are commutative?

☆ 3. Which of the operations in exercise 1 are associative?

4. Let $\mathcal{M} = \{A : A$ is an $m \times n$ matrix with real number entries$\}$.
    (a) If $\cdot$ is matrix multiplication, is $(\mathcal{M}, \cdot)$ an algebraic system? What if $m = n$?
    (b) If $+$ is matrix addition, is $(\mathcal{M}, +)$ an algebraic system? What if $m = n$?

5. The Cayley tables for operations $\circ$, $*$, $+$, $\times$ are listed below.

$\circ$	a	b	c	d
a	a	b	c	d
b	b	a	d	c
c	c	d	a	b
d	d	c	b	a

$*$	a	b	c
a	c	a	c
b	a	b	c
c	b	c	b

$+$	a	b
a	a	a
b	a	a

$\times$	a	b	c
a	a	c	b
b	c	b	a
c	b	a	c

   (a) Which of the operations are commutative?
   (b) Which of the operations are associative?
   (c) Which systems have an identity? What is the identity element?
   (d) For those systems that have an identity, which elements have inverses?

6. Prove that if $(A, *)$ is an algebraic system and $*$ is associative on A, then the product of any n elements from A can be grouped in any manner we wish (as long as the order of the factors is unchanged) without affecting the product. (*Hint:* Use complete induction to prove that the product of $a_1, a_2, a_3, \ldots, a_n$ is equal to $(\cdots ((a_1 * a_2) * a_3) \cdots) * a_n$.)

7. Prove Theorem 6.3.

☆ 8. Let $(A, \circ)$ be an algebraic structure. Prove that if e and f are both identities for $\circ$, then $e = f$.

9. Let $(A, \circ)$ be an algebraic structure, $a \in A$, and e the identity for $\circ$.
   ☆ (a) Prove that if $\circ$ is associative, and x and y are inverses of a, then $x = y$.
   (b) Give an example of a nonassociative structure in which inverses are not unique.

10. Let $(A, \circ)$ be an algebraic structure. An element $\ell \in A$ is a **left identity** for $\circ$ iff $\ell \circ a = a$ for every $a \in A$.
    (a) Give an example of a structure of order 3 with exactly two left identities.
    (b) Define a **right identity** for $(A, \circ)$.
    (c) Prove that if $(A, \circ)$ has a right identity r and left identity ℓ, then $r = \ell$, and that $r = \ell$ is an identity for $\circ$.

11. Define SQRT: $(\mathbf{R}^+, +) \rightarrow (\mathbf{R}^+, +)$ by SQRT $(x) = \sqrt{x}$. Is SQRT operation preserving?

12. Is SQRT: $(\mathbf{R}^+, \cdot) \rightarrow (\mathbf{R}^+, \cdot)$ operation preserving?

13. Is SQR: $(\mathbf{R}, +) \rightarrow (\mathbf{R}, +)$ defined by SQR $(x) = x^2$ operation preserving?

14. Is SQR: $(\mathbf{R}, \cdot) \rightarrow (\mathbf{R}, \cdot)$ operation preserving?

★ 15. Let $\mathcal{F}$ be the set of all real valued integrable functions defined on the interval $[a, b]$. Then $(\mathcal{F}, +)$ is an algebraic structure, where $+$ is the addition of functions. Define $I: (\mathcal{F}, +) \rightarrow (\mathbf{R}, +)$ by $I(f) = \int_a^b f(x)\,dx$. Use your knowledge of calculus to verify that I is an OP map.

16. Let $\mathcal{F}$ be as in exercise 15. Consider the algebraic structures $(\mathcal{F}, \cdot)$ and $(\mathbf{R}, \cdot)$ where $\cdot$ represents both function product and real number multiplication, respectively. Verify that $I: (\mathcal{F}, \cdot) \rightarrow (\mathbf{R}, \cdot)$, defined by $I(f) = \int_a^b f(x)\,dx$, is not operation preserving.

17. Let $f: (A, \circ) \rightarrow (B, *)$ and $g: (B, *) \rightarrow (C, \times)$ be OP maps.
    (a) Prove $g \circ f$ is an OP map.
    (b) Prove that if f is a one-to-one correspondence, then f^{-1} is an OP map.

18. Let $\mathcal{M}$ be the set of all 2×2 matrices with real entries. Define Det: $\mathcal{M} \rightarrow \mathbf{R}$ by

$$\text{Det} \begin{bmatrix} a & b \\ c & d \end{bmatrix} = ad - bc.$$

   (a) Let $(\mathcal{M}, \cdot)$ represent $\mathcal{M}$ with matrix multiplication and $(\mathbf{R}, \cdot)$ represent the reals with ordinary multiplication. Prove Det: $(\mathcal{M}, \cdot) \rightarrow (\mathbf{R}, \cdot)$ is operation preserving.
   (b) Let $(\mathcal{M}, +)$ represent $\mathcal{M}$ with matrix addition. Prove Det: $(\mathcal{M}, +) \rightarrow (\mathbf{R}, +)$ is not operation preserving.

19. (a) Let $\mathbf{C}$ denote the complex numbers with ordinary addition. Define Conj: $(\mathbf{C}, +) \rightarrow (\mathbf{C}, +)$ by Conj $(a + bi) = a - bi$. Prove that Conj is an OP map.
    (b) Prove that Conj: $(\mathbf{C}, \cdot) \rightarrow (\mathbf{C}, \cdot)$ is an OP map.

20. Let $t \in \mathbf{R}$. Define $S_t: (\mathbf{R}, +) \rightarrow (\mathbf{R}, +)$ by $S_t(x) = tx$. Prove that S_t is operation preserving.

21. Let $f: A \to B$.
    ★ (a)   Prove that the induced function $f: (\mathscr{P}(A), \cup) \to (\mathscr{P}(B), \cup)$ is an OP map.
    (b)   Prove that the induced function $f^{-1}: (\mathscr{P}(B), \cap) \to (\mathscr{P}(A), \cap)$ is an OP map.
    (c)   Prove that the induced function $f^{-1}: (\mathscr{P}(B), \cup) \to (\mathscr{P}(A), \cup)$ is an OP map.

22. **Proofs to Grade.**
    (a)   **Claim.**   Let $(A, \circ)$ be an algebraic structure. If e is an identity for $\circ$, and if x and y are both inverses of a, then $x = y$.
    "Proof." $x = x \circ e = x \circ (a \circ y) = (x \circ a) \circ y = e \circ y = y$. Thus $x = y$ and inverses are unique. ∎
    ★ (b)   **Claim.**   If every element of a structure $(A, \circ)$ has an inverse, then $\circ$ is commutative.
    "Proof."   Let x and y be in A. The element y has an inverse, which we will call y'. Then $y \circ y' = e$, so y is the inverse of y'. Now $x = x$, and multiplying both sides of this equation by the inverse of y', we have $y \circ x = x \circ y$. Therefore $\circ$ is commutative. ∎

6.2

Groups

We have considered algebraic structures $(A, \circ)$ with one binary operation. For the rest of this chapter we will give our attention to certain of these structures, the groups. The properties of associativity and existence of identities and inverses examined in section 6.1 are just the properties we need when we define a group. Our approach is **axiomatic.** That is, we shall list the desired properties (axioms) of a structure, and any system having these properties is called a group. There are some important observations to be made about such a method. Although stated for groups, the following comments apply generally to axiomatic studies.

First, a small set of axioms is advantageous (and may be challenging to produce) because a small set means that few properties need be checked to be sure a structure is a group. It may be best to leave a desired property out of the axioms if it can be deduced from the remaining axioms. This is so because every consequence of the axioms must be true of all structures satisfying the axioms. Finally, the fact that axiom systems may be altered by adding or deleting specific axioms does not mean that axioms are chosen at random, or that all axioms are equally worthy of study. The group axioms are chosen because the structures they describe are so important to modern mathematics and its applications.

It was the work of Evariste Galois (1811–1832) on solving polynomial equations that led to the study of groups. The concept of a group has influenced and enriched other areas of mathematics. In geometry, for example, the ideas of Euclidean and non-Euclidean geometries were unified by the notion of the group. Group theory has been applied outside mathematics, too, in fields such as nuclear physics and crystallography.

Definition.   A **group** is an algebraic structure $(G, \circ)$ such that

    (1) the operation $\circ$ is associative.
    (2) there is an identity element $e \in G$ for $\circ$.
    (3) every $x \in G$ has an inverse x^{-1} in G.

To prove that a structure is a group, we prove that axioms 1, 2, and 3 hold for the structure. Implicitly, we must also verify that $(G, \circ)$ is actually an algebraic structure. That is, we must verify that G is nonempty and that $\circ$ satisfies the closure property: if $x, y \in G$, then $x \circ y \in G$.

A fourth axiom may be considered. A group is called **abelian** iff

    (4) the operation $\circ$ is commutative.

The fourth axiom is an **independent** axiom because it cannot be proved from the other group axioms.

The systems $(\mathbf{R}, +)$, $(\mathbf{Q}, +)$, and $(\mathbf{Z}, +)$ are all abelian groups. The system $(\mathbf{R}, \cdot)$ is not a group because 0 has no inverse, but $(\mathbf{R} - \{0\}, \cdot)$ and $(\mathbf{Q} - \{0\}, \cdot)$ are (abelian) groups. The systems $(\mathbf{N}, +)$, $(\mathbf{Z}, -)$, and $(\mathbf{Z} - \{0\}, \cdot)$ are not groups. (Why?) However, $(\{0\}, +)$ is a group. Table 6.5 is the Cayley table for the abelian group $(\{1, -1\}, \cdot)$.

Table 6.5

$\cdot$	1	-1
1	1	-1
-1	-1	1

The group axioms do not require that the identity and the inverses of elements be unique. However, that the identity and the inverses are unique can be proved. (See exercises 8 and 9 in section 6.1.) Leaving the uniqueness condition out of the group axioms shortens the verification that a structure is a group.

Frequently we will simply say "the group G" meaning the set G where the product of elements x and y is denoted xy. Two elementary consequences of the group axioms facilitate calculations involving elements of a group.

Theorem 6.4 If G is a group and $a, b \in G$, then $(a^{-1})^{-1} = a$ and $(ab)^{-1} = b^{-1}a^{-1}$.

Proof.   Because a^{-1} is the inverse of a, $a^{-1}a = aa^{-1} = e$. Therefore a acts as the $\langle unique \rangle$ inverse of a^{-1}, so $(a^{-1})^{-1} = a$.

We know $(ab)^{-1}$ is the unique element x of G such that $(ab)x = x(ab) = e$. We see that $b^{-1}a^{-1}$ meets this criterion by computing

$$(ab)(b^{-1}a^{-1}) = a(bb^{-1})a^{-1} = a(e)a^{-1} = aa^{-1} = e.$$

Similarly, $(b^{-1}a^{-1})(ab) = e$, so $b^{-1}a^{-1}$ is the inverse of ab.   ∎

Theorem 6.5   Let G be a group. Then left and right cancellation both hold in G. That is, for elements x, y, z of G,

(a)   if $xy = xz$, then $y = z$.
(b)   if $yx = zx$, then $y = z$.

Proof.

(a)   Suppose $xy = xz$ in the group G. Then, using the fact that $x^{-1} \in G$, we have $x^{-1}(xy) = x^{-1}(xz)$. Therefore using the associative, inverse, and identity properties, we see that

$$x^{-1}(xy) = (x^{-1}x)y = ey = y \text{ and}$$
$$x^{-1}(xz) = (x^{-1}x)z = ez = z.$$

Therefore $y = z$. This proves (a), and the proof of (b) is similar (exercise 7).   ∎

It follows from (a) of Theorem 6.5 that no element can occur twice in any row of the operation table for a group. In fact, if a and b are in G, then $a(a^{-1}b) = b$, so every element b occurs in row a, for every a. Thus if G is a group, then every element of G occurs exactly once in every row and [using part (b) of the theorem] in every column of the operation table. The converse of this statement is false: It is possible to have each element occur exactly once in every row and in every column of an operation table for a structure that is not a group. (See exercise 5.)

If G is a group, it is convenient to have notation for powers of elements of G. Let $a \in G$ and $n \in \mathbf{N}$. We define inductively $a^0 = e$ and $a^{n+1} = a^n a$. Thus a^n is defined inductively for all $n \geqslant 0$. Define a^n for $n < 0$ by

$$a^n = (a^{-1})^{-n}.$$

Then we can prove that for $n > 0$, $(a^n)^{-1} = a^{-n}$. Now we have the usual laws of exponents. For m, $n \in \mathbf{Z}$:

$$a^n a^m = a^{n+m}$$
$$(a^n)^m = a^{nm}.$$

If and only if the group is abelian, we also have

$(ab)^n = a^n b^n$.

When the group operation is addition, we usually write the inverse of an element a as $-a$ and call it the **negative** of a. Rather than powers of a, we refer to **multiples** of a, but the only difference between these ideas is notation. Thus $0a = e$ and $(n + 1)a = na + a$ for $n > 0$, and $na = -n(-a)$ for $n < 0$.

Definition.   An operation preserving map from the group $(G, \circ)$ to the group $(H, *)$ is called a **homomorphism** from $(G, \circ)$ to $(H, *)$.

A homomorphism is just an OP map for groups. The range of a homomorphism is called a **homomorphic image.** The function f mapping $(\mathbf{Z}, +)$ to $(\mathbf{Z}, +)$ given by $f(x) = 3x$ is an example of a homomorphism. The corresponding homomorphic image is $\mathrm{Rng}(f) = \{3x : x \in \mathbf{Z}\} = \{. . ., -6, -3, 0, 3, 6, 9, . . .\}$, which is itself a group under the operation of addition.

Theorem 6.6   (a)   The homomorphic image of a group is a group.   (b)   The homomorphic image of an abelian group is an abelian group.

Proof.   The proof is nothing more than an observation that the image of a group under a homomorphism is an algebraic system ⟨*Theorem 6.1*⟩ that is associative, has an identity element, and has an inverse for every element ⟨*Theorem 6.3*⟩. Furthermore, if $(G, \circ)$ is abelian, then the image is also abelian ⟨*Theorem 6.2*⟩.

Let $(G, \circ)$ be any group with identity e. Define a mapping $F \colon G \to G$ by setting $F(x) = e$, for every $x \in G$. Then F is a homomorphism on G. We can verify this by observing that for x and $y \in G$, both $F(x \circ y) = e$ and $F(x) \circ F(y) = e \circ e = e$, so $F(x \circ y) = F(x) \circ F(y)$. In this case the group that is the image of the homomorphism is $(\{e\}, \circ)$.

Exercises 6.2

1.   Show that each of the following algebraic structures is a group.
★   (a) $(\{1, -1, i, -i\}, \cdot)$ where $i = \sqrt{-1}$ and $\cdot$ is complex number multiplication.
(b) $(\{1, \alpha, \beta\}, \cdot)$ where $\alpha = \dfrac{-1 + i\sqrt{3}}{2}$, $\beta = \dfrac{-1 - i\sqrt{3}}{2}$, and $\cdot$ is complex number multiplication.

    (c)   ($\{1, -1\}, \cdot$) where $\cdot$ is multiplication of integers

☆ (d) ($\mathcal{P}(X), \Delta$) where X is a nonempty set and Δ is the symmetric difference operation $A \Delta B = (A - B) \cup (B - A)$

★ 2. Given that $G = \{e, u, v, w\}$ is a group of order 4 with identity e and $u^2 = v$, $v^2 = e$, construct, if possible, the operation table for G.

3. Given that $G = \{e, u, v, w\}$ is a group of order 4 with identity e and $e^2 = u^2 = v^2 = w^2 = e$, construct, if possible, the operation table for G.

☆ 4. Which of the groups of exercise 1 are abelian?

5. Give an example of an algebraic system $(G, \circ)$ that is not a group such that in the operation table for $\circ$, every element of G appears exactly once in every row and once in every column. This can be done with as few as three elements in G.

6. By Theorem 6.4, $(ab)^{-1} = b^{-1}a^{-1}$.
    (a) Prove that $(abc)^{-1} = c^{-1}b^{-1}a^{-1}$.
    (b) Prove a similar result for $(a_1a_2a_3 \cdots a_n)^{-1}$ by induction.

7. Prove part (b) of Theorem 6.5. That is, prove that if G is a group, x, y, and z are elements of G, and if $yx = zx$, then $y = z$.

☆ 8. Let G be a group. Prove that if $g^2 = e$ for all $g \in G$, then G is abelian.

☆ 9. Give an example of an algebraic structure of order 4 that has both right and left cancellation but that is not a group.

10. Let G be an abelian group and let $a, b \in G$.
    (a) Prove that $a^2b^2 = (ab)^2$.
    (b) Prove that for all $n \in \mathbf{N}$, $a^nb^n = (ab)^n$.

11. Show that the structure $(\mathbf{R} - \{1\}, \circ)$, with operation $\circ$ defined by $a \circ b = a + b - ab$, is an abelian group. [You should first show that $(\mathbf{R} - \{1\}, \circ)$ is indeed a structure.]

☆ 12. Let $(G, *)$ be a group and $a, b \in G$. Show that there exist unique elements x and y in G such that $a * x = b$ and $y * a = b$.

13. Show that $(\mathbf{Z}, \#)$, with operation $\#$ defined by $a \# b = a + b + 1$, is a group. Find x such that $50 \# x = 100$.

14. **Proofs to Grade.**
    (a) **Claim.** If G is a group, then G is commutative.
        "Proof." Let a and b be elements of G. Then

$$
\begin{aligned}
ab &= a\,e\,b \\
&= a(ab)(ab)^{-1}b \\
&= a(ab)(b^{-1}a^{-1})b \\
&= (aa)(bb^{-1})a^{-1}b \\
&= (aa)a^{-1}b \\
&= (aa)(b^{-1}a)^{-1} \\
&= a(a(b^{-1}a)^{-1}) \\
&= a((b^{-1}a)a) \\
&= a((a^{-1}b)a) \\
&= (aa^{-1})(ba) \\
&= e(ba) \\
&= ba.
\end{aligned}
$$

    Therefore $ab = ba$, and G is commutative. ■

★ (b) **Claim.**   If G is a group with elements x, y, and z, and if $xz = yz$, then $x = y$.

"*Proof.*"   If $z = e$, then $xz = yz$ implies that $xe = ye$, so $x = y$. If $z \neq e$, then the inverse of z exists, and $xz = yz$ implies $xz/z = yz/z$ and $x = y$. Hence in all cases, if $xz = yz$, then $x = y$. ■

6.3

Examples of Groups

In this section we will see several examples where the concept of a group arises from another mathematical structure. The first example was mentioned briefly in section 6.1.

A function on a set A that is one-to-one and onto A is called a **permutation of A**. For example, let $A = \{1, 2, 3\}$. We shall adopt one of the common notations for a permutation. We think of a function f on A as follows:

$$\{1 \qquad 2 \qquad 3\}$$
$$\downarrow \qquad \downarrow \qquad \downarrow$$
$$\{f(1) \quad f(2) \quad f(3)\}$$

and write f as $(f(1)f(2)f(3))$. Then (123) represents the identity function on A, (213) represents the function g where $g(1) = 2$, $g(2) = 1$, and $g(3) = 3$, and (312) represents the function h where $h(1) = 3$, $h(2) = 1$, and $h(3) = 2$.

In section 2.5 we defined a permutation of a set as an arrangement of the elements of the set in some order. A permutation function has exactly the effect of arranging (or permuting) the elements of its domain.

As we have already observed in Section 6.1, we know from our study of functions that the set of all permutations of a set A is closed under function composition, that composition of functions is associative, that the identity permutation is an identity element, and that the inverse of a permutation is a permutation. This group of functions is called the **group of permutations of A**.

Definition.   For $n \in \mathbf{N}$, the group of all permutations of $\{1, 2, \ldots, n\}$ is called the **symmetric group on n symbols** and is designated by S_n.

The functions g and h above are elements of S_3. Recall that the operation $\circ$ here is function composition, so to compute the product of $g = (213)$ and $h = (312)$, we note that

$$(g \circ h)(1) = g(h(1)) = g(3) = 3$$
$$(g \circ h)(2) = g(h(2)) = g(1) = 2$$
$$(g \circ h)(3) = g(h(3)) = g(2) = 1$$

Therefore $g \circ h = (321)$. A similar computation will show that $h \circ g = (312) \circ (213) = (132)$. Thus S_3 is the first example we have seen of a group that is not abelian. We list all six elements of S_3 and, after extensive computation, fill in operation table 6.6.

Table 6.6: S_3

$\circ$	(123)	(213)	(321)	(132)	(231)	(312)
(123)	(123)	(213)	(321)	(132)	(231)	(312)
(213)	(213)	(123)	(312)	(231)	(132)	(321)
(321)	(321)	(231)	(123)	(312)	(213)	(132)
(132)	(132)	(312)	(231)	(123)	(321)	(213)
(231)	(231)	(321)	(132)	(213)	(312)	(123)
(312)	(312)	(132)	(213)	(321)	(123)	(231)

There are exactly $n!$ arrangements of the elements of a set A with n elements, so the number of permutation functions of A is $n!$ In particular the order of S_3 is $3! = 6$ and the order of S_n is $n!$

Groups whose elements are some (but not necessarily all) permutations of a set are called **permutation groups.** The reason for the importance of permutation groups is that, for every group with elements of any kind (numbers, sets, functions), there is a corresponding group of permutations with the same structure as the original group. (See Theorem 6.22.)

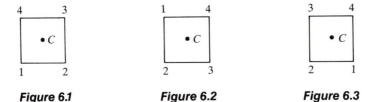

Figure 6.1   Figure 6.2   Figure 6.3

Our next example of a group involves using symmetric properties of familiar geometric figures. A symmetry of a geometric figure is a rigid motion (without bending or tearing) of the figure onto itself. More formally, a **symmetry** is a one-to-one distance-preserving transformation of the points of the figure. Every regular polygon and regular solid has an interesting group of symmetries. We consider the symmetries of a square by imagining a cardboard square with vertices 1, 2, 3, 4 and center C (figure 6.1). The square is carried onto itself by the following rigid motions:

R: a 90° rotation clockwise around center C; R transforms the square to the position shown in figure 6.2.

R^2:   a 180° rotation clockwise around C.

R^3:   a 270° rotation clockwise around C.

H: a reflection about the horizontal axis through C.

V: a reflection about the vertical axis through C; V transforms the square to the position shown in figure 6.3.

D: a reflection about the lower-left to upper-right diagonal.

D':   a reflection about the upper-left to lower-right diagonal.

I: the identity transformation.

We compute the product of two symmetries by performing them in succession. Thus VR is the result of a vertical axis reflection followed by a 90° clockwise rotation. By experimenting with a small cardboard square, we find that $VR = D \neq RV = D'$. This permutation group $\mathbb{O}$, called the **octic group,** is shown in the table.

	I	R	R^2	R^3	H	V	D	D'
I	I	R	R^2	R^3	H	V	D	D'
R	R	R^2	R^3	I	D	D'	V	H
R^2	R^2	R^3	I	R	V	H	D'	D
R^3	R^3	I	R	R^2	D'	D	H	V
H	H	D'	V	D	I	R^2	R^3	R
V	V	D	H	D'	R^2	I	R	R^3
D	D	H	D'	V	R	R^3	I	R^2
D'	D'	V	D	H	R^3	R	R^2	I

Other algebraic structures are based on the modulo m relation $\equiv_m$ on **Z**, which, you recall from section 3.2, is an equivalence relation on **Z**. We have seen that for each natural number m there are exactly m equivalence classes, $0/\equiv_m$, $1/\equiv_m$, . . ., $(m - 1)/\equiv_m$, and the set of all these classes is called $\mathbf{Z}_m$. When the modulus m is fixed throughout a discussion, we usually denote these classes by $\overline{0}, \overline{1}, . . ., \overline{m - 1}$. We define two operations $+_m$ and $\cdot_m$ as follows:

$$\overline{a} +_m \overline{b} = \overline{a + b}$$
$$\overline{a} \cdot_m \overline{b} = \overline{a \cdot b}$$

That is, the sum of the classes is the class of the sum, and so on.

The operation $+_{12}$ on $\mathbf{Z}_{12}$ is an abstraction of familiar "clock arithmetic" you have done many times: It is 11:00 in the evening and you settle in bed. At what time should your alarm be set so that you get 8 hours of rest? The answer, 7:00 in the morning, is arrived at by the computation

$$11/\!\equiv_{12} + 8/\!\equiv_{12} = 7/\!\equiv_{12}$$

because $11 + 8 = 19$ and $19 \equiv_{12} = 7$.

A difficulty in the definitions of $+_m$ and $\cdot_m$ must be overcome, for we must be sure that $+_m$ and $\cdot_m$ are operations on $\mathbf{Z}_m$. The problem is that $+_m$ and $\cdot_m$ must be functions from $\mathbf{Z}_m \times \mathbf{Z}_m$ to $\mathbf{Z}_m$. Take, for example, $m = 5$. Then $\overline{3} = \overline{-12}$ and $\overline{1} = \overline{31}$, so we must have $\overline{3} +_5 \overline{1} = \overline{-12} +_5 \overline{31}$ or else $+_5$ would not have unique images. We must be certain that addition and multiplication as defined in $\mathbf{Z}_m$ do not depend on which elements of the equivalence class are added. This is all taken care of by means of the following lemma.

Lemma 6.7   If $a \equiv_m b$ and $b \equiv_m d$, then

(i) $a + c \equiv_m b + d$.
(ii) $a \cdot c \equiv_m b \cdot d$.

Proof.   Exercise 3.   ∎

In light of the preceding discussion and Lemma 6.7, we now have that $(\mathbf{Z}_m, +_m)$ and $(\mathbf{Z}_m, \cdot_m)$ are algebraic systems for every $m \in \mathbf{N}$.

Theorem 6.8   For every natural number m, $(\mathbf{Z}_m, +_m)$ is an abelian group.

Proof.   The proof that $+_m$ is associative is a routine verification using the definition of $+_m$ and the associativity of $+$ for $\mathbf{Z}$. Commutativity is similar. The identity element is $\overline{0}$, and the negative of $\overline{a}$ is the element $\overline{(m - a)}$.   ∎

When the meaning is clear from the context, we simplify even further the notation $\overline{a}$ for the equivalence class of a modulo m and simply write a. At times we even use just $+$ for $+_m$. The tables for $+_2$, $+_6$, and $\cdot_6$ are shown.

$(\mathbf{Z}_2, +_2)$

$+_2$	0	1
0	0	1
1	1	0

$(\mathbf{Z}_6, +_6)$

$+_6$	0	1	2	3	4	5
0	0	1	2	3	4	5
1	1	2	3	4	5	0
2	2	3	4	5	0	1
3	3	4	5	0	1	2
4	4	5	0	1	2	3
5	5	0	1	2	3	4

$(\mathbf{Z}_6, \cdot_6)$

$\cdot_6$	0	1	2	3	4	5
0	0	0	0	0	0	0
1	0	1	2	3	4	5
2	0	2	4	0	2	4
3	0	3	0	3	0	3
4	0	4	2	0	4	2
5	0	5	4	3	2	1

Notice from its table that $(\mathbf{Z}_6, \cdot_6)$ is not a group and that it is possible to have $a \neq 0$, $b \neq 0$, but $a \cdot_6 b = 0$. In $\mathbf{Z}_m$, if $a \neq 0$ is an element such that

$a \cdot b = 0$ for some $b \neq 0$ in $\mathbf{Z}_m$, we say that a is a **divisor of zero.** It can be shown that $(\mathbf{Z}_m - \{0\}, \cdot_m)$ has no divisors of zero iff m is a prime.

Theorem 6.9 The structure $(\mathbf{Z}_m - \{0\}, \cdot_m)$ is an abelian group iff m is prime.

Proof. Exercise 4. ∎

Exercises 6.3

1. Construct the operation table for S_2, the symmetric group on 2 elements. Is S_2 abelian?

2. Construct the operation table for S_4, the symmetric group on 4 elements. What is the order of S_4? Is S_4 abelian?

☆ 3. Prove Lemma 6.7.

4. (a) Prove that $(\mathbf{Z}_m - \{0\}, \cdot_m)$ has no divisors of zero iff m is a prime.
   (b) Prove Theorem 6.9.

5. Let $e = (123), \alpha = (231), \beta = (132)$ be permutations in S_3.
   (a) Find α^2.
   (b) Show that $\beta\alpha = \alpha^2\beta$.
   (c) Show that $\{e, \alpha, \alpha^2, \beta, \alpha\beta, \alpha^2\beta\}$ is the set S_3.
   (d) Construct an operation table for S_3 using the symbols in part (c).

6. Let γ be the permutation on $\{1, 2, 3, 4, 5, 6\}$ given by $\gamma = (641253)$. Find
   (a) γ^2.   ★ (b) γ^3.   (c) γ^6.   (d) γ^{-1}.

7. For each of the following permutations α, find $\alpha^{-1}, \alpha^2, \alpha^3, \alpha^4, \alpha^{50}, \alpha^{51}$.
   (a) $\alpha = (21) \in S_2$ ★ (b) $\alpha = (231) \in S_3$ (c) $\alpha = (3412) \in S_4$

8. List the symmetries of an equilateral triangle and compute four typical products.

9. List the symmetries of a rectangle and give the group table.

★ 10. How many elements are there in the group of symmetries of a regular pentagon? a regular hexagon? a regular n-sided polygon?

11. Construct the operation table for each of the following groups.
   (a) $(\mathbf{Z}_8, +)$ (b) $(\mathbf{Z}_3 - \{0\}, \cdot)$ (c) $(\mathbf{Z}_7 - \{0\}, \cdot)$
   (d) $(\mathbf{Z}_9 - \{0, 3, 6\}, \cdot)$ (e) $(\mathbf{Z}_{15} - \{0, 3, 5, 6, 9, 10, 12\}, \cdot)$

12. Find all zero divisors in the following algebraic structures.
   ★ (a) $(\mathbf{Z}_{12}, \cdot)$ (b) $(\mathbf{Z}_{15}, \cdot)$ (c) $(\mathbf{Z}_m, \cdot)$

★ 13. If p is prime, show that $(p - 1)^{-1} = p - 1$ in $(\mathbf{Z}_p - \{0\}, \cdot)$.

14. Find all solutions in $(\mathbf{Z}_{20}, \cdot)$ for the following equations:
   ★ (a) $5 \cdot x = 0$.   (b) $3 \cdot x = 0$.
   (c) $x \cdot x = 0$.   (d) $x^2 - 6x + 5 = 0$.

15. **Proofs to Grade.**
    (a) **Claim.** If a and b are zero divisors in $(\mathbf{Z}_m, \cdot)$, then ab is a zero divisor.
       "Proof." If a and b are zero divisors, then $ab = 0$. Thus $(ab)(ab) = 0 \cdot 0 = 0$ and ab is a zero divisor. ∎

   ★ (b) **Claim.** If a and b are zero divisors in $(\mathbf{Z}_m, \cdot)$ and $ab \neq 0$, then ab is a zero divisor.
       "Proof." Since a is a zero divisor, $ax = 0$ for some $x \neq 0$ in $\mathbf{Z}_m$. Likewise, $by = 0$ for some $y \neq 0$ in $\mathbf{Z}_m$. Therefore $(ab)(xy) = (ax)(by) = 0 \cdot 0 = 0$. Thus ab is a zero divisor. ∎

6.4

Subgroups

In general, a substructure of an algebraic system $(A, *)$ consists of a subset of A together with all the operations and relations in the original structure, *provided* that this is an algebraic structure. This proviso is necessary, for it may happen that a subset of A is not closed under an operation. For example, the subset of $\mathbf{R}$ consisting of the irrationals is not closed under multiplication. Substructures are a natural idea, and they can be useful in describing a structure.

Definition. Let $(G, \circ)$ be a group and H a subset of G. Then $(H, \circ)$ is a **subgroup** of G iff $(H, \circ)$ is a group.

It is understood that the operation $\circ$ on H agrees with the operation $\circ$ on G. That is, the operation on H is the function $\circ$ restricted to $H \times H$.

If H is a nonempty subset of G and $(G, \circ)$ is a group, then to prove that H is a subgroup of G, *all three of the group properties and closure must be proved* for $(H, \circ)$. At least, that will be the situation until we find some way to shorten the work.

The set E of even numbers is a subgroup of $(\mathbf{Z}, +)$ because E is nonempty, E is closed under addition, the identity $0 \in E$, the negative of an even number is even, and $+$ is associative on E. In fact, for each integer m, the set $\{\ldots, -3m, -2m, -m, 0, m, 2m, 3m, \ldots\}$ of all multiples of m is a subgroup of $(\mathbf{Z}, +)$. All these groups and $(\mathbf{Z}, +)$ are subgroups of $(\mathbf{Q}, +)$ and $(\mathbf{R}, +)$.

Two subsets of $\mathbf{Z}_6$ that are closed under $+_6$ can be seen in the following tables. It is easy to check that both $H = \{0, 3\}$ and $K = \{0, 2, 4\}$ are subgroups of $(\mathbf{Z}_6, +)$.

$(\mathbf{Z}_6, +)$

+	0	1	2	3	4	5
0	**0**	1	2	**3**	4	5
1	1	2	3	4	5	0
2	2	3	4	5	0	1
3	**3**	4	5	**0**	1	2
4	4	5	0	1	2	3
5	5	0	1	2	3	4

$(\mathbf{Z}_6, +)$

+	0	1	2	3	4	5
0	**0**	1	**2**	3	**4**	5
1	1	2	3	4	5	0
2	**2**	3	**4**	5	**0**	1
3	3	4	5	0	1	2
4	**4**	5	**0**	1	**2**	3
5	5	0	1	2	3	4

$(H, +)$

+	0	3
0	0	3
3	3	0

$(K, +)$

+	0	2	4
0	0	2	4
2	2	4	0
4	4	0	2

For every group $(G, \circ)$ with identity e, $(\{e\}, \circ)$ is a group. This group is called the **identity subgroup**, or **trivial subgroup** of G. Also, every group is a subgroup of itself. All subgroups of G other than G and $\{e\}$ are called **proper subgroups.**

Important questions to be answered are whether the identity element in a subgroup can be different from the identity element of the original group, and whether the inverse of an element in H could be different from its inverse in G. The answers are "no" and "no."

Theorem 6.10 Let H be a subgroup of G. The identity of H is the identity e of G, and if $x \in H$, the inverse of x in H is its inverse in G.

Proof. If i is the identity element of H, then $ii = i$. But in G, $ie = i$, so $ii = ie$ and, by cancellation, $i = e$.

The proof for inverses is exercise 5. ∎

The next theorem is a labor-saving device for proving that a subset of a group is a subgroup. It is given in "iff" form for completeness, but the important result is that only two properties must be checked to show that H is a subgroup of G. The first is that H is nonempty. This is usually done by showing that the identity e of G is in H. The other is to show that $ab^{-1} \in H$ whenever a and b are in H. This is usually less work than showing both that H is closed under the group operation and that $b \in H$ implies $b^{-1} \in H$.

Theorem 6.11 Let G be a group. A subset H of G is a group iff H is nonempty and for all $a, b \in H$, $ab^{-1} \in H$.

Proof. First, suppose H is a subgroup of G. Then H is a group, so H contains the identity e. Therefore $H \neq \emptyset$. Also, if a and b are in H, then $b^{-1} \in H$ (by the inverse property) and $ab^{-1} \in H$ (by the closure property).

Now suppose $H \neq \emptyset$ and for all $a, b \in H$, $ab^{-1} \in H$. ⟨*We show that H is a subgroup of G by showing that the group axioms and closure hold for H. It is convenient to proceed in the order that follows.*⟩

(i)   Let x, y, and z be in H. Then x, y, and z are in G, so by associativity for G, $(xy)z = x(yz)$.

(ii) $H \neq \emptyset$, so there is some $a \in H$. Then $aa^{-1} = e \in H$.

(iii) Suppose $x \in H$. Then e and x are in H, so by hypothesis, $ex^{-1} = x^{-1} \in H$.

(iv) Let x and y be in H. Then by (iii), $y^{-1} \in H$. Then x and y^{-1} are in H, so by hypothesis, $x(y^{-1})^{-1} = xy \in H$.   ∎

There are subgroups associated with homomorphisms. We know that if f is a homomorphism from $(A, \circ)$ to $(B, *)$, then $(\mathrm{Rng}(f), *)$ is a subgroup of $(B, *)$. This is so because the homomorphic image of a group is a group, and $\mathrm{Rng}(f) \subseteq B$. The next definition and theorem show that for every homomorphism there is a subgroup on the domain side, too.

Definition.   Let $f: (A, \circ) \rightarrow (B, *)$ be a homomorphism and i be the identity element of B. The **kernel of f** is $\ker(f) = \{x \in A: f(x) = i\}$.

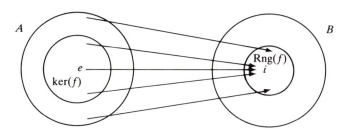

Figure 6.4

Theorem 6.12   The kernel of a homomorphism is a subgroup of the domain group.

Proof.   Let $f: (A, \circ) \rightarrow (B, *)$ be a homomorphism and let e and i be the identity elements of A and B, respectively (figure 6.4).

⟨*We apply Theorem 6.11.*⟩ First, $\ker(f)$ is nonempty, because $f(e) = i$, by Theorem 6.3; so $e \in \ker(f)$. Suppose a and b are in $\ker(f)$. Then $f(a) = f(b) = i$. By Theorem 6.3, $f(b^{-1}) = (f(b))^{-1}$. Therefore

$$f(a \circ b^{-1}) = f(a) \circ f(b^{-1}) = f(a) \circ (f(b))^{-1} = i \circ i^{-1} = i.$$

This shows that $a \circ b^{-1} \in \ker(f)$. Therefore $\ker(f)$ is a subgroup of A. ∎

For each natural number m, there is a particularly nice homomorphism from $(\mathbf{Z}, +)$ onto $(\mathbf{Z}_m, +_m)$. This is the canonical map defined by $f(a) = \bar{a}$ for each $a \in \mathbf{Z}$. By definition of $+_m$, $f(a + b) = \overline{a + b} = \bar{a} +_m \bar{b} = f(a) + f(b)$. The kernel of this homomorphism is $\bar{0} = \{0, \pm m, \pm 2m, \ldots\}$, which we have already said is a subgroup of $\mathbf{Z}$ under addition.

Consider the groups $(\mathbf{Z}_6, +_6)$ and $(\{a, b, c\}, \circ)$ with the operation table for $\circ$ as shown.

$\circ$	a	b	c
a	a	b	c
b	b	c	a
c	c	a	b

It can be verified (exercise 17) by checking all cases that the map $g: \mathbf{Z}_6 \xrightarrow{\text{onto}} \{a, b, c\}$ given by $g(0) = g(3) = a$, $g(1) = g(4) = b$, and $g(2) = g(5) = c$ is a homomorphism. The kernel of g is $\{0, 3\}$, which we know is a subgroup of $\mathbf{Z}_6$.

If a is a member of a group G, then all powers of a are in G by the closure property. This set $\{\ldots, a^{-2}, a^{-1}, a^0, a^1, a^2, \ldots\}$ of all powers of a produces a subgroup of G.

Definitions. Let a be an element of a group G. The **cyclic subgroup generated by** a is $(a) = \{a^n : n \in \mathbf{Z}\}$. If there is $b \in G$ such that $(b) = G$, then G is called a **cyclic group** and b is called a **generator** for G.

The name cyclic subgroup for (a) is justified by the fact that (a) is a subgroup of G. We verify this by observing that (a) is not empty since a^0 is the identity of G. Also, if $x, y \in (a)$, then $x = a^m$ and $y = a^n$ for some integers m and n, so $xy^{-1} = a^m(a^n)^{-1} = a^m a^{-n} = a^{m-n} \in (a)$. Thus (a) is a subgroup of G.

Every cyclic group is abelian, because $a^m a^n = a^{m+n} = a^n a^m$.

There are many examples of cyclic groups. With the operation of multiplication, the set of all powers of 2 is a cyclic group with generator 2. The element 2 (really $2/\equiv_5$) is also a generator of the group $(\mathbf{Z}_5 - \{0\}, \cdot_5)$. This is because with operation $\cdot_5$, $2^1 = 2$, $2^2 = 4$, $2^3 = 8 = 3$, and $2^4 = 16 = 1$, so every element of $\mathbf{Z}_5 - \{0\}$ is a power of 2.

Using additive notation, a group is cyclic when it consists of all multiples of some element. For example, $(\mathbf{Z}, +)$ is cyclic with generator 1 and every $(\mathbf{Z}_m, +_m)$ is cyclic with generator $1/\equiv_m$.

A cyclic group may have more than one generator: for the group $(\mathbf{Z}_4, +_4)$, $\mathbf{Z}_4 = (1) = (3)$. This is because

$$
\begin{aligned}
1 \cdot 1 &= 1 \\
2 \cdot 1 &= 1 + 1 = 2 \\
3 \cdot 1 &= 2 + 1 = 3 \\
4 \cdot 1 &= 3 + 1 = 0
\end{aligned}
\qquad \text{and} \qquad
\begin{aligned}
1 \cdot 3 &= 3 \\
2 \cdot 3 &= 3 + 3 = 2 \\
3 \cdot 3 &= 2 + 3 = 1 \\
4 \cdot 3 &= 1 + 3 = 0.
\end{aligned}
$$

The element 2 does not generate $\mathbf{Z}_4$; it generates the cyclic subgroup $\{0, 2\}$.

The group S_3 is not cyclic because none of its elements generates the entire group. For example

$$
\begin{aligned}
(312)^1 &= (312) \\
(312)^2 &= (231) \\
(312)^3 &= (123), \text{ the identity} \\
(312)^4 &= (312)^1 = (312) \\
(312)^5 &= (312)^2 = (231) \\
(312)^6 &= (312)^3 = (123), \text{ and so forth.}
\end{aligned}
$$

All other powers of (312) are equal to one of these three elements, so the cyclic subgroup generated by (312) is $\{(123), (312), (231)\}$. Similarly, the other elements of S_3 do not generate S_3 [exercises 11(a) and 12(a)]. Of course, the fact that S_3 is not abelian is sufficient to conclude that S_3 is not cyclic, because every cyclic group is abelian.

The **order of an element** $a \in G$ is the order of the cyclic subgroup (a) generated by a. If (a) is an infinite set, we say a has infinite order.

Example.   In the octic group, the element V has order 2 since $V^0 = I$, $V^1 = V$, $V^2 = VV = I$. Thus any power of V is either I or V; therefore $(V) = \{I, V\}$. Similarly, the element R has order 4 because $(R) = \{I, R, R^2, R^3\}$. The orders of the elements of the octic group are 1 (for I), 2 (for R^2, V, D, and D'), and 4 (for R and R^3). Thus the octic group is not cyclic.

Theorem 6.13   Let G be a group and a be an element of G with order r. Then r is the smallest positive integer such that $a^r = e$, the identity, and $(a) = \{e, a, a^2, \ldots, a^{r-1}\}$.

Proof.   Since (a) is finite, the powers of a are not all distinct. Let $a^m = a^n$ with $0 \leqslant m < n$. Then $a^{n-m} = e$ with $n - m > 0$. Therefore the set of positive integers p such that $a^p = e$ is nonempty.

Let k be the smallest such integer. ⟨*This k exists by the Well-Ordering Principle.*⟩ We prove that $k = r$ by showing that the elements of (a) are exactly $a^0 = e, a^1, a^2, \ldots, a^{k-1}$.

First, the elements $e, a^1, a^2, \ldots, a^{k-1}$ are distinct, for if $a^s = a^t$ with $0 \leqslant s < t < k$, then $a^{t-s} = e$ and $0 < t - s < k$, contradicting the definition of k.

Second, every element of (a) is one of $e, a^1, a^2, \ldots, a^{k-1}$. Consider a^t for $t \in \mathbf{Z}$. By the division algorithm, $t = mk + s$ with $0 \leqslant s < k$. Thus $a^t = a^{mk+s} = a^{mk}a^s = (a^k)^m a^s = e^m a^s = ea^s = a^s$, so that $a^t = a^s$ with $0 \leqslant s < k$.

We have shown that the elements a^s for $0 \leqslant s < k$ are all distinct and that every power of a is equal to one of these. Since (a) has exactly r elements, $r = k$ and $a^r = e$. ∎

If $a \in G$ has infinite order, then by the reasoning used in the proof of Theorem 6.13, all the powers of a are distinct and

$$(a) = \{\ldots, a^{-2}, a^{-1}, a^0 = e, a^1, a^2 \ldots\}.$$

Because homomorphisms preserve structure, you should not be surprised that a homomorphic image of a cyclic group is cyclic. In fact, the image of a generator of the group is a generator of the image group. Furthermore, every subgroup of a cyclic group is cyclic. This means that if $G = (a)$ and H is a subgroup of G, then $H = (a^s)$ for some $s \in \mathbf{N}$. The proofs are exercises 20 and 21.

Exercises 6.4

1. By looking for subsets closed under the group operation, then checking the group axioms, find all subgroups of
   ★ (a) $(\mathbf{Z}_8, +)$.
   (c) $(\mathbf{Z}_5, +)$.
   (b) $(\mathbf{Z}_7 - \{0\}, \cdot)$.
   ☆ (d) $(J, *)$, with table shown here.

*	a	b	c	d	e	f
a	a	b	c	d	e	f
b	b	a	f	e	d	c
c	c	e	a	f	b	d
d	d	f	e	a	c	b
e	e	c	d	b	f	a
f	f	d	b	c	a	e

2. In the group S_3, $(231) \circ (312) = (123)$. Is there a subgroup of S_3 that contains (231) but not (312)? Explain.

3. Find the smallest subgroup of S_3 that contains (312) and (321). (*Hint*: Use the closure property.)

4. For the octic group (symmetries of a square) in section 6.3, give the five different subgroups of order 2 and the three different subgroups of order 4.

5. Prove that if G is a group and H is a subgroup of G, then the inverse of an element $x \in H$ is the same as its inverse in G. (Theorem 6.10.)

6. Prove that if H and K are subgroups of a group G, then $H \cap K$ is a subgroup of G.

7. Prove that if $\{H_\alpha : \alpha \in \Delta\}$ is a family of subgroups of a group G, then $\bigcap_{\alpha \in \Delta} H_\alpha$ is a subgroup of G.

8. Give an example of a group G and subgroups H and K of G such that $H \cup K$ is not a subgroup of G.

9. Let G be a group and H be a subgroup of G.
    ★ (a)   If G is abelian, must H be abelian? Explain.
    (b)   If H is abelian, must G be abelian? Explain.

10. Let G be a group. If H is a subgroup of G and K is a subgroup of H, prove that K is a subgroup of G.

11. Find the order of each element of the group
    (a) S_3.   (b) $(\mathbf{Z}_7, +)$.
    ★ (c) $(\mathbf{Z}_8, +)$.   (d) $(\mathbf{Z}_7 - \{0\}, \cdot)$.

12. List all generators of each group in exercise 11.

★ 13. Let G be a group with identity e and let $a \in G$. Prove that the set $N_a = \{x \in G : xa = ax\}$, called the **normalizer** of a in G, is a subgroup of G.

14. Let G be a group and let $C = \{x \in G : \text{for all } y \in G, xy = yx\}$. Prove that C, the **center** of G, is a subgroup of G.

15. Prove that if G is a group and $a \in G$, then the center of G is a subgroup of the normalizer of a in G.

★ 16. Let G be a group and let H be a subgroup of G. Let a be a fixed element of G. Prove that $K = \{a^{-1}ha : h \in H\}$ is a subgroup of G.

17. Let $(\{a, b, c\}, \circ)$ be the group with operation table shown here.

$\circ$	a	b	c
a	a	b	c
b	b	c	a
c	c	a	b

Verify that the mapping $g : (\mathbf{Z}_6, +) \to (\{a, b, c\}, \circ)$ defined by $g(0) = g(3) = a$, $g(1) = g(4) = b$, and $g(2) = g(5) = c$ is a homomorphism.

18. Let $(\mathbf{C} - \{0\}, \cdot)$ be the group of complex numbers with ordinary complex number multiplication. Let $\alpha = (1 + i\sqrt{3})/2$.
    (a)   Find (α).
    (b)   Find a generator of (α) other than α.

19. Let x be an element of the group G with identity e. What are the possibilities for the order of x if
    ★ (a) $x^{15} = e$?　　　　(b) $x^{20} = e$?　　　　(c) $x^n = e$?

20. Prove that every subgroup of a cyclic group is cyclic.

21. Let $h: G \rightarrow K$ be a homomorphism.
    (a) Prove that if $x \in G$, then $h(x^k) = (h(x))^k$ for all $k \in \mathbf{N}$.
    (b) Prove that the homomorphic image of a cyclic group is cyclic.

22. Let $G = (a)$ be a cyclic group of order 30.
    (a) What is the order of a^6?　　　　(b) List all elements of order 2.
    ★ (c) List all elements of order 3.　　(d) List all elements of order 10.

23. Let $G = (a)$ be a cyclic group of order 9. Let G' be the group generated by the permutation (231). If Φ is a homomorphism from G to G' such that $\Phi(a) = $ (231), find all other images.

24. Let $\Phi: (G, *) \rightarrow (H, \circ)$ be a homomorphism. Let e be the identity of G. Prove that Φ is one-to-one iff ker$(\Phi) = \{e\}$.

☆ 25. Let $f: G \rightarrow H$ be a homomorphism. For $a \in G$, prove that the order of $f(a)$ divides the order of a.

6.5

Cosets and Lagrange's Theorem

The purpose of this section is to show that the order of a subgroup of a given finite group must be a divisor of the order of the group. Thus, given a group of order, say, 30, we will know it cannot have a subgroup with 8 elements because 8 does not divide 30. The way we prove that the order of a subgroup divides the order of the group is to use the subgroup to partition the group into disjoint subsets, one of which is the subgroup itself. Then because it turns out that all these subsets have the same number of elements as the subgroup, we know the group order is a multiple of the subgroup order. We first give an example.

　　　The group $(\mathbf{Z}_{12}, +)$ has 12 elements, which we denote simply by 0, 1, 2, . . . , 11. One subgroup of $\mathbf{Z}_{12}$ is the set H of all multiples of 4. The set H is a subset of $\mathbf{Z}_{12}$,

$$H = \{0, 4, 8\}.$$

If we add 1 to each element of H, we obtain another subset, which we will denote $1 + H$:

$$1 + H = \{1, 5, 9\}.$$

We note that $1 + H$ is not a subgroup of $\mathbf{Z}_{12}$ (closure fails), but $1 + H$ has the same number of elements as H. We call $1 + H$ a **coset** of H. Continuing, we have

$$2 + H = \{2, 6, 10\}$$
$$3 + H = \{3, 7, 11\}$$

We could go on and compute $4 + H$, $5 + H$, or any coset $x + H$ where $x \in \mathbf{Z}_{12}$, but these are not new sets of elements; $5 + H$, for example, is

$$5 + H = \{5, 9, 13\}$$
$$= \{5, 9, 1\}$$
$$= 1 + H$$

because 13 and 1 are the same equivalence class.

Finally we note that H, $1 + H$, $2 + H$, and $3 + H$ each have three elements, are disjoint, and their union is $\mathbf{Z}_{12}$ (figure 6.5). The number of elements in $\mathbf{Z}_{12}$ is the sum of the number of elements in the cosets H, $1 + H$, $2 + H$, and $3 + H$, which is 4 times the order of H. Thus we have an explanation for the fact that the order of H divides the order of $\mathbf{Z}_{12}$.

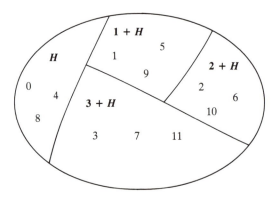

Figure 6.5

The result we seek, that the order of a subgroup must divide the order of the group, is Lagrange's Theorem, first proved by Joseph-Louis Lagrange (1736–1813). The first step is to give a formal definition of coset.

Definition.   Let $(G, \circ)$ be a group, H a subgroup of G, and $x \in G$. The set $x \circ H = \{x \circ h : h \in H\}$ is called the **left coset** of x and H, and $H \circ x = \{h \circ x : h \in H\}$ is called the **right coset** of x and H.

Table 6.7 shows a group T with identity a and subgroups $H = \{a, b\}$ and $K = \{a, e, f\}$. The cosets of H and K are also shown in tables.

Table 6.7   *Group T*

	a	*b*	*c*	*d*	*e*	*f*
a	*a*	*b*	*c*	*d*	*e*	*f*
b	*b*	*a*	*f*	*e*	*d*	*c*
c	*c*	*e*	*a*	*f*	*b*	*d*
d	*d*	*f*	*e*	*a*	*c*	*b*
e	*e*	*c*	*d*	*b*	*f*	*a*
f	*f*	*d*	*b*	*c*	*a*	*e*

Cosets of H	*Cosets of K*
$aH = \{a, b\} = Ha = \{a, b\}$	$aK = \{a, e, f\} = Ka = \{a, e, f\}$
$bH = \{b, a\} = Hb = \{b, a\}$	$bK = \{b, d, c\} = Kb = \{b, c, d\}$
$cH = \{c, e\} \neq Hc = \{c, f\}$	$cK = \{c, b, d\} = Kc = \{c, d, b\}$
$dH = \{d, f\} \neq Hd = \{d, e\}$	$dK = \{d, c, b\} = Kd = \{d, b, c\}$
$eH = \{e, c\} \neq He = \{e, d\}$	$eK = \{e, f, a\} = Ke = \{e, f, a\}$
$fH = \{f, d\} \neq Hf = \{f, c\}$	$fK = \{f, a, e\} = Kf = \{f, a, e\}$

Notice that the left coset fH and the right coset Hf are not equal. However, it *does* happen that for every x in T, $xK = Kx$.

As another example of cosets, consider the subgroup $H = \{0, \pm 5, \pm 10, \ldots\}$ of $(\mathbf{Z}, +)$. Then $2 + H = H + 2 = \{\ldots, -8, -3, 2, 7, \ldots\} = 2/\equiv_5$ and $0 + H = H + 0 = H$.

Theorem 6.14   If H is a subgroup of G and $x \in G$, then H and xH are equivalent (have the same number of elements).

Proof.   To show that H and xH are equivalent, we show that the function F given by $F(h) = xh$ maps H one-to-one and onto xH. If $w \in xH$, then $w = xh$ for some $h \in H$. Then for that h, $F(h) = w$. Therefore F maps onto xH. Suppose $F(h_1) = F(h_2)$ for $h_1, h_2 \in H$. Then $xh_1 = xh_2$. Then by cancellation, $h_1 = h_2$. Therefore F is one-to-one. ■

Because every left coset of H is equivalent to H, any two left cosets of H have the same number of elements. This result and others to follow also apply to right cosets.

Theorem 6.15   The set of left cosets of H in G is a partition of G.

Proof.   Every element x of G is a member of some left coset because $x = xe \in xH$. Therefore G is a union of left cosets. It remains to show that left cosets are either identical or disjoint.

Suppose $xH \cap yH \neq \varnothing$. Let $b \in xH \cap yH$. Then $b = xh_1 = yh_2$ for some $h_1, h_2 \in H$. Therefore $x = yh_2h_1^{-1}$ and

$xH = yh_2h_1^{-1} H$. Now let xh be in xH. Then $xh = yh_2h_1^{-1}h'$ for some $h' \in H$. But $h_2h_1^{-1}h' \in H$, so $xh \in yH$. This shows that $xH \subseteq yH$, and a similar argument shows $yH \subseteq xH$. Therefore $xH = yH$. ∎

Theorem 6.16. *(Lagrange)* Let G be a finite group and H a subgroup of G. Then the order of H divides the order of G.

Proof. Let the order of G be n and the order of H be m. Let the number of left cosets of H in G be k. Denote the distinct cosets of H by $x_1H, x_2H, \ldots, x_kH$. By Theorem 6.14, each coset contains exactly m elements. By Theorem 6.15, G may be written as $G = x_1H \cup x_2H \cup \cdots \cup x_kH$, and the cosets are pairwise disjoint. We have divided the elements of G into k classes, each class having m elements. Thus there are mk elements in G. Therefore $n = mk$, and so m divides n. ∎

For a group G with order n and subgroup H of order m, the integer n/m is called the **index** of H in G and is the number of distinct left cosets of H.

The converse of Lagrange's Theorem is false. There exists a group of order 12 that has no subgroup of order 6.

Corollary 6.17 Let G be a finite group and $a \in G$. Then the order of a divides the order of G.

Proof. Exercise 6. ∎

Corollary 6.18 If G is a finite group of order n and $a \in G$, then $a^n = e$, the identity.

Proof. Exercise 7. ∎

Corollary 6.19 A group of prime order is cyclic.

Proof. Let G have prime order p. Let a be an element of G other than e. By Corollary 6.17, the order of a divides the order of G, and so must be either 1 or p. But (a) contains both e and a, so the order of a must be p. Therefore $G = (a)$ and G is cyclic. ∎

Exercises 6.5

☆ 1. Let H be the subgroup of S_3 generated by the permutation (213). Find all left and right cosets of H in S_3.

2. Let H be the subgroup of S_3 generated by the permutation (231). Find all left and right cosets of H in S_3.

3. Find all the cosets of $H = (a^3)$ in a cyclic group $G = (a)$ of order 12.

★ 4. Show that a group G of order 4 is either cyclic or is such that $x^2 = e$ for all $x \in G$.

5. Show that a group G of order 27 is either cyclic or is such that $x^9 = e$ for all $x \in G$.

6. Prove Corollary 6.17.

7. Prove Corollary 6.18.

8. Let H be a subgroup of G.
   (a) Prove that $aH = bH$ iff $a^{-1}b \in H$.
   (b) If $aH = bH$, must $b^{-1}a$ be in H? Explain.
   (c) If $aH = bH$, must ab^{-1} be in H? Explain.

☆ 9. Prove that if G is a cyclic group of order n and if m divides n, then G has a subgroup of order m.

10. **Proofs to Grade.**
   ★ (a) **Claim.** If H is a subgroup of G and $aH = bH$, then $a^{-1}b \in H$.
      "Proof." Suppose $aH = bH$. Then $ah = bh$ for some $h \in H$, and so $h = a^{-1}bh$. Thus $hh^{-1} = a^{-1}b$ and since $hh^{-1} \in H$, we conclude that $a^{-1}b \in H$. ∎

   (b) **Claim.** If H is a subgroup of G and $a^{-1}b \in H$, then $aH \subseteq bH$.
      "Proof." Suppose $a^{-1}b \in H$. Then $a^{-1}b = h$ for some $h \in H$. Thus $a^{-1} = hb^{-1}$, so that $a = (hb^{-1})^{-1} = bh^{-1}$. Thus $a \in bH$. Suppose that $x \in aH$. Then $x = ah'$ for some $h' \in H$. Thus $x = (bh^{-1})h' = b(h^{-1}h')$, so $x \in bH$. Therefore $aH \subseteq bH$. ∎

   ★ (c) **Claim.** If H is a subgroup of G and $aH = bH$, then $b^{-1}a \in H$.
      "Proof." Suppose $aH = bH$. Then $b^{-1}aH = b^{-1}bH = eH = H$. Therefore $b^{-1}aH = H$. Since $b^{-1}a \in b^{-1}aH$, $b^{-1}a \in H$. ∎

6.6

Quotient Groups

Let G be a group and let H be a subgroup of G. The set of left cosets of H in G is a partition of G, by Theorem 6.15. Therefore there is a corresponding equivalence relation on G with the property that a, $b \in G$ are related iff they belong to the same left coset of H. (See chapter 3, section 3.) The equivalence classes under this relation are exactly the left cosets of H. The set of all equivalence classes (cosets) is denoted G/H (read "G mod H").

Example.   In $(\mathbf{Z}_6, +)$, let H be the subgroup $\{0, 3\}$. The left cosets of H are

$$0 + H = 3 + H = \{0, 3\}$$
$$1 + H = 4 + H = \{1, 4\}$$
$$2 + H = 5 + H = \{2, 5\}$$

Here $\mathbf{Z}_6/H = \{0 + H, 1 + H, 2 + H\}$.

It is our intention to impose a structure—an operation we hope will satisfy the group axioms—on G/H. If we try to define a "product" for cosets xH and yH, nothing could be more natural than $xH \cdot yH = xyH$. In order to try out this "structure" $(G/H, \cdot)$, we consider as examples the sets T/H and T/K where T, H, and K are the groups presented in table 6.7.

The left cosets of H in T are $\{a, b\}$, $\{c, e\}$, and $\{d, f\}$. Applying the definition for $\cdot$ on T/H, we have

$$\{c, e\} \cdot \{d, f\} = cH \cdot dH = cdH = fH = \{d, f\}$$

$$\{c, e\} \cdot \{d, f\} = eH \cdot fH = efH = aH = \{a, b\}.$$

The fact that $\cdot$ yields two different images means that it is not an operation on T/H.

The left cosets of K in T are $\{a, e, f\} = aK = eK = fK$ and $\{b, c, d\} = bK = cK = dK$. In this case $\cdot$ is indeed an operation on T/K because the product $xK \cdot yK$ does not depend on the representatives for the cosets. For example, $eK = fK$, and $bK = cK$, and $eK \cdot bK = ebK = cK$, and $fK \cdot cK = fcK = bK$, but $cK = bK$. Table 6.8 is the operation table for T/K.

Table 6.8

	aK	bK
aK	aK	bK
bK	bK	aK

That the proposed operation $\cdot$ works for T/K and $xK = Kx$ for all $x \in T$ is more than coincidence. We are led to the following definition.

Definition.   Let G be a group and H a subgroup of G. Then H is called **normal** in G iff for all $x \in G$, $xH = Hx$.

Thus K is a normal subgroup of T but H is not. If G is any group, then both G and $\{e\}$ are normal in G (see exercise 4). If G is an abelian group, then every subgroup of G is normal. Notice that $xK = Kx$ does not imply that

$xk = kx$ for every $k \in K$. For the subgroup K of T considered above, $f \in K$ and $dK = Kd$, but $df \neq fd$. (See table 6.7.)

Theorem 6.20   If H is a normal subgroup of G with $xH = yH$ and $wH = vH$, then $xwH = yvH$.

Proof.   Let xwh be an element of xwH. Then $wh \in wH = vH = Hv$ ⟨*because H is normal*⟩, so $wh = h_1v$ for some $h_1 \in H$. Thus $xwh = xh_1v$. Now $xh_1 \in xH = yH = Hy$, so $xh_1 = h_2y$ for some $h_2 \in H$. Thus $xwh = h_2yv \in Hyv = yvH$. Since cosets are either identical or disjoint, $xwH = yvH$.   ∎

If H is a normal subgroup of G and if $xH = yH$ and $wH = vH$, then $xH \cdot wH = xwH = yvH = yH \cdot vH$, by Theorem 6.20. This establishes that $(G/H, \cdot)$ is an algebraic structure.

Theorem 6.21   If H is a normal subgroup of G, then $(G/H, \cdot)$, with $\cdot$ as defined above, is a group called the **quotient group of G modulo H** (G mod H). The identity element is the coset $H = eH$, and the inverse of xH is $x^{-1}H$.

Proof.   Exercise 3.   ∎

Let $m \in \mathbf{N}$ and let H_m be the subgroup of $(\mathbf{Z}, +)$ consisting of all multiples of m. Then H_m is normal because $(\mathbf{Z}, +)$ is abelian. Elements of $\mathbf{Z}/H_m$ are cosets of the form $x + H_m$, which is the equivalence class $x/\equiv_m$. The operation on cosets in $\mathbf{Z}/H_m$ is the same as the operation $+_m$ on $\mathbf{Z}_m$. Therefore $\mathbf{Z}/H_m$ is the group $(\mathbf{Z}_m, +_m)$.

The group $H = (\{1, 6\}, \cdot_7)$ is a normal subgroup of the abelian group $(\mathbf{Z}_7 - \{0\}, \cdot_7)$. The cosets in $(\mathbf{Z}_7 - \{0\})/H$ are H, $2H = \{2, 5\}$, and $3H = \{3, 4\}$. Table 6.9 shows $(\mathbf{Z}_7 - \{0\}, \cdot_7)$ and the quotient group.

Table 6.9 $(\mathbf{Z}_7 - \{0\}, \cdot_7)$

$\cdot_7$	1	2	3	4	5	6
1	1	2	3	4	5	6
2	2	4	6	1	3	5
3	3	6	2	5	1	4
4	4	1	5	2	6	3
5	5	3	1	6	4	2
6	6	5	4	3	2	1

$((\mathbf{Z}_7 - \{0\})/H, \cdot)$

$\cdot$	H	$2H$	$3H$
H	H	$2H$	$3H$
$2H$	$2H$	$3H$	H
$3H$	$3H$	H	$2H$

Exercises 6.6

★   1.   Let G be a group and H be a subgroup of G. Prove that H is normal iff $a^{-1}Ha \subseteq H$ for all $a \in G$, where $a^{-1}Ha = \{a^{-1}ha : h \in H\}$.

☆ 2. Prove that a subgroup H of G is normal in G iff $Ha \subseteq aH$ for every $a \in G$.

3. Prove Theorem 6.21.

4. Let G be a group with identity e.
    (a) Prove that $\{e\}$ and G are normal subgroups of G.
    (b) Describe the quotient group $G/\{e\}$.
    (c) Describe the quotient group G/G.

5. Let H be a normal subgroup of G. Prove that
    (a) if G is abelian, then G/H is abelian.
    ☆ (b) if G is cyclic, then G/H is cyclic.

6. Let G be the group $(\mathbf{R} - \{0\}, \cdot)$ and let H be the subgroup consisting of all positive real numbers. Show that H is a normal subgroup of G of index 2, and construct the operation table for G/H.

7. For the group $\mathcal{O}$ of symmetries of a square (section 6.3), let $J = \{I, R, R^2, R^3\}$, $K = \{I, R^2, H, V\}$, and $L = \{I, H\}$.
    (a) Construct the operation table for $\mathcal{O}/J$.
    (b) Construct the operation table for $\mathcal{O}/K$.
    (c) Is L normal in $\mathcal{O}$?
    (d) Is L normal in K?

☆ 8. If H_1 is a normal subgroup of G, and H_2 is a normal subgroup of H_1, must H_2 be a normal subgroup of G? Explain.

9. Consider the subgroup $C = \{I, R^2\}$ of the octic group $\mathcal{O}$.
    (a) Prove that C is normal in the octic group and hence that the quotient group $\mathcal{O}/C$ exists.
    (b) Construct the operation table for $\mathcal{O}/C$.
    (c) Is $\mathcal{O}/C$ abelian?
    (d) Is $\mathcal{O}/C$ cyclic?

10. Let f be a homomorphism of G onto H. Let N be a normal subgroup of G. Prove that $f(N) = \{f(n): n \in N\}$ is a normal subgroup of H.

11. Let f be a homomorphism of G onto H. Let M be a subgroup of H. Prove that $N = \{g: f(g) \in M\}$ is a subgroup of G that includes $\ker(f)$. Show that if M is normal in H, then N is normal in G.

12. **Proofs to Grade.**
    (a) **Claim.**   If H is a normal subgroup of G, and K is a normal subgroup of H, then K is a normal subgroup of G.
    "Proof."   We must show $xK = Kx$ for all $x \in G$. Since H is normal in G, $xH = Hx$ for all $x \in G$. Since K is a normal subgroup of H, $xK = Kx$ for all $x \in H$ and hence for all $x \in G$. Therefore K is a normal subgroup of G. ■
    ★ (b) **Claim.**   If H and K are normal subgroups of G, then $H \cap K$ is a normal subgroup of G.
    "Proof."   Let $x \in G$. Then $x(H \cap K) = xH \cap xK = Hx \cap Kx$ ⟨*since H and K are normal*⟩ $= (H \cap K)x$. Therefore $H \cap K$ is normal. ■
    (c) **Claim.**   If H is a normal subgroup of G and G/H is abelian, then G is abelian.
    "Proof."   Let $x, y \in G$. Then $xH, yH \in G/H$. Since G/H is abelian, $(xH)(yH) = (yH)(xH)$. Therefore $xyH = yxH$. Therefore $(xy)h = (yx)h$ for some $h \in H$. By cancellation, $xy = yx$. ■

6.7

Isomorphism; The Fundamental Theorem of Group Homomorphisms

A homomorphism that is one-to-one is called an **isomorphism.** If there is an isomorphism $f: (A, \circ) \xrightarrow{\text{onto}} (B, *)$, then $(A, \circ)$ is said to be **isomorphic** to $(B, *)$. Inverses and composites of isomorphisms are also isomorphisms, so the relation of being isomorphic is an equivalence relation on the class of all groups.

The word *isomorphic* comes from the Greek words *isos* (equal) and *morphe* (form). Isomorphic groups are literally "equal form" because they differ only in the names or nature of their elements; *all their algebraic properties are identical.* For example, the groups $(\mathbf{Z}_2, +_2)$, $(\{1, 6\}, \cdot_7)$, and $(\{\varnothing, A\}, \Delta)$ are isomorphic, where A is any nonempty set and Δ is the symmetric difference operation defined by $X \Delta Y = (X - Y) \cup (Y - X)$. The three groups are shown in table 6.10.

Table 6.10

$+_2$	0	1
0	0	1
1	1	0

$\cdot_7$	1	6
1	1	6
6	6	1

Δ	$\varnothing$	A
$\varnothing$	$\varnothing$	A
A	A	$\varnothing$

In fact, any two groups of order 2 are isomorphic. This observation can be extended: Every cyclic group of order m is isomorphic to $(\mathbf{Z}_m, +_m)$. Every infinite cyclic group is isomorphic to $(\mathbf{Z}, +)$ (see exercises 6–9). Because of these results, we can say that $(\mathbf{Z}_m, +_m)$ and $(\mathbf{Z}, +)$ are the only cyclic groups, up to isomorphism. The noncyclic group T of section 6.5 is isomorphic to the group $(S_3, \circ)$ under the isomorphism

$$f = \{(a, (123)), (b, (213)), (c, (321)), (d, (132)), (e, (231)), (f, (321))\}.$$

We claimed in section 6.3 that for every group there is a permutation group with the same structure. We now prove this result, due to Arthur Cayley (1821–1895).

Theorem 6.22   Every group G is isomorphic to a permutation group.

Proof.   As the set whose elements are to be permuted, we choose the set G. If $a \in G$, then $\{ax: x \in G\} = G$. Therefore the function $\theta_a: G \to G$ defined by $\theta_a(x) = ax$ is a mapping from G onto G that *(by the cancellation property of the group)* is one-to-one. Therefore θ_a is a permutation of G associated with the element a of G. The function f sending a to θ_a will be our isomorphism from G to the set $H = \{\theta_a: a \in G\}$.

Let $x \in G$. Then $\theta_{ab}(x) = (ab)x = a(bx) = \theta_a(\theta_b(x)) = (\theta_a \circ \theta_b)(x)$, so $\theta_{ab} = \theta_a \circ \theta_b$. Therefore f is a homomorphism. Also, f maps onto H by definition of H. Thus H ⟨*the homomorphic image*⟩ is a permutation group. Now suppose $\theta_a = \theta_b$. Then $ax = bx$ for every $x \in G$, so $a = b$. Therefore f is one-to-one; G and H are isomorphic. ∎

Example.   Let G be the group $(\{1, -1, i, -i\}, \cdot)$, where $\cdot$ is the usual multiplication of complex numbers. As elements of the permutation group H, we take the functions that are left translations by elements of G (that is, $\theta_a(x) = a \cdot x$). For example,

$$\theta_i(1) \quad = i \cdot 1 = i$$
$$\theta_i(-1) = i \cdot (-1) = -i$$
$$\theta_i(i) \quad = i \cdot i = -1,$$
$$\theta_i(-i) = i \cdot (-i) = 1,$$

so $\theta_i = (i, -i, -1, 1)$. Similar computations show that

$$\theta_1 \quad = (1, -1, i, -i),$$
$$\theta_{-1} = (-1, 1, -i, i),$$
and $\theta_{-i} = (-i, i, 1 -1)$.

Thus $H = \{\theta_1, \theta_{-1}, \theta_i, \theta_{-i}\}$ and the operation on H is composition. The tables for G and H clearly have the same structure.

·	1	-1	i	-i
1	1	-1	i	-i
-1	-1	1	-i	i
i	i	-i	-1	1
-i	-i	i	1	-1

$(G, \cdot)$

∘	θ_1	θ_{-1}	θ_i	θ_{-1}
θ_1	θ_1	θ_{-1}	θ_i	θ_{-i}
θ_{-1}	θ_{-1}	θ_1	θ_{-i}	θ_i
θ_i	θ_i	θ_{-i}	θ_{-1}	θ_1
θ_{-i}	θ_{-i}	θ_i	θ_1	θ_{-1}

$(H, \circ)$

The main purpose of this section is to establish the connection between homomorphisms defined on a group and normal subgroups. First, for every homomorphism, there is a corresponding normal subgroup. This subgroup determines a quotient group that is isomorphic to the image group.

Theorem 6.23a   The kernel K of a homomorphism f from a group $(G, \circ)$ onto $(B, *)$ is a normal subgroup of G. Furthermore, $(G/K, \cdot)$ is isomorphic to $(B, *)$.

Proof.   We saw in Theorem 6.12 that $K = \{y \in G: f(y) = e_B\}$ ⟨*where e_B is the identity of B*⟩ is a subgroup of G. We first prove that for all x and $a \in G$, $x \in a \circ K$ iff $f(x) = f(a)$. Suppose

$x \in a \circ K$. Then $x = a \circ k$ for some $k \in K$. Then $f(x) = f(a \circ k) = f(a) * f(k) = f(a) * e_B = f(a)$, so $f(x) = f(a)$. Now suppose $f(x) = f(a)$. Then $f(a^{-1} \circ x) = f(a^{-1}) * f(x) = (f(a))^{-1} * f(a) = e_B$. Thus $a^{-1} \circ x \in K$. Therefore $x = a \circ (a^{-1} \circ x) \in aK$.

Similarly, $x \in K \circ a$ iff $f(x) = f(a)$. Hence $x \in a \circ K$ iff $f(x) = f(a)$ iff $x \in K \circ a$, so $a \circ K = K \circ a$ for all $a \in G$. Therefore K is normal in G.

This argument shows that for every left coset $a \circ K$ of the kernel K, all the elements of $a \circ K$ are mapped by f to the same element $f(a)$ in B. Therefore we may define a mapping $h: G/K \rightarrow B$ by setting $h(a \circ K) = f(a)$. Further, h is one-to-one. $\langle$*If $h(a_1 \circ K) = h(a_2 \circ K)$ then $f(a_1) = f(a_2)$, so $a_2 \in a_1 \circ K$. As $a_2 = a_2 \circ e$ is also in $a_2 \circ K$, and cosets form a partition of G, $a_1 \circ K = a_2 \circ K$.*$\rangle$

The mapping h is onto B: If $b \in B$, then $b = f(a)$ for some $a \in G$, and $b = f(a) = h(a \circ K)$.

Finally, h is operation preserving:

$$
\begin{aligned}
h((a_1 \circ K)(a_2 \circ K)) &= h(a_1 \circ a_2 \circ K) \\
&= f(a_1 \circ a_2) \\
&= f(a_1) * f(a_2) \\
&= h(a_1 \circ K) * h(a_2 \circ K).
\end{aligned}
$$

Therefore h is an isomorphism from $(G/K, \cdot)$ onto $(B, *)$.   ∎

The other aspect of the connection between homomorphisms and normal subgroups is that it is possible to start with any normal subgroup and construct a homomorphic image of the group. This image is the quotient group.

Theorem 6.23b   Let H be a normal subgroup of $(G, \circ)$. Then the mapping $g: G \rightarrow G/H$ given by $g(a) = a \circ H$ is a homomorphism from G onto G/H with kernel H.

Proof.   We have seen the mapping g before (section 4.1) and called it the canonical map from G to the set of cosets $\langle$*equivalence classes*$\rangle$. First, g maps onto G/H because, if $a \circ K$ is a left coset, then $a \circ K = g(a)$. To verify that g is operation preserving, we compute

$$
g(a \circ b) = a \circ b \circ H = a \circ H \cdot b \circ H = g(a) \cdot g(b).
$$

It remains to show that H is the kernel of g. The identity of $(G/H, \cdot)$ is the coset H, so $x \in \ker(g)$ iff $g(x) = x \circ H = H$ iff $x \in H$. Therefore $\ker(g) = H$.   ∎

The two preceding theorems are summarized in the next one.

Theorem 6.24 *(The Fundamental Theorem of Group Homomorphisms)* Every homomorphism of a group G determines a normal subgroup of G. Every normal subgroup of G determines a homomorphic image of G. Up to isomorphism the only homomorphic images of G are the quotient groups of G modulo normal subgroups. Finally, the diagram in figure 6.6 "commutes"; that is $f = h \circ g$, where h and g are the mappings of Theorems 6.23 (a) and (b).

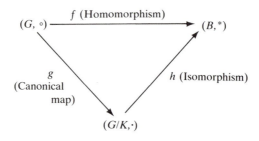

Figure 6.6

Proof. What remains to be proved is that f is the composite of h with g. Let $a \in G$. Then $g(a) = a \circ K$ and $h(g(a)) = h(a \circ K) = f(a)$. Therefore $f = h \circ g$. ∎

Finally, we note that the theorem's primary assertion is that every homomorphic image can be identified with (is isomorphic to) the quotient group $G/\ker(f)$. Furthermore, with this identification, the homomorphism f acts like the canonical homomorphism. Thus we can think of homomorphic images as though they are quotient groups, and homomorphisms as though they are canonical maps. The Fundamental Theorem is used to conclude that two groups are isomorphic to each other without actually constructing the isomorphism.

As an example of the use of the Fundamental Theorem, we describe all homomorphic images of $(\mathbf{Z}_6, +)$. Every such image must be isomorphic to a quotient group of $\mathbf{Z}_6$. The subgroups of $\mathbf{Z}_6$ have orders 1, 2, 3, and 6, and every subgroup is normal in $\mathbf{Z}_6$. The corresponding quotient groups have orders 6, 3, 2, and 1. These quotient groups are cyclic [see exercise 5(b) of section 6.6] and therefore isomorphic to $\mathbf{Z}_6$, $\mathbf{Z}_3$, $\mathbf{Z}_2$, and $\{0\}$, respectively.

Exercises 6.7

---

1. Show that the mapping $f: (\mathbf{R}^+, \cdot) \to (\mathbf{R}, +)$ given by $f(x) = \log_{10} x$ is an isomorphism.

2. Is $(\mathbf{Z}_4, +)$ isomorphic to $(\{1, -1, i, -i\}, \cdot)$? Explain.

3. Is S_3 isomorphic to $(\mathbf{Z}_6, +)$? Explain.

4. Give an example of two groups of order 8 that are not isomorphic.

☆ 5. Let $f: (G, \cdot) \rightarrow (H, *)$ be an isomorphism. For $a \in G$, prove that the order of a equals the order of $f(a)$.

6. Prove that any two groups of order 2 are isomorphic.

7. Prove that any two groups of order 3 are isomorphic.

8. Show that every cyclic group of order m is isomorphic to $(\mathbf{Z}_m, +)$.

9. Show that every infinite cyclic group is isomorphic to $(\mathbf{Z}, +)$.

10. Use the method of proof of Cayley's Theorem to find a group of permutations isomorphic to
    (a) $(\mathbf{Z}_5 - \{0\}, \cdot_5)$.
    (b) the subgroup $\{I, R^2, D, D'\}$ of the octic group.
    (c) $(\mathbf{R}, +)$.

11. (a) Define a homomorphism from $(\mathbf{Z}_{2n}, +)$ onto $(\mathbf{Z}_n, +)$.
    (b) What is the kernel K of the homomorphism you defined in (a)?
    (c) Prove that $\mathbf{Z}_{2n}/K$ is isomorphic to $\mathbf{Z}_n$.
    (d) Construct the operation tables for the groups described in (a), (b), and (c) in the case $n = 3$.

★ 12. Describe all homomorphic images of $\mathbf{Z}_8$.

13. Describe all homomorphic images of $\mathbf{Z}_{11}$.

14. Describe all homomorphic images of S_3.

7

Real Analysis

In this chapter we give an introduction to the analyst's point of view of the real numbers. Section 7.1 discusses the reals as an example of a complete ordered field. To fully understand the formal definitions in this section requires some familiarity with the first concepts of algebraic structure presented in section 1 of chapter 6. Sections 7.2 and 7.3 deal with topological properties of **R**, and section 7.4 is about sequences of real numbers. Each section presents another aspect of the concept of completeness for the real number system. Section 7.5 ties together all the other sections and gives an application of completeness.

7.1

Field Properties of the Real Numbers

In this section we describe the real numbers as a complete ordered field. We shall give general definitions of the words "complete," "ordered," and "field," but you should think of the real numbers as your principal example of each concept. We assume a knowledge of rational, real, and complex number systems commonly obtained in elementary calculus.

Definition. A **field** is an algebraic structure $(F, +, \cdot)$ where $+$ and $\cdot$ are binary operations on F such that

(1) $(F, +)$ is an abelian group with identity denoted by 0; that is, $+$ is an operation on F so that for all $x, y, z \in F$,
   (a) $(x + y) + z = x + (y + z)$.
   (b) $x + 0 = 0 + x = x$.
   (c) For every $x \in F$, there exists $-x \in F$ such that
      $x + (-x) = (-x) + x = 0$.
   (d) $x + y = y + x$.

(2) $(F - \{0\}, \cdot)$ is an abelian group with identity denoted by 1; that is, $\cdot$ is an operation on $F - \{0\}$ so that for all $x, y, z \in F - \{0\}$,
   (a) $x \cdot (y \cdot z) = (x \cdot y) \cdot z$.
   (b) $x \cdot 1 = 1 \cdot x = x$.
   (c) For every $x \in F - \{0\}$, there exists $x^{-1} \in F - \{0\}$ such that $x \cdot x^{-1} = x^{-1} \cdot x = 1$.
   (d) $x \cdot y = y \cdot x$.

(3) For all x, y, z in F, $x \cdot (y + z) = x \cdot y + x \cdot z$.

(4) $0 \neq 1$.

The rationals, the reals, and the complex numbers are all fields. If you have read section 6.3 you are familiar with the modular arithmetic structures $(\mathbf{Z}_m, +, \cdot)$. If p is a prime, then $(\mathbf{Z}_p, +, \cdot)$ is a field with p elements. By property (4), every field must have at least two elements, and thus the smallest field is $\mathbf{Z}_2 = \{0, 1\}$ with the operations shown in the tables.

+	0	1
0	0	1
1	1	0

$\cdot$	0	1
0	0	0
1	0	1

We now develop properties that collectively will distinguish the real numbers from all other fields. To begin, we consider a property of the fields **Q** and **R** not shared by the other fields we have named.

Definition. A field $(F, +, \cdot)$ is **ordered** iff there is a relation $<$ on F such that for all $x, y, z \in F$,

(1) $x \not< x$ (irreflexivity).
(2) If $x < y$ and $y < z$, then $x < z$ (transitivity).
(3) Either $x < y$, $x = y$, or $y < x$ (trichotomy).
(4) If $x < y$, then $x + z < y + z$.
(5) If $x < y$ and $0 < z$, then $xz < yz$.

Properties (4) and (5) ensure the compatibility of the order relation $<$ with the field operations $+$ and $\cdot$. Familiar properties (such as $0 < 1$, $-1 < 0$, and $a < b$ implies $-b < -a$) can be derived from the order axioms. Therefore, these properties must be true in every ordered field. It can also be shown that in an ordered field, if $x < y$, and $z < 0$, then $xz > yz$. The notation $x \leqslant y$ means $x < y$ or $x = y$, while $x > y$ stands for $y < x$.

Both the rational numbers and the real numbers are ordered fields under their usual orderings. It can be shown that the complex numbers do not have an ordering that satisfies properties (1)–(5). Consider, for example, the situation if we assume $0 < i$. Then $0 = 0 \cdot i < i \cdot i = i^2 = -1$ and thus $0 < -1$. On the other hand, if $i < 0$, then $i^2 > 0 \cdot i$, so again $-1 > 0$. By trichotomy, either $0 < i$ or $i < 0$ must be true, but both lead us to the false statement $0 < -1$. Therefore the field of complex numbers cannot be an ordered field.

The finite fields also cannot be ordered. To show, for example, that $\mathbf{Z}_2$ is not ordered, suppose $0 < 1$. Then $1 = 0 + 1 < 1 + 1 = 0$, so $1 < 0$. Thus, by transitivity, $0 < 0$, contradicting property (1).

Developing a distinction between $\mathbf{R}$ and $\mathbf{Q}$ takes a bit more care. We first note that every ordered field has the property that between any two elements there is a third element. If $a < b$ in an ordered field, then using $\frac{1}{2}$ as the symbol for the inverse of the element $1 + 1$, we have $a < \frac{1}{2}(a + b) < b$. Thus ordered fields do not have "gaps" between any two given elements.

However, some ordered fields, like the rationals, seem to be missing some points. For instance, if we look at a sequence of rationals such as 1.4, 1.41, 1.414, 1.4142, 1.41421, 1.414213, . . ., formed by approximating $\sqrt{2}$, we see the sequence consists entirely of rationals, yet the value it approaches ($\sqrt{2}$) is not a rational number. In this respect the rationals appear to be incomplete. We shall see that this situation does not happen for sequences of real numbers; if there is a limiting value for a sequence of reals, it must be a real number.

We will now make precise the idea of being incomplete or having missing points.

Definition.   Let A be a subset of an ordered field F. We say $u \in F$ is an **upper bound** for A iff $a \leqslant u$ for all $a \in A$. If A has an upper bound, A is **bounded above**. Likewise, $\ell \in F$ is a **lower bound** for A iff $\ell \leqslant a$ for all $a \in A$, and A is **bounded below** iff any lower bound for A exists. The set A is **bounded** iff A is both bounded above and bounded below.

In $\mathbf{R}$, the half-open interval $[0, 3)$ has 3 as an upper bound. In fact, π, 18, and 206 are also upper bounds for $[0, 3)$. Both -0.5 and 0 are lower bounds. We note that a bound for a set might or might not be an element of the set. The set $\mathbf{N}$ is bounded below but not above.

In $\mathbf{Q}$, the set $A = \{x \in \mathbf{Q} : x < \sqrt{2}\}$ has many upper bounds: 8, 1.42, 1.4146, and so on. A has no lower bounds.

The best possible upper bound for a set A is called the supremum of A.

Definition.   Let A be a subset of an ordered field F. We say $s \in F$ is a **least upper bound** or **supremum of A in F** iff

     (1) s is an upper bound for A.
     (2) $s \leqslant t$ for all upper bounds t of A.

Likewise, $i \in F$ is the **greatest lower bound** or **infimum of A in F** iff

     (3) i is a lower bound for A.
     (4) $\ell \leqslant i$ for all lower bounds ℓ of A.

While several numbers may serve as an upper bound for a given set A, when the supremum of A exists it is unique (see exercise 4). We shall denote the supremum and infimum of A by $\sup(A)$ and $\inf(A)$, respectively.

> **Example.**   In the field **R**, $\sup([0, 3)) = 3$ and $\inf([0, 3)) = 0$. If $A = \{2^{-k}: k \in \mathbf{N}\}$, then $\sup(A) = \frac{1}{2}$ and $\inf(A) = 0$. For $B = \{x \in \mathbf{R}: x^2 < 2\}$, $\sup(B) = \sqrt{2}$ and $\inf(B) = -\sqrt{2}$. Also, $\inf(\mathbf{N}) = 1$ and $\sup(\mathbf{N})$ does not exist.

> **Example.**   In the field of rationals, if $A = \{2^{-k}: k \in \mathbf{N}\}$, then $\sup(A) = \frac{1}{2}$ and $\inf(A) = 0$. For $B = \{x \in \mathbf{Q}: x^2 < 2\}$, even though B is bounded in **Q**, B has no supremum or infimum in **Q**.

The following theorem provides a characterization of the supremum of a set.

Theorem 7.1   Let A be a subset of an ordered field F. Then $s = \sup(A)$ iff

(1)   for all $\epsilon > 0$, if $x \in A$, then $x < s + \epsilon$.
(2)   for all $\epsilon > 0$, there exists $y \in A$ such that $y > s - \epsilon$.

Proof.   First, suppose $s = \sup(A)$. Let $\epsilon > 0$ be given. Then $x \leqslant s < s + \epsilon$ for all $x \in A$, which establishes property (1).

     To verify (2), suppose there were no y such that $y > s - \epsilon$. Then $s - \epsilon$ is an upper bound for A less than the least upper bound of A, a contradiction.

     Suppose now that s is a number that satisfies conditions (1) and (2). To show that $s = \sup(A)$, we must first show that s is an upper bound for A. Suppose there is $y \in A$ such that $y > s$. If we let $\epsilon = (y - s)/2$, then $y > s + \epsilon$, which violates condition (1). Hence $y \leqslant s$ for all $y \in A$, and s is an upper bound.

   To show s is the least of all upper bounds, suppose that there is another upper bound t such that $t < s$. If we let $\epsilon = s - t$, then by condition (2), there is a number $y \in A$ such that $y > s - \epsilon$. Thus $y > s - \epsilon = s - (s - t) = t$. Hence $y > t$ and t is not an upper bound. Therefore s is indeed the least upper bound of A. ∎

We have seen that the field of rational numbers contains bounded subsets with no supremums in **Q**, and therefore **Q** is not complete in this sense. On the other hand every bounded subset of the reals does have a supremum; a proof of this requires considerable preliminary study of exactly what is a real number and will be omitted. The distinction we seek between **R** and **Q** is given in the next definition.

Definition.   An ordered field F is **complete** iff every nonempty subset of F that has an upper bound in F has a supremum that is in F.

The rationals are not complete. We state without proof that the reals are a complete ordered field. In fact, with much work we could show that if F and F' are any two complete ordered fields, then they are essentially the same in both number of elements and structure: the fields are isomorphic. This means there is a one-to-one correspondence from F to F' that preserves both operations, the order relations, and all supremums. Thus the real numbers form essentially the only complete ordered field, so you should think of the terms "real number system" and "complete ordered field" as synonymous.

Exercises 7.1

1. Let F be an ordered field and x, y, $z \in F$. Prove that
   (a) exactly *one* of $x < y$, $x = y$, or $y < x$ is true.
   (b) if $x < 0$, then $-x > 0$.
   (c) $0 < 1$.
   (d) $-1 < 0$.
   (e) if $x < y$, then $-y < -x$.
   (f) if $x < y$ and $z < 0$, then $xz > yz$.

2. Find the supremum and infimum in the real numbers, if they exist, of each of the following:
   ★ (a) $\left\{ \dfrac{1}{n} : n \in \mathbf{N} \right\}$ (b) $\left\{ \dfrac{n + 1}{n} : n \in \mathbf{N} \right\}$

* (c) $\{2^x : x \in \mathbf{Z}\}$

(d) $\left\{(-1)^n\left(1 + \dfrac{1}{n}\right) : n \in \mathbf{N}\right\}$

* (e) $\left\{\dfrac{n}{n + 2} : n \in \mathbf{N}\right\}$

(f) $\{x \in \mathbf{Q} : x^2 < 10\}$

* (g) $\{x : -1 \leqslant x \leqslant 1\} \cup \{5\}$

(h) $\{x : -1 \leqslant x \leqslant 1\} - \{0\}$

(i) $\left\{\dfrac{x}{2^y} : x, y \in \mathbf{N}\right\}$

(j) $\{x : |x| > 2\}$

3. Let x be an upper bound for $A \subseteq \mathbf{R}$. Prove that
    (a) if $x < y$, then y is an upper bound for A.
    (b) if $x \in A$, then $x = \sup(A)$.

4. Let $A \subseteq \mathbf{R}$.
* (a) Prove that $\sup(A)$ is unique. That is, prove that if x and y are both least upper bounds for A, then $x = y$.
    (b) Prove that $\inf(A)$ is unique.

5. Let $A \subseteq B \subseteq \mathbf{R}$.
    (a) Prove that if $\sup(A)$ and $\sup(B)$ both exist, then $\sup(A) \leqslant \sup(B)$.
    (b) Prove that if $\inf(A)$ and $\inf(B)$ both exist, then $\inf(A) \geqslant \inf(B)$.

6. Formulate and prove a characterization of greatest lower bounds similar to that in Theorem 7.1 for least upper bounds.

7. If possible, give an example of
    (a) a set $A \subseteq \mathbf{R}$ such that $\sup(A) = 4$ and $4 \notin A$.
    (b) a set $A \subseteq \mathbf{Q}$ such that $\sup(A) = 4$ and $4 \notin A$.
    (c) a set $A \subseteq \mathbf{N}$ such that $\sup(A) = 4$ and $4 \notin A$.
    (d) a set $A \subseteq \mathbf{N}$ such that $\sup(A) > 4$ and $4 \notin A$.

8. Give an example of a set of rational numbers that has a rational lower bound but no rational greatest lower bound.

9. Let $A \subseteq \mathbf{R}$. Prove that
* (a) if $\sup(A)$ exists, then $\sup(A) = \inf\{u : u$ is an upper bound of $A\}$.
    (b) if $\inf(A)$ exists, then $\inf(A) = \sup\{\ell : \ell$ is a lower bound of $A\}$.

10. Prove that if $B \subseteq A \subseteq \mathbf{R}$ and A is bounded, then B is bounded.

11. For $A \subseteq \mathbf{R}$, let $A^- = \{x : -x \in A\}$. Prove that if $\sup(A)$ exists, then $\inf(A^-)$ exists and $\inf(A^-) = -\sup(A)$.

12. Let A and B be subsets of $\mathbf{R}$.
* (a) Prove that if $\sup(A)$ and $\sup(B)$ exist, then $\sup(A \cup B)$ exists and $\sup(A \cup B) = \max\{\sup(A), \sup(B)\}$.
    (b) State and prove a similar result for $\inf(A \cup B)$.
    (c) Give an example where $A \cap B \neq \varnothing$, $\sup(A \cap B) < \sup(A)$, and $\sup(A \cap B) < \sup(B)$.
    (d) For $A \cap B \neq \varnothing$, state and prove a relationship between $\sup(A)$, $\sup(B)$, and $\sup(A \cap B)$.
    (e) Give an example where $A \cap B \neq \varnothing$, $\inf(A \cap B) > \inf(A)$, and $\inf(A \cap B) > \inf(B)$.
    (f) For $A \cap B \neq \varnothing$, state and prove a relationship between $\inf(A)$, $\inf(B)$, and $\inf(A \cap B)$.

13. **Proofs to Grade.**

★ (a)   **Claim.**   Let $A \subseteq \mathbf{R}$. If $i = \inf(A)$ and $\epsilon > 0$, then there is $y \in A$ such that
$y < i + \epsilon$.
        "Proof."   Let $y = i + \epsilon/2$. Then $i < y$ so $y \in A$. By construction of y,
$y < i + \epsilon$. ■

   (b)   **Claim.**   Let $(F, +, \cdot)$ be an ordered field. Then $0 < 1$.
        "Proof."   By property (4) of fields, $0 \ne 1$. Therefore by property (3)
of ordered fields, $0 < 1$ or $1 < 0$. If $1 < 0$, then $-1 > 0$. Thus
$(-1)(-1) > 0(-1)$. Therefore $1 > 0$, a contradiction. Thus $0 < 1$. ■

   (c)   **Claim.**   If $f: \mathbf{R} \to \mathbf{R}$ and A is a bounded subset of $\mathbf{R}$, then $f(A)$ is
bounded.
        "Proof."   Let m be an upper bound for A. Then $a \le m$ for all $a \in A$.
Therefore $f(a) \le f(m)$ for all $a \in A$. Thus $f(m)$ is an upper bound for
$f(A)$. ■

7.2

The Heine-Borel Theorem

In the next three sections we shall limit our discussion to the field of real
numbers. It is not necessary to do so, but this provides a familiar model of
an ordered field in which to work.

Definition.   For a real number a, if δ is a positive real, the **δ-neigh-
borhood** of a is the set $\mathcal{N}(a, \delta) = \{x \in \mathbf{R}: |x - a| < \delta\}$.

Because $|x - a| < \delta$ is equivalent to $a - \delta < x < a + \delta$, the δ-neigh-
borhood of a is the open interval $(a - \delta, a + \delta)$ centered at a with radius δ.
Thus, for example, $\mathcal{N}(3, 0.5) = (2.5, 3.5)$ and $\mathcal{N}(1, 0.01) = (0.99, 1.01)$.

Definition.   For a set $A \subseteq \mathbf{R}$, a point x is an **interior point of A** iff there
exists $\delta > 0$ such that $\mathcal{N}(x, \delta) \subseteq A$. The set A is **open** in $\mathbf{R}$ iff every point
of A is an interior point of A. The set A is **closed** in $\mathbf{R}$ iff its complement
$\tilde{A}$ is open in $\mathbf{R}$.

For the interval $[2, 5)$, 3 is an interior point since $\mathcal{N}(3, 0.5) \subseteq [2, 5)$.
Also, 4.98 is an interior point because $\mathcal{N}(4.98, 0.01) \subseteq [2, 5)$. See figure 7.1.
In fact, every point in $(2, 5)$ is an interior point of $[2, 5)$. The point 2 is not
interior to $[2, 5)$ since every δ-neighborhood of 2 contains points that are less

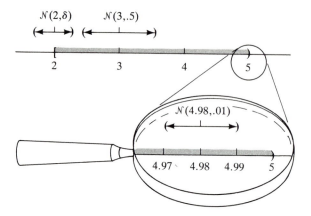

Figure 7.1

than 2 and hence not in [2, 5). Thus the set $\overline{[2, 5)}$ is not open. Should we conclude that [2, 5) is closed? Its complement $\overline{[2, 5)} = (-\infty, 2) \cup [5, \infty)$ contains 5 but not as an interior point. Thus $\overline{[2, 5)}$ is not open, so [2, 5) is not closed.

A set is open if about each point you can find a δ-neighborhood that lies entirely within the set. This means no point of the set can be on the "boundary" or outer edges of the set (see exercise 7). With experience you should come to recognize open subsets of **R**. The next two theorems will help.

Theorem 7.2   Every open interval is an open set.

Proof.   Let (a, b) be an open interval. To show (a, b) is open, we let $x \in (a, b)$ and show that x is an interior point of (a, b). ⟨*That is, show $\mathcal{N}(x, \delta) \subseteq (a, b)$ for some $\delta > 0$.*⟩ We choose $\delta = \min\{x - a, b - x\}$. ⟨*This minimum is the largest possible δ we can use. See figure 7.2.*⟩ Then $\delta > 0$. To show that $\mathcal{N}(x, \delta) \subseteq (a, b)$, let $y \in \mathcal{N}(x, \delta)$. Then $a \leq x - \delta < y < x + \delta \leq b$; and so $y \in (a, b)$. ∎

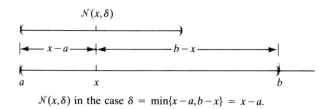

$\mathcal{N}(x, \delta)$ in the case $\delta = \min\{x - a, b - x\} = x - a$.

Figure 7.2

Theorem 7.3

(a) Both $\varnothing$ and **R** are open sets.

(b) If $\mathscr{A}$ is a collection of open sets, then $\underset{A \in \mathscr{A}}{\cup} A$ is open.

(c) If $\mathscr{A}$ is a finite collection of open sets, then $\underset{A \in \mathscr{A}}{\cap} A$ is open.

Proof.

(a) Since $\mathscr{N}(x, 1) \subseteq \mathbf{R}$ for all $x \in \mathbf{R}$, **R** is open. Also, $\varnothing$ is open, vacuously.

(b) Suppose $x \in \underset{A \in \mathscr{A}}{\cup} A$. Then there exists $B \in \mathscr{A}$ such that $x \in B$. Since B is in the collection $\mathscr{A}$, B is open and thus x is an interior point of B. Therefore there exists $\delta > 0$ such that $\mathscr{N}(x, \delta) \subseteq B$. Since $B \subseteq \underset{A \in \mathscr{A}}{\cup} A$, $\mathscr{N}(x, \delta) \subseteq \underset{A \in \mathscr{A}}{\cup} A$. Therefore x is an interior point of $\underset{A \in \mathscr{A}}{\cup} A$, which proves $\underset{A \in \mathscr{A}}{\cup} A$ is open.

(c) Suppose $x \in \underset{A \in \mathscr{A}}{\cap} A$. Then $x \in A$ for all $A \in \mathscr{A}$, and so for each open set $A \in \mathscr{A}$ there corresponds $\delta_A > 0$ such that $\mathscr{N}(x, \delta_A) \subseteq A$. By letting $\delta = \min\{\delta_A : A \in \mathscr{A}\}$, we have $\delta > 0$ and $\mathscr{N}(x, \delta) \subseteq A$ for all $A \in \mathscr{A}$. Thus $\mathscr{N}(x, \delta) \subseteq \underset{A \in \mathscr{A}}{\cap} A$, which proves $\underset{A \in \mathscr{A}}{\cap} A$ is open. ∎

In part (c) it is necessary to assume the collection $\mathscr{A}$ is finite to conclude that $\delta = \min\{\delta_A : A \in \mathscr{A}\}$ is positive. Indeed, the infinite collection $\mathscr{A} = \{(2 - \frac{1}{n}, 5) : n \in \mathbf{N}\}$ of open sets has intersection

$$\underset{n \in \mathbf{N}}{\cap} (2 - \tfrac{1}{n}, 5) = [2, 5),$$

which we have seen is not open. In this case for $A = (1, 5)$, we might choose $\delta_A = 0.75$, for $A = (1\frac{1}{2}, 5)$ we could choose $\delta_A = .41$, and so forth, but there is no minimum for the set $\{\delta_A : A \in \mathscr{A}\}$.

Theorems 7.2 and 7.3 can be combined to produce many examples of open subsets of **R**. For example,

$(5, 7) \cup (-3, 4) \cup (10, 20)$

$(2, \infty) = \underset{A \in \mathscr{A}}{\cup} A$, where $\mathscr{A} = \{(2, x) : x > 2\}$

$(-\infty, 2) = \underset{A \in \mathscr{A}}{\cup} A$, where $\mathscr{A} = \{(x, 2) : x < 2\}$

$(-5, 0) \cup (2, \infty)$

and

$$\mathbf{R} - \{2\} = (-\infty, 2) \cup (2, \infty)$$

are open sets.

In exercise 8 you are invited to state and prove a theorem similar to Theorem 7.3 for closed sets. Some examples of closed sets are **R** and $\emptyset$, any closed interval, and every finite subset of **R**. (See exercise 4.) The set **Z** is closed in **R** because its complement, $\mathbf{R} - \mathbf{Z} = \underset{a \in \mathbf{Z}}{\cup} (a, a + 1)$, is a union of open sets and hence is open.

The remainder of this section will concentrate on closed and bounded sets. Observe that the sets $(\pi, 4)$ and $(2, \infty)$ do not contain their suprema or infima. Each of the sets $[-1, 0.83]$ and $\{6, 10\}$ does contain its supremum and infimum, and this is true of all closed and bounded subsets of **R**.

Theorem 7.4   If A is a nonempty closed and bounded subset of **R**, then $\sup(A) \in A$ and $\inf(A) \in A$.

Proof.   Denote $\sup(A)$ by s and suppose $s \notin A$. Then $s \in \widetilde{A}$, which is open, since A is closed. Thus $\mathcal{N}(s, \delta) \subseteq \widetilde{A}$ for some positive δ. This implies $s - \delta$ is an upper bound for A, since the interval $(s - \delta, s + \delta)$ is a subset of $\widetilde{A}$. $\langle$*Notice that no element of A is greater than or equal to $s + \delta$, because s is an upper bound for A.*$\rangle$ This contradicts Theorem 7.1. Therefore $s \in A$. The proof that $\inf(A) \in A$ is similar.   ∎

Theorem 7.5   Let A be a closed set and $x \in \mathbf{R}$. If $A \cap \mathcal{N}(x, \delta) \neq \emptyset$ for all $\delta > 0$, then $x \in A$.

Proof.   Exercise 9.   ∎

Definition.   Let C be a set. A collection $\mathcal{A}$ of sets is a **cover** for C iff $C \subseteq \underset{A \in \mathcal{A}}{\cup} A$. A **subcover of** $\mathcal{A}$ for C is a subcollection $\mathcal{B}$ of $\mathcal{A}$ that is also a cover for C.

Example.   If we let $H_a = (-\infty, a)$ for each $a \in \mathbf{R}$, then $\mathcal{H} = \{H_a : a \in \mathbf{R}\}$ is a cover for the reals, since $\underset{a \in \mathbf{R}}{\cup} H_a = \mathbf{R}$.

The collection $\mathcal{W} = \{H_a : a \in \mathbf{Z}\}$ is a subcover of $\mathcal{H}$; and $\mathcal{L} = \{H_a : a \in \mathbf{N}\}$ is a subcover of $\mathcal{W}$ and of $\mathcal{H}$. We note in passing that there is no finite subset of $\mathcal{H}$ that covers **R**.

Example.   For any set C, $\{C\}$ is a cover. Also $\{\{c\} : c \in C\}$ is a cover with no proper subcover at all.

Example.   Let $A_n = (n - \frac{1}{n}, n + \frac{1}{n})$ for each $n \in \mathbf{N}$. The collection $\mathcal{A} = \{A_n : n \in \mathbf{N}\}$ is a cover for **N** that has no subcover other than itself (figure 7.3).

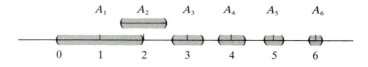

Figure 7.3

Example.   The collection $\mathscr{C} = \{(1/x, 1): x \in (1, \infty)\}$ is a cover of the interval $(0, 1)$ by open sets, and $\mathscr{D} = \{(1/x, 1): x \in \mathbf{N} - \{1\}\}$ is a subcover of $\mathscr{C}$. The cover $\mathscr{C}$ of $(0, 1)$ has no finite subcover. It also happens that $\mathscr{C}$ is a cover of the set $[\frac{1}{5}, 1)$. Among the many finite subcovers of $\mathscr{C}$ for $[\frac{1}{5}, 1)$ is $\mathscr{E} = \{(\frac{1}{2}, 1), (\frac{1}{11}, 1)\}$.

Definition.   A subset A of $\mathbf{R}$ is **compact** iff every cover of A by open sets has a finite subcover.

The concept of compactness may be difficult to imagine initially. You should think of set A being compact as meaning that any time $A \subseteq \underset{\alpha \in \Delta}{\cup} O_\alpha$, a union of open sets, then there are a finite number of $O_{\alpha_i}, i = 1, \ldots, n$ such that $A \subseteq \underset{i=1}{\overset{n}{\cup}} O_{\alpha_i}$. Thus any cover of A with open sets needs to use only finitely many of its sets to cover A.

Example.   We have seen that $\mathscr{H} = \{(-\infty, a): a \in \mathbf{R}\}$ is an open cover of $\mathbf{R}$ with no finite subcover. Thus $\mathbf{R}$ is not compact. Likewise, $(0, 1)$ is not compact, since $\mathscr{C} = \{(1/x, 1): x \in (1, \infty)\}$ has no finite subcover. What kind of sets are compact? Certainly, any finite set F is compact; if $F \subseteq \underset{\alpha \in \Delta}{\cup} O_\alpha$ where each O_α is open, simply select one covering set O_{α_x}, such that $x \in O_{\alpha_x}$ for each $x \in F$; then $\{O_{\alpha_x}: x \in F\}$ is a finite subcover.

We close this section with an elegant characterization of the compact subsets of $\mathbf{R}$. The next theorem can be traced to the works of Edward Heine (1821–1881) and Emile Borel (1871–1938). Watch how heavily the proof depends upon the completeness of $\mathbf{R}$.

Theorem 7.6 (*The Heine-Borel Theorem*)   A subset A of $\mathbf{R}$ is compact iff A is closed and bounded.

Proof.

(i)   Suppose A is compact. We first show that A is bounded. We note that $\mathbf{R} = \underset{n \in \mathbf{N}}{\cup} (-n, n)$; thus $A \subseteq \underset{n \in \mathbf{N}}{\cup} (-n, n)$.

Therefore $\mathcal{H} = \{(-n, n): n \in \mathbf{N}\}$ is an open cover of A. By compactness, $\mathcal{H}$ has a finite subcover $\mathcal{H}' = \{(-n, n): n \in \{n_1, n_2, \ldots, n_k\}\}$. If we choose $N = \max\{n_1, n_2, \ldots, n_k\}$, then $A \subseteq (-N, N)$. Therefore A is bounded.

We next show that A is closed by proving $\widetilde{A}$ is open. Suppose $y \in \widetilde{A}$. ⟨*We must show y is an interior point of $\widetilde{A}$*⟩ For each $x \in A$, $x \neq y$ and thus $\delta_x = \frac{1}{2}|x - y|$ is a positive number. The collection $\{\mathcal{N}(x, \delta_x): x \in A\}$ is a family of open sets that covers A. Hence by compactness,

$$A \subseteq \mathcal{N}(x_1, \delta_{x_1}) \cup \mathcal{N}(x_2, \delta_{x_2}) \cup \cdots \cup \mathcal{N}(x_k, \delta_{x_k}\}$$

for some $x_1, x_2, \ldots, x_k \in A$.
By choosing $\delta = \min\{\delta_{x_1}, \delta_{x_2}, \ldots, \delta_{x_k}\}$, we have $\mathcal{N}(y, \delta) \subseteq \widetilde{A}$. ⟨*If $z \in A$, then $|z - x_i| < \delta_{x_i}$ for some i. Thus if $z \in \mathcal{N}(y, \delta)$, then $|z - y| < \delta \leq \delta_{x_i}$ and $|x_i - y| \leq |x_i - z| + |z - y| < 2\delta_{x_i} = |x_i - y|$.*⟩ Thus $\widetilde{A}$ is open. Hence A is closed (figure 7.4).

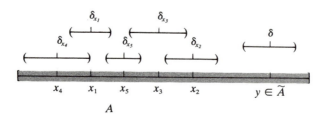

Figure 7.4

(ii)   Conversely, suppose A is a closed and bounded set and $\mathcal{C}$ is an open cover for A. ⟨*We show $\mathcal{C}$ has a finite subcover.*⟩ For each $x \in \mathbf{R}$, let $A_x = \{a \in A: a \leq x\}$. Also, let $D = \{x \in \mathbf{R}: A_x$ is included in a union of finitely many sets from $\mathcal{C}\}$.

Since A is bounded, $\inf(A)$ exists. Thus if $x < \inf(A)$, $A_x = \varnothing$ and it follows that $x \in D$. Therefore $(-\infty, \inf(A)) \subseteq D$ and so D is nonempty.

We claim D has no upper bound. If D is bounded above, then $x_0 = \sup(D)$ exists ⟨*by the completeness of $\mathbf{R}$*⟩. Let $\delta > 0$ and choose $t \in D$ such that $x_0 - \delta < t \leq x_0$ ⟨*applying Theorem 7.1*⟩. If $A \cap \mathcal{N}(x_0, \delta) = \varnothing$, then $A_t = \{a \in A: a \leq t\} = \{a \in A: a \leq x_0 + \frac{\delta}{2}\} = A_{x_0 + \frac{\delta}{2}}$. Since

$t \in D$, $x_0 + \frac{8}{2} \in D$, which is a contradiction to $x_0 = \sup(D)$. But t is in D, so $x_0 + \frac{8}{2}$ is in D. This is a contradiction to $x_0 = \sup(D)$. Therefore for all $\delta > 0$, we have $A \cap \mathcal{N}(x_0, \delta) \neq \emptyset$. Since A is closed, $x_0 \in A$. Let C^* be an element of $\mathscr{C}$ such that $x_0 \in C^*$. Since C^* is open, there exists $\epsilon > 0$ such that $\mathcal{N}(x_0, \epsilon) \subseteq C^*$. Choose $x_1 \in D$ such that $x_0 - \epsilon < x_1 \leq x_0$. Since $x_1 \in D$, there are open sets $C_1, C_2, \ldots, C_n$ in $\mathscr{C}$ such that $A_{x_1} \subseteq C_1 \cup C_2 \cup \cdots \cup C_n$. Now let $x_2 = x_0 + \frac{\epsilon}{2}$. Then $x_2 \in C^*$ and $A_{x_2} \subseteq C_1 \cup C_2 \cup \cdots \cup C_n \cup C^*$. Thus $x_2 \in D$, a contradiction, since $x_2 > x_0$ and $x_0 = \sup(D)$. We conclude that D has no upper bound.

Finally, since D has no upper bound, choose $x \in D$ such that $x > \sup(A)$ $\langle \sup(A)$ *exists because* **R** *is complete*$\rangle$. Thus $A_x = A$ and since $x \in D$, A is included in a union of finitely many sets from $\mathscr{C}$. Therefore A is compact.   ∎

Exercises 7.2

1. Classify each of the following subsets of **R** as open, closed, both open and closed, or neither open nor closed.
   (a) $(-\infty, -3)$ ★ (b) $\mathcal{N}(a, \delta) - \{a\}$, $a \in$ **R**
   (c) $(5, 8) \cup \{9\}$ (d) **Q**
   ★ (e) **R** − **N**   (f) $\{x : |x - 5| = 7\}$
   (g) $\{x : |x - 5| > 7\}$ (h) $\{x : |x - 5| \neq 7\}$
   ★ (i) $|x : |x - 5| \leq 7\}$ (j) $\{x : 0 < |x - 5| \leq 7\}$

☆ 2. Prove that if A and B are closed subsets of **R** then $A \cup B$ is closed.

3. Give an example of a family of sets $\mathscr{A}$ such that each $A \in \mathscr{A}$ is closed but $\underset{A \in \mathscr{A}}{\cup} A$ is not closed.

★ 4. Prove that any finite set $A \subseteq$ **R** is closed.

5. Prove that if A is open and B is closed, then $A - B$ is open.

6. Let A be a subset of **R**. Prove that the set of all interior points of A is an open set.

7. A point x is a **boundary point** of the set A iff for all $\delta > 0$, $\mathcal{N}(x, \delta) \cup A \neq \emptyset$ and $\mathcal{N}(x, \delta) \cap \tilde{A} \neq \emptyset$.
   (a) Find all boundary points of $(2, 5]$, $(0, 1)$, $[3, 5] \cup \{6\}$, **Q**.
   (b) Prove that x is a boundary point of A iff x is not an interior point of A and not an interior point of $\tilde{A}$.
   (c) Prove that A is open iff A contains none of its boundary points.
   (d) Prove that A is closed iff A contains all of its boundary points.

8. State and prove a theorem similar to Theorem 7.3 for closed sets.

9. Prove Theorem 7.5.

10. Prove that if A and B are both compact subsets of $\mathbf{R}$, then
    ☆ (a) $A \cup B$ is compact.   (b) $A \cap B$ is compact.

11. Give an example of an open bounded subset of $\mathbf{R}$ and an open cover of that set that has no finite subcover.

12. Give an example of a closed subset of $\mathbf{R}$ and an open cover of that set that has no finite subcover.

13. Which of the following subsets of $\mathbf{R}$ are compact?
    ★ (a) $\mathbf{Z}$ (b) $[0, 10] \cup [20, 30]$
       (c) $[\pi, \sqrt{10}]$ ★ (d) $\mathbf{R} - A$, where A is any finite set

       (e) $\{1, 2, 3, 4, 9, 12, 18\}$ (f) $\{0\} \cup \left\{\dfrac{1}{n} : n \in \mathbf{N}\right\}$

    ★ (g) $(-3, 5]$ (h) $[0, 1] \cap \mathbf{Q}$

☆ 14. Give two different proofs that any finite set is compact.

15. Let $S = (0, 1]$ and let

$$\mathscr{C} = \left\{\left(\frac{n + 2}{2^n}, 2^{1/n}\right) : n \in \mathbf{N}\right\}.$$

   (a) Prove that $\mathscr{C}$ is an open cover of S.
   (b) Is there a finite subcover of $\mathscr{C}$ that covers S?
   (c) What does the Heine-Borel Theorem say about S?

16. Use the Heine-Borel Theorem to prove that if $\{A_\alpha : \alpha \in \Delta\}$ is a collection of compact sets, then $\bigcap_{\alpha \in \Delta} A_\alpha$ is compact.

17. Give an example of a collection $\{A_\alpha : \alpha \in \Delta\}$ of compact sets such that $\bigcup_{\alpha \in \Delta} A_\alpha$ is not compact.

18. **Proofs to Grade.**
    (a) **Claim.** If A and B are compact then $A \cup B$ is compact.
    "Proof." If A and B are compact. Then for any open cover $\{O_\alpha : \alpha \in \Delta\}$ of A, there exists a finite subcover $O_{\alpha_1}, O_{\alpha_2}, \ldots, O_{\alpha_n}$, and for any open cover $\{U_\beta : \beta \in \Gamma\}$ of B, there exists a finite subcover $U_{\beta_1}, U_{\beta_2}, \ldots, U_{\beta_m}$. Thus $A \subseteq O_{\alpha_1} \cup O_{\alpha_2} \cup \cdots \cup O_{\alpha_n}$ and $B \subseteq U_{\beta_1} \cup U_{\beta_2} \cup \cdots \cup U_{\beta_m}$. Therefore $A \cup B \subseteq O_{\alpha_1} \cup O_{\alpha_2} \cup \cdots \cup O_{\alpha_n} \cup U_{\beta_1} \cup U_{\beta_2} \cup \cdots \cup U_{\beta_m}$, a union of a finite number of open sets. Thus $A \cup B$ is compact. ∎
    ★ (b) **Claim.** If A is compact, $B \subseteq A$, and B is closed, then B is compact.
    "Proof." Let $\{O_\alpha : \alpha \in \Delta\}$ be an open cover of B. If $\{O_\alpha : \alpha \in \Delta\}$ is an open cover of A, then there is a finite subcover of $\{O_\alpha : \alpha \in \Delta\}$ that covers A and hence covers B. If $\{O_\alpha : \alpha \in \Delta\}$ is not an open cover of A, add one more open set $O^* = \mathbf{R} - B$ to the collection to obtain an open cover of A. This open cover of A has a finite subcover that is a cover of B. In either case B is covered by a finite number of open sets. Therefore B is compact. ∎
    (c) **Claim.** If A is compact, $B \subseteq A$, and B is closed, then B is compact.
    "Proof." B is closed by assumption. Since A is compact, A is bounded. Since $B \subseteq A$, B is also bounded. Thus B is closed and bounded. Therefore B is compact. ∎

7.3

The Bolzano-Weierstrass Theorem

In the previous section we used the completeness of **R** to prove the Heine-Borel Theorem. In this section we use the Heine-Borel Theorem to prove another classical result of modern analysis, the Bolzano-Weierstrass Theorem.

Bernard Bolzano (1781–1848) was a leader in developing highly rigorous standards of mathematics. Karl Weierstrass (1815–1897) has been called the father of modern analysis because of his many deep contributions and his absolute insistence upon rigor. The theorem that bears their names is easy to visualize if you think of the terms of a sequence "piling up" around some number. Consider the set $A = \{(-1)^{n+1}/n: n \in \mathbf{N}\}$, which is bounded above by 1 and below by $-\frac{1}{2}$. Since $A = \{1, -\frac{1}{2}, \frac{1}{3}, -\frac{1}{4}, \frac{1}{5}, \ldots\}$ is infinite, it would seem that its points must pile up or "accumulate" around some number or numbers between $-\frac{1}{2}$ and 1. The number 0 is a point about which points of A accumulate (figure 7.5). The fact that such a number must exist in **R** is a consequence of the Bolzano-Weierstrass Theorem. We first define what it means for points to "accumulate."

$$-\frac{1}{2} \qquad -\frac{1}{4} \quad -\frac{1}{6} \quad \frac{1}{8} \qquad 0 \qquad \frac{1}{7}\,\frac{1}{5} \quad \frac{1}{3} \qquad\qquad\qquad 1$$

Figure 7.5

Definition.   The number x is an **accumulation point for the set A** iff for all $\delta > 0$, $\mathcal{N}(x, \delta)$ contains a point of A distinct from x.

The idea is that when x is an accumulation point for A, there are so many elements of A near x that every neighborhood of x, no matter how small, must contains points of A.

An accumulation point might or might not belong to the set in question. The number 0 is the only accumulation point of $\{(-1)^{n+1}/n: n \in \mathbf{N}\}$, although 0 is not in the set. On the other hand, the closed interval [0, 1] has every one of its points as an accumulation point and no others. The set $\{(-1)^n(n + 1)/n: n \in \mathbf{N}\}$ has exactly two accumulation points, -1 and 1.

It is important to remember that every δ-neighborhood of an accumulation point x of a set A contains points of A *other than* x. This means that no finite set A can have an accumulation point since $\mathcal{N}(x, \delta) \cap A = \{x\}$ for all $x \in A$ if we choose $\delta = \min\{|x - y|: x, y \in A, x \neq y\}$.

The set of accumulation points of A is called the **derived set of A,** denoted A'. As examples

$$(0, 1)' = [0, 1],$$
$$[4, 6]' = [4, 6],$$

and

$$\mathbf{N}' = \varnothing.$$

If $B = \{1.4, 1.41, 1.414, 1.4142, 1.41421, \ldots\}$ is the set of successive decimal approximations to $\sqrt{2}$, then $B' = \{\sqrt{2}\}$. The following theorem relates derived sets and closed sets.

Theorem 7.7   A set A is closed iff $A' \subseteq A$.

Proof. Suppose A is closed and $x \in A'$. We prove $x \in A$ by contradiction. If $x \notin A$, then $x \in \widetilde{A}$, an open set. Thus $\mathcal{N}(x, \delta) \subseteq \widetilde{A}$ for some positive δ. But then $\mathcal{N}(x, \delta)$ can contain no points of A except perhaps for x. Thus $x \notin A'$, a contradiction. We conclude $x \in A$.

Suppose now $A' \subseteq A$. To show that A is closed, we show $\widetilde{A}$ is open by contradiction. If $\widetilde{A}$ is not open, there is at least one $x \in \widetilde{A}$ that is not an interior point of $\widetilde{A}$. Therefore no δ-neighborhood of x is a subset of $\widetilde{A}$; that is, each δ-neighborhood of x contains a point of A. This point must be different from x, since $x \in \widetilde{A}$. Thus $x \in A'$. But $A' \subseteq A$, so $x \in A$, a contradiction. ∎

At the beginning of this section we suggested that the set $A = \{(-1)^{n+1}/n : n \in \mathbf{N}\}$, and by inference all other bounded infinite sets, must have at least one accumulation point in $\mathbf{R}$. We are now in a position to prove this is so, using the Heine-Borel Theorem.

Theorem 7.8   *(The Bolzano-Weierstrass Theorem)*
Every bounded infinite subset of real numbers has an accumulation point in $\mathbf{R}$.

Proof.   Suppose the set A is bounded and infinite but has no accumulation points. Then A is closed by Theorem 7.7. $\langle A' = \varnothing$, hence $A' \subseteq A.\rangle$ Thus by the Heine-Borel Theorem, A is compact.

Since A has no accumulation points for each $x \in A$, there exists $\delta_x > 0$ such that $\mathcal{N}(x, \delta_x) \cap A = \{x\}$. Thus for $x \neq y$ in A, $\mathcal{N}(x, \delta_x) \neq \mathcal{N}(y, \delta_y)$. But then the family $\mathcal{A} = \{\mathcal{N}(x, \delta_x) : x \in A\}$ is an infinite collection of open sets that covers A with no subcover other than itself; hence it has no finite subcover. This contradicts the fact that A is compact. Therefore A must have an accumulation point. ∎

Exercises 7.3

1. Find the derived set of each of the following sets.

   ★ (a) $\left\{\dfrac{n + 1}{2n} : n \in \mathbf{N}\right\}$ (b) $\{2^n : n \in \mathbf{N}\}$

   ★ (c) $\{6n : n \in \mathbf{N}\}$ (d) $\{m/2^n : n, m \in \mathbf{N}\}$

      (e) $(0, 1]$ (f) $(3, 7) \cup \{4, 6, 8\}$

   ★ (g) $\left\{1 + \dfrac{(-1)^n n}{n + 1} : n \in \mathbf{N}\right\}$ (h) $\mathbf{Z}$

      (i) $\mathbf{Q}$ (j) $\left\{\dfrac{1 + n^2(1 + (-1)^n)}{n} : n \in \mathbf{N}\right\}$

2. Find an example of an infinite subset of $\mathbf{R}$ that has
   (a) no accumulation points.
   (b) exactly one accumulation point.
   (c) exactly two accumulation points.
   (d) denumerably many accumulation points.
   (e) an uncountable number of accumulation points.

☆ 3. Prove that if $A \subseteq \mathbf{R}$, $z = \sup(A)$, and $z \notin A$, then z is an accumulation point of A.

4. (a) Prove that if $A \subseteq B \subseteq \mathbf{R}$, then $A' \subseteq B'$.
   (b) Is the converse of part (a) true? Explain.

5. Prove that $(A \cup B)' = A' \cup B'$. (The operation of finding the derived set preserves unions.)

6. Prove that $(A \cap B)' \subseteq A' \cap B'$. Show by example that equality need not hold.

7. Prove that if B is closed and $A \subseteq B$, then $A' \subseteq B$.

8. Define the **closure** of a set A to be $c(A) = A \cup A'$.
   (a) Prove that $c(A)$ is a closed set.
   (b) Prove that $A \subseteq c(A)$.
   ★ (c) Prove that $c(A)$ is the smallest closed set containing A. That is, if B is closed and $A \subseteq B$, show that $c(A) \subseteq B$.
   (d) Prove that the derived set of the closure of A is the same as the derived set of A.

9. Which of the following have at least one accumulation point?
   ★ (a) An infinite subset of $\mathbf{N}$ (b) An infinite subset of $(-10, 10)$
   ★ (c) An infinite subset of $[0, 100]$ (d) An infinite subset of $\mathbf{Q}$

10. Prove that if a bounded subset of the real numbers has no accumulation points, the set is finite.

11. Let x be an accumulation point of the set A and let F be a finite set. Prove that x is an accumulation point of $A - F$.

★ 12. Let $S = (0, 1]$. Find $S' \cap (\widetilde{S})'$.

13. Prove that if $S \subseteq \mathbf{R}$ is open, then every point of S is an accumulation point of S.

14. **Proofs to Grade.**
    ☆ (a) **Claim.** For $A, B \subseteq \mathbf{R}, (A \cup B)' = A' \cup B'$.
       Proof.''
       (i) Since $A \subseteq A \cup B$, $A' \subseteq (A \cup B)'$ by exercise 4(a). Likewise $B' \subseteq (A \cup B)'$. Therefore $A' \cup B' \subseteq (A \cup B)'$.
       (ii) To show that $(A \cup B)' \subseteq A' \cup B'$, let $x \in (A \cup B)'$. Then for all $\delta > 0$, $\mathcal{N}(x, \delta)$ contains a point of $A \cup B$ distinct from x. Restating this, we have, for all $\delta > 0$, that $\mathcal{N}(x, \delta)$ contains a point of A distinct from x or a point of B distinct from x. Thus for all $\delta > 0$, $\mathcal{N}(x, \delta)$ contains a point of A distinct from x, or, for all $\delta > 0$, $\mathcal{N}(x, \delta)$ contains a point of B distinct from x. But this means $x \in A'$ or $x \in B'$. Therefore $x \in A' \cup B'$. ∎

    (b) **Claim.** If A is a set with an accumulation point, $B \subseteq A$, and B is infinite, then B has an accumulation point.
       "Proof." First, A is infinite because $B \subseteq A$ and B is infinite. Since A has an accumulation point, by the Bolzano-Weierstrass Theorem, A must be bounded. Since $B \subseteq A$, this means B is bounded. Hence by the Bolzano-Weierstrass Theorem again, B has an accumulation point. ∎

    ★ (c) **Claim.** For $A, B \subseteq \mathbf{R}, (A - B)' \subseteq A' - B'$.
       "Proof." $(A - B)' = (A \cap \widetilde{B})'$ ⟨*definition of* $A - B$⟩
       $\subseteq A' \cap (\widetilde{B})'$ ⟨*exercise 6*⟩
       $\subseteq A' \cap (\widetilde{B'})$ ⟨*since* $(\widetilde{B})' \subseteq (\widetilde{B'})$⟩
       $= A' - B'$ ∎

7.4

The Bounded Monotone Sequence Theorem

In this section we shall use the Bolzano-Weierstrass Theorem to prove an important result about sequences. The concept of a sequence should be familiar to you from calculus.

A **sequence** is a function x whose domain is $\mathbf{N}$. A sequence of real numbers, or **real sequence,** has codomain $\mathbf{R}$. Every sequence we consider will be a real sequence. For $n \in \mathbf{N}$, the image of n under the sequence x is written as x_n rather than the customary functional notation $x(n)$. We call x_n the **nth term of the sequence.**

Suppose x is the sequence given by $x_n = 2^n$. The first few terms are $x_1 = 2, x_2 = 4, x_3 = 8$, and so on. The sequence y, where $y_n = (-1)^n$, has the alternating terms -1 and 1.

A sequence x is **bounded** iff its range $\{x_n : n \in \mathbf{N}\}$ is a bounded subset of $\mathbf{R}$. Boundedness may be described in terms of absolute value.

Theorem 7.9 A sequence x of real numbers is bounded iff there exists a number B such that $|x_n| \leq B$ for all $n \in \mathbf{N}$.

Proof.   Exercise 7.   ∎

The sequence given by $x_n = 2^n$ is bounded below but not bounded, whereas the sequence y defined by $y_n = (-1)^n$ is bounded, because it is bounded above by 1 and below by -1. The sequence whose nth term is $z_n = e^{-n}$ is bounded above by e^{-1} and below by 0. Thus we may select any $B \geq e^{-1}$ to satisfy Theorem 7.9.

The bounded sequence x given by $x_n = (n+1)/n$ has for its first few terms, $2, \frac{3}{2}, \frac{4}{3}, \frac{5}{4}, \ldots$. Furthermore, for large values of n, the values of x_n are close to the number 1. This observation gives rise to the concept of the limit of x.

Definition.   A sequence x **has the limit L,** or x **converges to L,** iff, for every real $\epsilon > 0$, there exists a natural number N such that if $n > N$, then $|x_n - L| < \epsilon$. If x converges to L, we write $\lim\limits_{n \to \infty} x_n = L$ or $x_n \to L$. If no such L exists, the sequence **diverges.**

To prove that a given sequence converges involves first deciding the value of the limit L and then verifying the definition of convergence. For example, suppose $x_n = 3n^2/(n^2 + 1)$. After calculating x_n for several large values of n, we make an educated guess that x converges to 3. Next we must show that the terms approach 3 by showing that, for every $\epsilon > 0$, there is a natural number N such that $n > N$ implies $|(3n^2/(n^2 + 1)) - 3| < \epsilon$. Since $|(3n^2/(n^2 + 1)) - 3| = |(-3/(n^2 + 1))| = 3/(n^2 + 1)$, we require an integer N such that $n > N$ implies $3/(n^2 + 1) < \epsilon$ or, equivalently, $n^2 + 1 < 3/\epsilon$. By selecting N to be any natural number greater than $3/\epsilon$, we have that $n > N$ implies (since $n^2 + 1 > n$) $n^2 + 1 > N > 3/\epsilon$. The above is scratchwork for the formal proof that follows.

Example.   Prove that the sequence x given by $x_n = 3n^2/(n^2 + 1)$ converges.

Proof.   We will show $\lim\limits_{n \to \infty} 3n^2/(n^2 + 1) = 3$. Let $\epsilon > 0$. Let N be a natural number greater than $3/\epsilon$. Then since $n^2 + 1 > n$ for all n, if $n > N$ then $n > 3/\epsilon$, and thus $n^2 + 1 > 3/\epsilon$. Therefore if $n > N$, then $3/(n^2 + 1) < \epsilon$. Thus $|(3n^2/(n^2 + 1)) - 3| < \epsilon$. Therefore the definition is satisfied and $3n^2/(n^2 + 1) \to 3$.   ∎

Intuitively, a sequence x converges if the terms x_n get closer to a single number L for larger values of n. For this reason the sequence $y_n = (-1)^n$ must diverge. A proof of this must show the negation of the definition of a limit: If $\lim\limits_{n \to \infty} y_n \neq L$, then there exists a real $\epsilon > 0$ such that for all $N \in \mathbf{N}$ there exists $n > N$ such that $|y_n - L| \geq \epsilon$.

Example.   Prove that the sequence given by $y_n = (-1)^n$ has no limit.

Proof.   Suppose $\lim_{n \to \infty} y_n = L$ for some number L. Let $\epsilon = 1$. We must show that for all $N \in \mathbf{N}$, there exists $n > N$ such that $|y_n - L| \geq 1$. Let $N \in \mathbf{N}$. If $L > 0$, let n be any odd integer greater than N. Then $y_n = -1$. For $L \leq 0$, let n be any even integer greater than N; here $y_n = 1$. In each of these cases n was selected so that $n > N$ and $|y_n - L| \geq 1$. Therefore $\lim_{n \to \infty} y_n \neq L$. ∎

Theorem 7.10   If a sequence x converges, then its limit is unique.

Proof.   Suppose $x_n \to L$ and $x_n \to M$ and $L \neq M$. Let $\epsilon = \frac{1}{3}|L - M|$. ⟨*The idea of the proof is to suppose there are two different limits and select ϵ so small that the terms cannot simultaneously be within ϵ of each limit.*⟩

Since $x_n \to L$ and $x_n \to M$, there are natural numbers N_1 and N_2 such that $n > N_1$ implies $|x_n - L| < \epsilon$ and $n > N_2$ implies $|x_n - M| < \epsilon$. Let $N = N_1 + N_2$. Then since N is greater than both N_1 and N_2,

$$
\begin{aligned}
|L - M| &= |(L - x_n) + (x_n - M)| \\
&\leq |L - x_n| + |x_n - M| \\
&= |x_n - L| + |x_n - M| \\
&< \epsilon + \epsilon \\
&= 2\epsilon \\
&= \tfrac{2}{3}|L - M|.
\end{aligned}
$$

Thus the assumption $L \neq M$ leads to $|L - M| < \frac{2}{3}|L - M|$, which is a contradiction since $|L - M| > 0$. We conclude the limit of x is unique. ∎

The next theorem is useful for determining limits without resorting to the definition. Sometimes dubbed the "sandwich" or "squeeze" theorem, it states that if a sequence b is "sandwiched" between two sequences a and c, both of which converge to L, then b must also converge to L.

Theorem 7.11   Suppose a, b, and c are real sequences such that $a_n \leq b_n \leq c_n$ for all $n \in \mathbf{N}$. If $a_n \to L$ and $c_n \to L$, then $b_n \to L$.

Proof.   Suppose $a_n \to L$ and $c_n \to L$. Let $\epsilon > 0$. There are natural numbers N_1 and N_2 such that $n > N_1$ implies $|a_n - L| < \epsilon$ and

$n > N_2$ implies $|c_n - L| < \epsilon$. Let N be the larger of N_1, N_2. Since $a_n \leq b_n \leq c_n$, $a_n - L \leq b_n - L \leq c_n - L$. Therefore for $n > N$, $-\epsilon < a_n - L \leq b_n - L \leq c_n - L < \epsilon$. Thus $|b_n - L| < \epsilon$ for all $n > N$, so $b_n \to L$. ∎

Example.   To illustrate Theorem 7.11, let's look at the sequence whose nth term is $x_n = \dfrac{\sin n}{n}$. Since sine is a function with range $[-1, 1]$, $-\dfrac{1}{n} \leq \dfrac{\sin n}{n} \leq \dfrac{1}{n}$, for all $n \in \mathbf{N}$. Because both $-\dfrac{1}{n} \to 0$ and $\dfrac{1}{n} \to 0$, we conclude that $\dfrac{\sin n}{n} \to 0$.

The sequence with nth term $x_n = 2^n$ diverges because the terms increase without bound and cannot stay "close" to any one real number. In general, a sequence whose terms form an unbounded set must diverge. In other words, a convergent sequence must be bounded because all but the first few terms will be close to the limit. In this setting, closeness is determined by the choice of ϵ, and the first few terms are the first to the Nth, where N depends on ϵ.

Theorem 7.12   Every sequence that converges is bounded.

Proof.   Suppose x is a sequence convergent to a number L. For $\epsilon = 1$, there is a natural number N such that if $n > N$, then $|x_n - L| < 1$. Since $||x_n| - |L|| \leq |x_n - L|$, we have for all $n > N$, $|x_n| - |L| < 1$. Thus for all $n > N$, $|x_n| < |L| + 1$. ⟨*All but the first N terms are bounded by $|L| + 1$. We now take care of the first terms as well.*⟩ Let $B = \max\{|x_1|, |x_2|, \ldots, |x_N|, |L| + 1\}$. Then $|x_n| \leq B$ for all $n \in \mathbf{N}$, and x is bounded. ∎

Example.   Let x be the sequence

$$x_n = \begin{cases} (-2)^n & 1 \leq n \leq 1000 \\ \dfrac{15n}{n + 1} & n > 1000. \end{cases}$$

This sequence converges to 15 and therefore must be bounded. In this case the first "few" (1000) terms hop around a bit before the terms settle in close to 15. A bound for the sequence is given by $|x_n| \leq 2^{1000}$ for all $n \in \mathbf{N}$.

Definition.   A sequence x is **increasing** iff for all n, $m \in \mathbf{N}$, $x_n \leq x_m$ whenever $n < m$. A **decreasing** sequence y requires $y_n \geq y_m$ for $n < m$. A **monotone** sequence is one that is either increasing or decreasing.

A constant sequence t, where $t_n = c$ for some fixed $c \in \mathbf{R}$ and for all n, is monotone since it is both increasing and decreasing. The sequence given by $x_n = 2^n$ is increasing while $z_n = e^{-n}$ is decreasing. The sequence y defined by $y_n = (-1)^n$ is not monotone.

The next theorem relates all the concepts of this section: a sequence that is both bounded and monotone must converge. Its proof makes use of the Bolzano-Weierstrass Theorem.

Theorem 7.13   (*The Bounded Monotone Sequence Theorem*)   For every bounded monotone sequence x, there is a real number L such that $x_n \to L$.

Proof.   Assume x is a bounded and increasing sequence. The proof in the case where x is decreasing is similar.

If $\{x_n : n \in \mathbf{N}\}$ is finite, then let $L = \max\{x_n : n \in \mathbf{N}\}$. For some $N \in \mathbf{N}$, $x_N = L$ and, since x is increasing, $x_n = L$ for all $n > N$. Therefore $x_n \to L$.

Suppose $\{x_n : n \in \mathbf{N}\}$ is infinite. Then by the Bolzano-Weierstrass Theorem, the bounded infinite set $\{x_n : n \in \mathbf{N}\}$ must have at least one accumulation point. Let L be an accumulation point. We claim $x_n \leq L$ for all $n \in \mathbf{N}$. If there exists N such that $x_N > L$, then $x_n > L$ for all $n \geq N$. By exercise 11 of section 7.3, L is an accumulation point of $\{x_n : n \geq N\}$. But for $\delta = |x_N - L|$, $\mathcal{N}(L, \delta)$ contains no points of $\{x_n : n \geq N\}$. This is a contradiction. Thus $x_n \leq L$ for all n. We claim that the sequence x converges to L.

Let $\epsilon > 0$. Since L is an accumulation point of $\{x_n : n \in \mathbf{N}\}$, there exists $M \in \mathbf{N}$ such that $x_M \in \mathcal{N}(L, \epsilon)$. Thus $L - \epsilon < x_M$, and so, for $n > M$,

$$L - \epsilon < x_M \leq x_n \leq L < L + \epsilon$$

Therefore for $n > M$, $|x_n - L| < \epsilon$. Thus $x_n \to L$.   ∎

Example.   The Bounded Monotone Sequence Theorem can be used to ensure the existence of several important real numbers. For instance, let x be the sequence whose nth term is $x_n = (1 + \frac{1}{n})^n$. By the Binomial Theorem,

$$\left(1 + \frac{1}{n}\right)^n = 1 + \frac{n}{1!}\frac{1}{n} + \frac{n(n-1)}{2!}\frac{1}{n^2}$$

$$+ \frac{n(n - 1)(n - 2)}{3!} \frac{1}{n^3} + \cdots + \frac{1}{n^n}$$

$$= 1 + 1 + \frac{1}{2!} \frac{n(n - 1)}{n^2} + \frac{1}{3!} \frac{n(n - 1)(n - 2)}{n^3}$$

$$+ \cdots + \frac{1}{n!} \frac{n!}{n^n}.$$

$$\leq 1 + 1 + \frac{1}{2!} + \frac{1}{3!} + \cdots + \frac{1}{n!}$$

$$\leq 1 + 1 + \frac{1}{2} + \frac{1}{4} + \cdots + \frac{1}{2^{n-1}}$$

$$= 1 + \frac{2^n - 1}{2^{n-1}}$$

$$< 1 + 2$$

$$= 3.$$

Thus the sequence $x_n = (1 + \frac{1}{n})^n$ is bounded by 3.

We next show that x is an increasing sequence. We again use the Binomial Theorem to compare x_n and x_{n+1}.

$$x_n = \left(1 + \frac{1}{n}\right)^n = 1 + 1 + \frac{1}{2!} \frac{n(n - 1)}{n^2}$$

$$+ \frac{1}{3!} \frac{n(n - 1)(n - 2)}{n^3} + \cdots + \frac{1}{n!} \frac{n!}{n^n}$$

and

$$x_{n+1} = \left(1 + \frac{1}{n + 1}\right)^{n+1}$$

$$= 1 + 1 + \frac{1}{2!} \frac{(n + 1)n}{(n + 1)^2}$$

$$+ \frac{1}{3!} \frac{(n + 1)(n)(n - 1)}{(n + 1)^3} + \cdots$$

$$+ \frac{1}{n!} \frac{(n + 1)(n)(n - 1) \cdots 3 \cdot 2}{(n + 1)^n}$$

$$+ \frac{1}{(n + 1)!} \frac{(n + 1)!}{(n + 1)^{n+1}}.$$

We leave it as exercise 10 to show that each term in the expansion of x_n is less than or equal to the corresponding term in the expansion of x_{n+1}. Additionally, x_{n+1} has one more term, $\frac{1}{(n + 1)!} \frac{(n + 1)!}{(n + 1)^{n+1}}$, in its binomial expansion. Thus $x_n \leq x_{n+1}$ for all $n \in \mathbf{N}$.

Because $x_n = (1 + \frac{1}{n})^n$ is bounded and increasing, it must converge, by Theorem 7.13.

Definition.   We define the real number *e* as $e = \lim\limits_{n \to \infty} (1 + \frac{1}{n})^n$.

The constant *e* should be familiar to you from your study of exponential functions and logarithms.

Exercises 7.4

1. For each sequence *x*, prove that *x* converges or diverges.

   ☆ (a) $x_n = \dfrac{(-1)^n}{n}$ (b) $x_n = \dfrac{n+1}{n}$

   ☆ (c) $x_n = n^2$ (d) $x_n = \dfrac{(-1)^n n}{2n+1}$

   (e) $x_n = \dfrac{\cos n}{n}$ ☆ (f) $x_n = \sqrt{n+1} - \sqrt{n}$

   (g) $x_n = \left(\dfrac{n}{2}\right)^n$ (h) $x_n = \dfrac{6}{2^n}$

   (i) $x_n = \dfrac{5000}{n!}$ (j) $x_n = \sin\left(\dfrac{n\pi}{2}\right)$

   (k) $x_n = \dfrac{n!}{n^n}$ (l) $x_n = n^{1/n}$

   (m) $\begin{cases} x_1 = 1 \\ x_n = \sqrt{1 + x_{n-1}}, \ n \geq 2 \end{cases}$ (n) $\begin{cases} x_1 = 1 \\ x_n = 2\sqrt{x_{n-1}}, \ n \geq 2 \end{cases}$

2. Give an example of a bounded sequence that is not convergent.

3. Give an example of an increasing sequence that is not convergent.

4. Prove that if $x_n \to L$ and $y_n \to M$ and $r \in \mathbf{R}$, then
   ★ (a) $x_n + y_n \to L + M$.   (b) $x_n - y_n \to L - M$.
   (c) $rx_n \to rL$.   (d) $x_n y_n \to LM$.

☆ 5. Prove that if $x_n \to 0$ and *y* is a bounded sequence, then $x_n y_n \to 0$.

6. ☆ (a) Prove that if $x_n \to L$, then $|x_n| \to |L|$.
   (b) Give an example of a sequence *x* such that $|x_n| \to |L|$ but *x* does not converge to *L*.

7. Prove Theorem 7.9.

8. Give a proof of the Bounded Monotone Sequence Theorem for the case in which the sequence *x* is assumed bounded and decreasing.

9. ☆ (a) Prove that if $x_n \to L$ and $L \neq 0$, then there is a number *N* such that if $n \geq N$, then $|x_n| > (|L|/2)$.
   (b) Prove that if $x_n \to L$, $x_n \neq 0$ for all *n*, $L \neq 0$, and if $y_n \to M$, then $(y_n/x_n) \to (M/L)$.

10. Complete the proof that $x_n = (1 + \frac{1}{n})^n$ is an increasing sequence by showing that for all $k \leq n$,

$$\frac{1}{k!} \frac{n(n-1)(n-2)\cdots[n-(k-1)]}{n^k}$$

$$\le \frac{1}{k!} \frac{(n+1)(n)(n-1)\cdots[n-(k-2)]}{(n+1)^k}$$

11. **Proofs to Grade.**

   (a) **Claim.** Every bounded decreasing sequence converges.
   "Proof." Let x be a bounded decreasing sequence. Then $y_n = -x_n$ defines a bounded increasing sequence. By the proof of Theorem 7.13, $y_n \to L$ for some L. Thus $x_n \to -L$. ∎

   ★ (b) **Claim.** If the sequence x converges and the sequence y diverges, then $x + y$ diverges.
   "Proof." Suppose $x_n + y_n \to K$ for some real number K. Since $x_n \to L$ for some number L, $(x_n + y_n) - x_n \to K - L$; that is, $y_n \to K - L$. This is a contradiction. Thus $x_n + y_n$ diverges. ∎

   (c) **Claim.** If two sequences x and y both diverge, then $x + y$ diverges.
   "Proof." Suppose $\lim_{n\to\infty} (x_n + y_n) = L$. Since x diverges, there exists $\epsilon_1 > 0$ such that for all $N \in \mathbf{N}$ there exists $n > N$ such that $|x_n - (L/2)| \ge \epsilon_1$. Since y diverges, there exists $\epsilon_2 > 0$ such that, for all $N \in \mathbf{N}$, there exists $n > N$ such that $|y_n - (L/2)| \ge \epsilon_2$. Let $\epsilon = \frac{1}{2} \min\{\epsilon_1, \epsilon_2\}$. Then for all $N \in \mathbf{N}$, there exists $n > N$ such that

$$
\begin{aligned}
|(x_n + y_n) - L| &= |[x_n - (L/2)] + [y_n - (L/2)]| \\
&\ge |x_n - (L/2)| + |y_n - (L/2)| \\
&\ge \epsilon_1 + \epsilon_2 \\
&\ge \tfrac{1}{2}\epsilon + \tfrac{1}{2}\epsilon \\
&= \epsilon.
\end{aligned}
$$

   Therefore $\lim_{n\to\infty} (x_n + y_n) \ne L$. ∎

7.5

Equivalents of Completeness

In the first section of this chapter we stated without proof that the ordered field of real numbers is complete. Completeness was then used to prove the Heine-Borel Theorem in section 7.2. The Heine-Borel Theorem was the main result used to prove the Bolzano-Weierstrass Theorem in section 7.3, which in turn justified the Bounded Monotone Sequence Theorem of section 7.4. In this section we complete a circle of arguments by showing that if the ordered field of reals is assumed to have every bounded monotone sequence convergent, then the reals are complete. Thus completeness and the three properties described by the theorems are equivalent for **R**. What has been omitted is a proof of the (correct) statement that $\mathbf{R}$ is complete. Furthermore, the theorems stated for the reals could be generalized to prove that the properties are all equivalent for any ordered field. Our proof of completeness from the Bounded Monotone Sequence Theorem will use the following two lemmas about convergent sequences. The proofs are exercises 1 and 2.

Lemma 7.14 If x and y are two sequences such that $\lim_{n \to \infty} y_n = s$ and $\lim_{n \to \infty} (x_n - y_n) = 0$, then $\lim_{n \to \infty} x_n = s$.

Lemma 7.15 If x is a sequence with $\lim_{n \to \infty} x_n = s$ and t is a real number such that $t < s$, then there exists $N \in \mathbf{N}$ such that $x_n > t$ for all $n \geq N$.

Theorem 7.16 Suppose the real number system has the property that every bounded monotone sequence must converge. Then $\mathbf{R}$ is complete.

Proof. Let A be a nonempty subset of $\mathbf{R}$ that is bounded above by a real number b. To prove completeness, we must show $\sup(A)$ exists and is a real number. Since $A \neq \varnothing$, choose $a \in A$. If a is an upper bound for A, then $a = \sup(A)$ ⟨*exercise 3, section 7.1*⟩, and we are done. Assume that a is not an upper bound for A. If $(a + b)/2$ is an upper bound for A, let $x_1 = a$ and $y_1 = (a + b)/2$; if not, let $x_1 = (a + b)/2$ and $y_1 = b$. In either case $y_1 - x_1 = (b - a)/2$, x_1 is not an upper bound, and y_1 is an upper bound.

Now if $(x_1 + y_1)/2$ is an upper bound for A, let $x_2 = x_1$ and $y_2 = (x_1 + y_1)/2$; otherwise, let $x_2 = (x_1 + y_1)/2$ and $y_2 = y_1$. In either case the result is $y_2 - x_2 = (b - a)/4$, $x_2 \geq x_1$, x_2 is not an upper bound for A, while $y_2 \leq y_1$, and y_2 is an upper bound for A.

Continuing in this manner, we inductively define an increasing sequence x such that no x_n is an upper bound for A, and a decreasing sequence y such that every y_n is an upper bound for A. In addition, $y_n - x_n = (b - a)/2^n$ and y is bounded below by a. Therefore by hypothesis, y converges to a point $s \in \mathbf{R}$. Furthermore, since $\lim_{n \to \infty} (y_n - x_n) = \lim_{n \to \infty} (b - a)/2^n = 0$, $\lim_{n \to \infty} x_n = s$ by Lemma 7.14.

We claim that s is an upper bound for A. If $z > s$ for some $z \in A$, then $z > y_N$ for some N ⟨*because $y_n \to s$*⟩. This contradicts the fact that y_N is an upper bound for A.

Finally, if t is a real number and $t < s$, then $t < x_N$ for some $N \in \mathbf{N}$, by Lemma 7.15. Since x_N is not an upper bound for A, t is not an upper bound. Thus s is a real number that is an upper bound for A, and no number less than s is an upper bound; that is, $s = \sup(A)$. Therefore **R** is complete. ∎

We saw in Section 7.1 that the field of rationals is not complete. Since the major theorems of this chapter are equivalent to completeness in a field, all three must fail for the rationals.

The set $A = \{x \in \mathbf{Q}: x^2 \leqslant 2\} = [-\sqrt{2}, \sqrt{2}] \cap \mathbf{Q}$ is a closed subset of **Q**. A is also bounded but not compact because $\{(-x, x): x \in A \text{ and } x \neq 0\}$ is an open cover of A with no finite subcover. This example shows that the Heine-Borel Theorem fails for **Q**.

The set $B = \{1, 4, 1.41, 1.414, 1.4142, \ldots\}$ of (rational) decimal approximations of $\sqrt{2}$ is a bounded and infinite subset of **Q** whose only accumulation point in **R** is not in **Q**. Thus the Bolzano-Weierstrass Theorem fails in **Q**.

A counterexample to the Bounded Monotone Sequence Theorem in **Q** is the sequence whose terms are the successive decimal expansions of $\sqrt{2}$. This bounded and increasing sequence fails to converge to any rational number.

Why is completeness such a crucial property of the real number system? One possible answer lies in the last sequence considered in section 7.4. The sequence $x_n = (1 + \frac{1}{n})^n$ is bounded and increasing, so by the completeness of **R** (via the Bounded Monotone Sequence Theorem), the terms x_n must approach a unique real number, which is the number e. The fact that e must exist in **R** is a consequence of completeness.

Not only must the limit of a bounded monotone sequence of rational numbers be in **R**, but the same is true for a bounded monotone sequence of irrational numbers, or of rationals and irrationals. The completeness property and its equivalents assure us that every number that can be approached by reals is in fact a real number.

Exercises 7.5

1. Prove Lemma 7.14.

2. Prove Lemma 7.15.

3. Give an example of a closed subset A of **Q** such that $A \subseteq [7, 8]$ and A is not compact. What open cover of A has no finite subcover?

4. Give a sequence x of distinct rationals such that $\{x_n: n \in \mathbf{N}\} \subseteq [5, 6]$ and x has no accumulation point in **Q**.

Answers to Selected Exercises

1. (a)

P	$\sim P$	$P \wedge \sim P$
T	F	F
F	T	F

(c)

P	Q	R	$Q \vee R$	$P \wedge (Q \vee R)$
T	T	T	T	T
F	T	T	T	F
T	F	T	T	T
F	F	T	T	F
T	T	F	T	T
F	T	F	T	F
T	F	F	F	F
F	F	F	F	F

(e)

P	Q	$\sim Q$	$P \wedge \sim Q$
T	T	F	F
F	T	F	F
T	F	T	T
F	F	T	F

253

(g)

P	*Q*	*~Q*	*P* ∧ *Q*	(*P* ∧ *Q*) ∨ *~Q*
T	T	F	T	T
F	T	F	F	F
T	F	T	F	T
F	F	T	F	T

2.   (a)   equivalent
     (c)   equivalent
     (e)   equivalent
     (g)   not equivalent
     (i)   not equivalent
     (k)   equivalent
   (m)   equivalent

3.   (a)   false
     (c)   true
     (e)   false
     (g)   false
     (i)   true
     (k)   false

4.   (a) x is not a positive integer.
     (c) $5 < 3$.
     (e)   Roses are not red or violets are not blue.
     (g) T is green and T is not yellow.

5.   (a)   Since P is equivalent to Q, P has the same truth table as Q. Therefore Q has the same truth table as P, so Q is equivalent to P.

7.   (a)

P	*Q*	*P* ⓥ *Q*
T	T	F
F	T	T
T	F	T
F	F	F

Exercises 1.2

1.   (a)   Antecedent: squares have three sides.
            Consequent: triangles have four sides.
     (d)   Antecedent: f is differentiable.
            Consequent: f is continuous.
     (f)   Antecedent: f is integrable.
            Consequent: f is bounded.

2.   (a)   Converse: If triangles have four sides, then squares have three sides.
            Contrapositive: If triangles do not have four sides, then squares do not have three sides.
     (d)   Converse: If f is continuous, then f is differentiable.
            Contrapositive: If f is not continuous, then f is not differentiable.
     (f)   Converse: If f is bounded, then f is integrable.
            Contrapositive: If f is not bounded, then f is not integrable.

3.   (a)   true
     (c)   true
     (e)   true

4. (a) true
   (c) true

5. (b)

P	Q	$\sim P$	$\sim P \Rightarrow Q$	$Q \Leftrightarrow P$	$(\sim P \Rightarrow Q) \vee (Q \Leftrightarrow P)$
T	T	F	T	T	T
F	T	T	T	F	T
T	F	F	T	F	T
F	F	T	F	T	T

(c)

P	Q	$\sim Q$	$Q \Rightarrow P$	$\sim Q \Rightarrow (Q \Leftrightarrow P)$
T	T	F	T	T
F	T	F	F	T
T	F	T	T	T
F	F	T	T	T

6. (f)

P	Q	R	$Q \vee R$	$P \wedge (Q \vee R)$	$P \wedge Q$	$P \wedge R$	$(P \wedge Q) \vee (P \wedge R)$
T	T	T	T	T	T	T	T
F	T	T	T	F	F	F	F
T	F	T	T	T	F	T	T
F	F	T	T	F	F	F	F
T	T	F	T	T	T	F	T
F	T	F	T	F	F	F	F
T	F	F	F	F	F	F	F
F	F	F	F	F	F	F	F

Since the fifth and eighth columns are the same, the propositions
$P \wedge (Q \vee R)$ and $(P \wedge Q) \vee (P \wedge R)$ are equivalent.

7. (a) $((f$ has a relative minimum at $x_0) \wedge (f$ is differentiable at $x_0)) \Rightarrow (f'(x_0) = 0)$.
   (d) $((x = 1) \vee (x = -1)) \Rightarrow (|x| = 1)$.
   (e) $(x_0$ is a critical point for $f) \Leftrightarrow ((f'(x_0) = 0) \vee (f'(x_0)$ does not exist$))$.

8. (b)

P	Q	R	$P \wedge Q$	$P \wedge Q \Rightarrow R$	$\sim R$	$\sim Q$	$P \wedge \sim R$	$P \wedge \sim R \Rightarrow \sim Q$
T	T	T	T	T	F	F	F	T
F	T	T	F	T	F	F	F	T
T	F	T	F	T	F	T	F	T
F	F	T	F	T	F	T	F	T
T	T	F	T	F	T	F	T	F
F	T	F	F	T	T	F	F	T
T	F	F	F	T	T	T	T	T
F	F	F	F	T	T	T	F	T

Since the fifth and ninth columns are the same, the propositions
$P \wedge Q \Rightarrow R$ and $P \wedge \sim R \Rightarrow \sim Q$ are equivalent.

9. (a) If x is an even integer, then $x + 1$ is an odd integer.
   (c) not possible.

Exercises 1.3

1. (a) $\sim(\forall x)(x$ is precious $\Rightarrow x$ is beautiful$)$.
      Or, $(\exists x)(x$ is precious and x is not beautiful$)$.
   (b) This question is not the same as 1(a).

(c) ($\exists x$)(x is a positive integer and x is smaller than all other positive integers).
Or, ($\exists x$)(x is a positive integer and ($\forall y$)(y is a positive integer $\Rightarrow x \leqslant y$)).
(f) ($\forall x$)($\sim$($\forall y$)(x loves y)). Or $\sim$($\exists x$)($\forall y$)(x loves y).
(h) ($\exists x$)(x cares about me).

2. (a) ($\forall x$)(x is precious $\Rightarrow x$ is beautiful). All precious stones are beautiful.
(c) ($\forall x$)(x is a positive integer $\Rightarrow$ ($\exists y$)(y is a positive integer) $\wedge x > y$)). For every positive integer there is a smaller positive integer.
Or, $\sim$($\exists x$)(x is a positive integer $\wedge$ ($\forall y$)(y is a positive integer $\Rightarrow x \leqslant y$)).
There is no smallest positive integer.

3. *Hint:* To use part (a), note that $\sim$($\exists x$)($\sim A(x)$) is equivalent to ($\forall x$)($\sim\sim A(x)$).

4. (a) true
(b) false
(f) false
(i) false

5. (d) This statement is not a denial. It implies the negation of ($\exists! x$)$P(x)$, but if ($\forall x$)$\sim P(x)$, then both the statement and ($\exists! x$)$P(x)$ are false.

7. For every backwards E, there exists an upside down A!

Exercises 1.4

2. (a) tautology
(b) neither
3. (a) E. The converse rather than the statement is proved.
(d) A.
4. (a) Suppose $(G, *)$ is a cyclic group.

　·
　·
　·

Thus $(G, *)$ is abelian.
Therefore, if $(G, *)$ is a cyclic group, then $(G, *)$ is abelian.
(c) Suppose $(G, *)$ is not abelian.

　·
　·
　·

Thus $(G, *)$ is not a cyclic group.
Therefore, if $(G, *)$ is a cyclic group, then $(G, *)$ is abelian.

Exercises 1.5

1. (a) E. The false statement referred to is not a denial of the claim.
(b) C. Uniqueness has not been shown.
(d) A.
5. Let x, y, z be three consecutive odd primes. Suppose $x \neq 3$. Since x is prime, when x is divided by 3 the remainder is 1 or 2. In case the remainder is 1, then $x = 3k + 1$ for some integer $k \geqslant 1$. But then $y = x + 2 = 3k + 3 = 3(k + 1)$, so y is not prime. In case the remainder is 2, then $x = 3k + 2$ for some integer $k \geqslant 1$. But then $z = x + 4 = 3k + 2 + 4 = 3(k + 2)$, so z is not prime. In either case we reach the contradiction that y or z is not prime. Thus $x = 3$ and $y = 5, z = 7$.

7. (c) Let n be a natural number. Then both $2n$ and $2n + 1$ are natural numbers. Let $M = 2n + 1$. Then M is a natural number greater than $2n$.

(g) Let $\epsilon > 0$ be a real number. Then $\frac{1}{\epsilon}$ is a positive real number and so has a decimal expression as an integer part plus a decimal part. Let M be the integer part of $\frac{1}{\epsilon}$, plus 1. Then M is an integer and $M > \frac{1}{\epsilon}$. To prove for all natural numbers $n > M$ that $\frac{1}{n} < \epsilon$, let n be a natural number and $n > M$. Since $M > \frac{1}{\epsilon}$, we have $n > \frac{1}{\epsilon}$. Thus $\frac{1}{n} < \epsilon$. Therefore, for every real number $\epsilon > 0$, there is a natural number M such that for all natural numbers $n > M$, $\frac{1}{n} < \epsilon$.

Exercises 2.1

1. (a) $\{x \in \mathbf{N}: x < 6\}$ or $\{x: x \in \mathbf{N} \text{ and } x < 6\}$.
    (c) $\{x \in \mathbf{R}: 2 \leqslant x \leqslant 6\}$ or $\{x: x \in \mathbf{R} \text{ and } 2 \leqslant x \leqslant 6\}$.
2. (a) $\{1, 2, 3, 4, 5\}$
    (c) not possible
3. (a) true
    (c) true
    (e) false
    (g) true
    (i) false
4. (a) $\{\{0\}, \{\triangle\}, \{\square\}, \{0, \triangle\}, \{0, \square\}, \{\triangle, \square\}, X, \varnothing\}$
    (c) $\{\{\varnothing\}, \{\{a\}\}, \{\{b\}\}, \{\{a, b\}\}, \{\varnothing, \{a\}\}, \{\varnothing, \{b\}\}, \{\varnothing, \{a, b\}\}, \{\{a\}, \{b\}\},$
         $\{\{a\}, \{a, b\}\}, \{\{b\}, \{a, b\}\}, \{\varnothing, \{a\}, \{b\}\}, \{\varnothing, \{a\}, \{a, b\}\}, \{\varnothing, \{b\}, \{a, b\}\},$
         $\{\{a\}, \{b\}, \{a, b\}\}, X, \varnothing\}$.
5. (a) no proper subsets
    (c) $\{1\}, \{2\}$
6. (a) true
    (c) true
    (e) false
    (g) false
    (i) false
    (k) true
7. (a) $A = \{1, 2\}, B = \{1, 2, 4\}, C = \{1, 2, 5\}$
    (c) $A = \{1, 2, 3\}, B = \{1, 4\}, C = \{1, 2, 3, 5\}$
    (e) not possible
8. (a) true
    (c) true
    (e) true
    (g) true
    (i) true
    (k) true
9. (a) Let x be any object. Suppose $x \in A$. Then $P(x)$ is true.
         Since $(\forall x)(P(x) \Rightarrow Q(x))$, $Q(x)$ is true.
         Thus $x \in B$. Therefore $A \subseteq B$.　∎
12. *Hint:* To prove $A = B$, we are given $A \subseteq B$. To prove $B \subseteq A$, use Theorem 2.3.
16. Both $X \in X$ and $X \notin X$ are false because they lead to a contradiction. Thus the collection of all ordinary sets *is not a set.*
17. (a) A. Every statement in the proof is correct. No reasons are given, but no explanation of the steps is required for correctness.

(c)   C. The "proof" asserts that $x \in C$, but fails to justify this assertion with a definite statement that $x \in A$ or that $x \in B$. This problem could be corrected by inserting a second sentence "Suppose $x \in A$" and a fourth sentence "Then $x \in B$."

Exercises 2.2

1. (a) $\{0, 1, 2, 3, 4, 5, 6, 7, 8, 9\}$
    (c) $\{1, 3, 5, 7, 9\}$
    (e) $\{3, 9\}$
    (g) $\{1, 5, 7\}$
    (i) $\{1, 5, 7\}$
2. (a) $\{0, -2, -4, -6, -8, -10, \ldots\}$
    (c) D
    (e) $N \cup \{0\}$
    (g) D
    (i) $\{0, 2, 4, 6, 8, \ldots\}$
    (k) $\varnothing$
3. (a) $\{\{1\}, \{2\}, \{1, 2\}, \varnothing\}$
    (d) $\{\varnothing, \{1\}, \{3\}\}$
    (f) $\{\{3\}\}$
4. A and B are disjoint.
8. (a) *Hint:* To prove that $A \subseteq B$ implies $A - B = \varnothing$, assume that $A \subseteq B$ and show that $x \in A - B$ is always false. To prove the converse, begin by assuming $A - B = \varnothing$ and that some object $x \in A$.
9. (c) We show that if C and D are not disjoint, then A and B are not disjoint. If C and D are not disjoint then there is an object $x \in C \cap D$. But then $x \in C$ and $x \in D$. Since $C \subseteq A$ and $D \subseteq B$, $x \in A$ and $x \in B$. Therefore $x \in A \cap B$ so A and B are not disjoint.
11. (a) $S \in \mathcal{P}(A \cap B)$ iff $S \subseteq A \cap B$
     $\qquad\qquad\qquad$ iff [by exercise 8(c)] $S \subseteq A$ and $S \subseteq B$
     $\qquad\qquad\qquad$ iff $S \in \mathcal{P}(A)$ and $S \in \mathcal{P}(B)$
     $\qquad\qquad\qquad$ iff $S \in \mathcal{P}(A) \cap \mathcal{P}(B)$
     (c) Let A and B be any sets. Since $\varnothing$ is a subset of every set we have $\varnothing \in \mathcal{P}(A - B)$ and $\varnothing \in \mathcal{P}(A), \varnothing \in \mathcal{P}(B)$ so $\varnothing \notin \mathcal{P}(A) - \mathcal{P}(B)$. This shows that $\mathcal{P}(A - B) \not\subseteq \mathcal{P}(A) - \mathcal{P}(B)$.
12. (a) $A = \{1, 2\}, B = \{1, 3\}, C = \{2, 3, 4\}$.
     (c) $A = \{1, 2\}, B = \{1, 3\}$.
     (e) $A = \{1, 2\}, B = \{1, 3\}, C = \{1\}$.
14. (a) C. The proof that $A \cap B = A$ is incomplete.
     (b) E. The claim is false. The statement "$x \in A$ and $x \in \varnothing$ iff $x \in A$" is false.
     (d) A.
     (f) E. A picture often helps to bring forth ideas around which a correct proof may be built. A proof by picture alone is not sufficient.

Exercises 2.3

1. (a) $\bigcup_{A \in \mathcal{A}} A = \{1, 2, 3, 4, 5, 6, 7, 8\}$; $\bigcap_{A \in \mathcal{A}} A = \{4, 5\}$

(c) $\underset{n \in \mathbf{N}}{\cup} A_i = \mathbf{N}$; $\underset{n \in \mathbf{N}}{\cap} A_n = \{1\}$

(e) $\underset{A \in \mathcal{A}}{\cup} A = \mathbf{Z}$; $\underset{A \in \mathcal{A}}{\cap} A = \{10\}$

(g) $\underset{n \in \mathbf{N}}{\cup} A_n = (0, 1)$; $\underset{n \in \mathbf{N}}{\cap} A_n = \varnothing$

(i) $\underset{r \in \mathbf{R}}{\cup} A_r = [0, \infty)$; $\underset{r \in \mathbf{R}}{\cap} A_r = \varnothing$

2.   The family in exercise 1(a) is not pairwise disjoint. The family in 1(b) is pairwise disjoint.

4.   (a)   Let $\beta \in \Delta$. Suppose $x \in \underset{\alpha \in \Delta}{\cap} A_\alpha$. Then $x \in A_\alpha$ for each $\alpha \in \Delta$.

   Since $\beta \in \Delta$, $x \in A_\beta$. Therefore $\underset{\alpha \in \Delta}{\cap} A \subseteq A_\beta$.

5.   (a) $x \in B \cap \underset{\alpha \in \Delta}{\cup} A_\alpha$ iff $x \in B$ and $x \in \underset{\alpha \in \Delta}{\cup} A_\alpha$

   iff $x \in B$ and $x \in A_\alpha$ for some $\alpha \in \Delta$

   iff $x \in B \cap A_\alpha$ for some $\alpha \in \Delta$

   iff $x \in \underset{\alpha \in \Delta}{\cup} (B \cap A_\alpha)$.

6.   (a) $(\underset{\alpha \in \Delta}{\cup} A_\alpha) \cap (\underset{\beta \in \Gamma}{\cup} B_\beta) = \underset{\beta \in \Gamma}{\cup} ((\underset{\alpha \in \Delta}{\cup} A_\alpha) \cap B_\beta)$

   $= \underset{\beta \in \Gamma}{\cup} (\underset{\alpha \in \Delta}{\cup} (A_\alpha \cap B_\beta))$.

8.   (a)   Suppose $x \in \underset{\alpha \in \Gamma}{\cup} A_\alpha$. Then $x \in A_\alpha$ for some $\alpha \in \Gamma$.

   Since $\Gamma \subseteq \Delta$, $x \in A_\alpha$ for some $\alpha \in \Delta$.

   Therefore, $x \in \underset{\alpha \in \Delta}{\cup} A_\alpha$.   ∎

12.   Suppose B is a set such that $B \subseteq A_\beta$ for all $\beta \in \Delta$.
   Suppose $x \in B$. Then $x \in A_\beta$ for all $\beta \in \Delta$.
   Therefore $x \in \underset{\beta \in \Delta}{\cap} A_\beta$.

   Thus $B \subseteq \underset{\beta \in \Delta}{\cap} A_\beta$.

14. $A_r = (0, r)$, $\mathcal{A} = \{A_r : r \in \mathbf{R}^+\}$

16.   (a)   A. Note that if we allowed $\Delta = \varnothing$, the claim would be false.
   (b)   C. No connection is made between the first and second sentences.

Exercises 2.4

1.   Only $\varnothing$ is inductive.
2.   (b)   must be true
   (e)   might be false
3.   (b) $f_4 = 3, f_7 = 13, f_{n+3} - f_{n+1} = f_{n+2}$.
6.   (a)   Let S be a subset of $\mathbf{N}$ such that $1 \in S$ and S is inductive. We wish to show that $S = \mathbf{N}$. Assume that $S \neq \mathbf{N}$ and let $T = \mathbf{N} - S$. By the WOP, the nonempty set T has a least element. This least element is not 1, because $1 \in S$. If the least element is n, then $n \in T$ and $n - 1 \in S$. But by the inductive property of S, $n - 1 \in S$ implies that $n \in S$. This is a contradiction. Therefore, $S = \mathbf{N}$.
12.   (b)   *Hint:* For each player x, consider the set W_x of all players who win against player x.
13.   (a)   E. Let $S = \{n \in \mathbf{N}:$ all horses in every set of n horses have the same color$\}$. Then $1 \in S$. In fact, $S = \{1\}$. The statement $n \in S \Rightarrow n + 1 \in S$ is correct

for $n \geq 2$. The only counterexample to $(\forall n)(n \in S \Rightarrow n + 1 \in S)$ occurs when $n = 1$. The "proof" fails to consider that a special argument would be necessary when $n = 1$. In this case there would be no way to remove a different horse from a set of n horses.

(d)   A.

Exercises 2.5

2. (b)   16
3. *Hint:* Since $1,000,000 = (10^3)^2 = (10^2)^3 = 10^6$, there are 10^3 squares less than or equal to $1,000,000$; 10^2 cubes less than or equal to $1,000,000$; and 10 natural numbers that are both squares and cubes (sixth powers) less than or equal to $1,000,000$.
9. (a) $3! \cdot 4!$
    (b)   *Hint:* Think of arranging 5 objects, where each boy and the group of girls are the objects. Then consider that the girls can be arranged in 3! ways.
15. (e)   *Hint:* Consider two disjoint sets containing n and m elements.

Exercises 3.1

2. (b) $(a, b) \in A \times (B \cap C)$ iff $a \in A$ and $b \in B \cap C$
                              iff $a \in A$ and $b \in B$ and $b \in C$
                              iff $a \in A$ and $b \in B$ and $a \in A$ and $b \in C$
                              iff $(a, b) \in A \times B$ and $(a, b) \in A \times C$
                              iff $(a, b) \in (A \times B) \cap (A \times C)$.   ∎
5. (a)   domain **R**, range **R**.
    (c)   domain $[1, \infty)$, range $[0, \infty)$.
    (e)   domain **R**, range **R**.
6. (a) $R_1^{-1} = R_1$.
    (c) $R_3^{-1} = \{(x, y) \in \mathbf{R} \times \mathbf{R}: y = \tfrac{1}{7}(x + 10)\}$.
    (e) $R_5^{-1} = \{(x, y) \in \mathbf{R} \times \mathbf{R}: y = \pm\sqrt{(5 - x)/4}\}$.
    (g) $R_7^{-1} = \left\{(x, y) \in \mathbf{R} \times \mathbf{R}: y < \dfrac{x + 4}{3}\right\}$.
    (i) $R_9^{-1} = \{(x, y) \in \mathbf{P} \times \mathbf{P}: y$ is a child of x, and x is male$\}$.
7. (a) $R_1 \circ R_1 = \{(x, z): x = z\} = R_1$.
    (d) $R_2 \circ R_3 = \{(x, z) \in \mathbf{R} \times \mathbf{R}: z = -35x + 52\}$.
    (g) $R_4 \circ R_5 = \{(x, z) \in \mathbf{R} \times \mathbf{R}: z = 16x^4 - 40x^2 + 27\}$.
    (j) $R_6 \circ R_6 = \{(x, z) \in \mathbf{R} \times \mathbf{R}: z < x + 2\}$.
    (m) $R_3 \circ R_8 = \{(x, z) \in \mathbf{R} \times \mathbf{R}: z = \dfrac{14x}{x - 2} - 10\}$.
    (o) $R_9 \circ R_9$ is *not* $\{(x, z): z$ is a grandfather of $x\}$.
15. (a)   E. The statements "$x \in A \times B$" and "$x \in A$ and $x \in B$" are not equivalent.
    (b)   C. The only error is that $(a, c) \notin B \times D$ implies $a \notin B$ or $c \notin D$.
    (d)   A.

Exercises 3.2

1. (a) not reflexive, not symmetric, transitive
   (e) reflexive, not symmetric, transitive
   (k) not reflexive, symmetric, not transitive (*Note:* Sibling means "a brother or sister.")
2. (a) $\{(1, 1), (2, 2), (2, 3), (3, 1)\}$.
   (d) $\{(1, 1), (2, 2), (3, 3), (1, 3), (2, 3), (3, 1), (3, 2)\}$.
3. (f)   This is the graph of the relation $\{(x, y): y \leq x\}$.

4. (a) To show that R is reflexive, let a be a natural number. All prime factorizations of a have the same number of 2's. Thus $a\,R\,a$. It must also be shown that R is symmetric and transitive. Three elements of $4/R$ are $4 = 2 \cdot 2$, $28 = 2 \cdot 2 \cdot 7$, and $300 = 2 \cdot 2 \cdot 3 \cdot 5 \cdot 5$.
   (d) To show transitivity, begin by supposing $(x, y)\,R\,(z, w)$ and $(z, w)\,R\,(u, v)$. Then $xw = yz$ and $zv = wu$. Multiply the first equation by v and the second by y. Comparison of the resulting equations will show that $(x, y)\,R\,(u, v)$. It must also be shown that R is reflexive and symmetric. The equivalence class of $(2, 3)$ contains pairs (x, y) such that $x/y = 3/2$.
5. (a) $0/\equiv_5 = \{\ldots, -15, -10, -5, 0, 5, 10, \ldots\}$
        $1/\equiv_5 = \{\ldots, -9, -4, 1, 6, 11, \ldots\}$
        $2/\equiv_5 = \{\ldots, -8, -3, 2, 7, 12, \ldots\}$
        $3/\equiv_5 = \{\ldots, -7, -2, 3, 8, 13, \ldots\}$
        $4/\equiv_5 = \{\ldots, -6, -1, 4, 9, 14, \ldots\}$
7. (b) Assume R is symmetric. Then $(x, y) \in R$ iff $(y, x) \in R$ iff $(x, y) \in R^{-1}$. Thus $R = R^{-1}$. Now, suppose $R = R^{-1}$. Then $(x, y) \in R$ implies $(x, y) \in R^{-1}$, which implies $(y, x) \in R$. Thus R is symmetric. ■
10. One part of the proof is to show that R is symmetric. Suppose $x\,R\,y$. Then $x\,L\,y$ and $y\,L\,x$, so $y\,L\,x$ and $x\,L\,y$. Therefore, $y\,R\,x$.
11. (b) Assume R is asymmetric. If $(x, y) \in R$ then by asymmetry $(y, x) \notin R$. Thus $(x, y) \in R$ and $(y, x) \in R$ is false. Therefore if (x, y) and $(y, x) \in R$, then $x = y$. That is, R is antisymmetric.
12. (c) E. The last sentence confuses $R \cap S$ with $R \circ S$. A correct proof requires a more complete second sentence.

Exercises 3.3

2. *Hint:* The partition has four elements. One of them is the set $\{(1, -1), (-1, 1), (i, i), (-i, -i)\}$.
4. (a) $\{(1, 1), (1, 2), (2, 1), (2, 2), (3, 3), (3, 4), (3, 5), (4, 3), (4, 4), (4, 5), (5, 3), (5, 4), (5, 5)\}$.
7. No. Let R be the relation $\{(1, 1), (2, 2), (3, 3), (1, 2), (1, 3), (2, 1), (3, 1)\}$ on the set $A = \{1, 2, 3\}$. Then $R(1) = \{1, 2, 3\}$, $R(2) = \{1, 2\}$, and $R(3) = \{1, 3\}$. The set $\mathcal{A} = \{\{1, 2, 3\}, \{1, 2\}, \{1, 3\}\}$ is not a partition of A.

10. (b) Yes, $\{\widetilde{B}_1, \widetilde{B}_2\}$ is a partition of A, because $\{\widetilde{B}_1, \widetilde{B}_2\} = \{B_2, B_1\}$. If $B_1 = B_2$ then $B_1 = B_2 = A$ and $\{\widetilde{B}_1 = \widetilde{B}_2 = \emptyset$, so $\{\widetilde{B}_1, \widetilde{B}_2\}$ is not a partition.
11. (b) C. This is a tough one because the ideas are all there and every statement is true. We give it C because the ideas are not well connected.

Exercises 3.4

3. (b) All such graphs are isomorphic.
13. Suppose the graph G has order $n \geq 2$, and all vertices have different degrees. Consider what these degrees must be. Can one vertex have degree $n - 1$ and another have degree 0?
24. (a) E. The claim is false. To discover the error in the proof, consider exercise 14.

Exercises 4.1

1. (a) R_1 is a function. $\text{Dom}(R_1) = \{0, \triangle, \square, \cap, \cup\}$.
A possible codomain is $\text{Rng}(R_1) = \{0, \triangle, \square, \cap, \cup\}$.
   (g) R_7 is a function. $\text{Dom}(R_7) = \mathbf{R}$. Possible codomains are $\mathbf{R}$ and $[0, \infty)$.
2. (a) Domain $= \mathbf{R} - \{1\}$. Range $= \{y \in \mathbf{R}: y \neq 0\}$. A possible codomain is $\mathbf{R}$.
   (d) Domain $= \mathbf{R} - \{\frac{\pi}{2} + k\pi: k \in \mathbf{Z}\}$. A possible codomain is $\mathbf{R}$.
3. (c) $5, -5$
4. (a) $\text{Dom}(f) = \mathbf{R} - \{3\}$, $\text{Rng}(f) = \mathbf{R} - \{-1\}$.
8. (a) A.
11. (a) $f(3) = 3/\equiv_6 = \{\ldots, -9, -3, 3, 9, \ldots\}$.
12. (a) $\text{Dom}(S)$.
13. (a) Let $x, y, z \in \mathbf{N}$.
   (i) By definition of absolute value, $d(x, y) = |x - y| \geq 0$ for all $x, y \in \mathbf{N}$.
   (ii) $d(x, y) = |x - y| = 0$ iff $x - y = 0$ iff $x = y$.
   (iii) $d(x, y) = |x - y| = |y - x| = d(y, x)$.
   (iv) By the triangle property of absolute value,
   $|x - y| + |y - z| \geq |x - z|$.
   Thus $d(x, y) + d(y, z) \geq d(x, z)$.
14. (c) $\binom{m}{2} n^2$. We choose 2 elements from A for first coordinates; each may then be assigned any element of B as image.

Exercises 4.2

1. (a) $(f \circ g)(x) = 17 - 14x$, $(g \circ f)(x) = -29 - 14x$
   (c) $(f \circ g)(x) = \sin(2x^2 + 1)$, $(g \circ f)(x) = 2 \sin^2 x + 1$
2. (a) $\text{Dom}(f \circ g) = \mathbf{R} = \text{Rng}(f \circ g) = \text{Dom}(g \circ f) = \text{Rng}(g \circ f)$.
   (c) $\text{Dom}(f \circ g) = \mathbf{R}$, $\text{Rng}(f \circ g) = [-1, 1]$.
   $\text{Dom}(g \circ f) = \mathbf{R}$, $\text{Rng}(g \circ f) = [1, 3]$.
3. (a) $f^{-1}(x) = \dfrac{x - 2}{5}$

(c) $f^{-1}(x) = \dfrac{1 - 2x}{x - 1}$

(e) $f^{-1}(x) = -3 + \ln x$

7. (a) $\{(x, y) \in \mathbf{R} \times \mathbf{R}\colon y = 0 \text{ if } x < 0 \text{ and } y = x^2 \text{ if } x \geqslant 0\}$
    $\{(x, y) \in \mathbf{R} \times \mathbf{R}\colon y = x^2\}$.

11. (a) $h \cup g$ is a function.

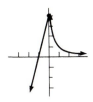

12. (a) *Hint:* Write $A \cup C$ as $A \cup (C - E)$. Then show $h \cup g = h \cup (g|_{C - E})$ and use Theorem 4.6.

14. We show that $f_1 + f_2$ is a function with domain **R**. First $f_1 + f_2$ is by definition a relation. For all $x \in \mathbf{R}$ there is some $u \in \mathbf{R}$ such that $(x, y) \in f_1$ because $f_1\colon \mathbf{R} \to \mathbf{R}$ and there exists $v \in \mathbf{R}$ such that $(x, v) \in f_2$ because $f_2\colon \mathbf{R} \to \mathbf{R}$. Then $(x, u + v) \in f_1 + f_2$, so $x \in \mathrm{Dom}(f_1 + f_2)$. It is clear from the definition of $f_1 + f_2$ that $x \in \mathrm{Dom}(f_1 + f_2)$ implies $x \in \mathbf{R}$, so $\mathrm{Dom}(f_1 + f_2) = \mathbf{R}$.

    Let $x \in \mathbf{R}$. Then because f_1 and f_2 are functions there are unique real numbers c and d such that $(x, c) \in f_1$ and $(x, d) \in f_2$. Then there is a unique real number $c + d$ and by construction of $f_1 + f_2$ this number is the unique second coordinate associated with x in $f_1 + f_2$. Thus $f_1 + f_2$ is a function.

18. (a) A.

Exercises 4.3

1. (a) Onto **R**. Let $w \in \mathbf{R}$. Then for $x = 2(w - 6)$, we have $f(x) = \frac{1}{2}[2(w - 6)] + 6 = w$. Thus $w \in \mathrm{Rng}(f)$. Therefore, f maps onto **R**. ∎
    (c) Not onto $\mathbf{N} \times \mathbf{N}$. Since $(5, 8) \in \mathbf{N} \times \mathbf{N}$ and $(5, 8) \notin \mathrm{Rng}(f)$, f does not map onto $\mathbf{N} \times \mathbf{N}$.

2. (a) One-to-one. Suppose $f(x) = f(y)$. Then $\frac{1}{2}x + 6 = \frac{1}{2}y + 6$. Then $\frac{1}{2}x = \frac{1}{2}y$, so $x = y$.
    (g) Not one-to-one, because $\sin(\frac{\pi}{6}) = \sin(\frac{13\pi}{6}) = \frac{1}{2}$.

3. (a) $B = \{0, 3\}$, $f = \{(1, 0), (2, 3), (3, 0), (4, 0)\}$.

8. (a) Let $f\colon \mathbf{R} \to \mathbf{R}$ be given by $f(x) = 2x$ and $g\colon \mathbf{R} \to \mathbf{R}$ be given by $g(x) = x^2$. Then f maps onto **R** but $g \circ f$ is not onto **R**.
    (e) Let $A = \{a, b, c\}$, $B = \{1, 2, 3\}$, $C = \{x, y, z\}$, $f = \{(a, 2), (b, 2), (c, 3)\}$, and $g = \{(1, x), (2, y), (3, z)\}$. Then g is one-to-one, but $g \circ f = \{(a, y), (b, y), (c, z)\}$ is not.

10. (a) Let $A = B = \mathbf{R}$ and $S = \{(x, y) \in \mathbf{R} \times \mathbf{R}\colon x^2 + y^2 = 25\}$. Then $(3, 4) \in S$ and $(3, -4) \in S$ and $\pi_1(3, 4) = 3 = \pi_1(3, -4)$, so π_1 is not one-to-one.

11. (a) Yes.
    (c) Not necessarily. The projection π_2 is one-to-one iff S is one-to-one.

12. (b) None.

(c) $\binom{m}{n}$ $n!$ n^{m-n}. We choose n elements of A to be the pre-images of elements of B and then assign images to arguments in $n!$ ways. Finally each of the remaining $m - n$ elements of B may be assigned any element of B as its image.

13.   (c)   A.

(d)   E. The "proof" does not show that every $w \in (-\frac{\pi}{2}, \frac{\pi}{2})$ is in the range.

Exercises 4.4

1.   (a) $(\varnothing, \varnothing)$, $(\{1\}, \{4\})$, $(\{2\}, \{4\})$, $(\{3\}, \{5\})$, $(\{1, 2\}, \{4\})$, $(\{1, 3\}, \{4, 5\})$, $(\{2, 3\}, \{4, 5\})$, $(A, \{4, 5\})$

2.   (a) $[2, 10]$

(c) $\{0\}$

4.   (a) $\{(1, 1)\}$

5.   (b) $\{p, s, t\}$

(d) $\{1, 2, 3, 5\}$

6.   (c) $\left[2 - \sqrt{3}, \dfrac{3 - \sqrt{5}}{2}\right) \cup \left(\dfrac{3 + \sqrt{5}}{2}, 2 + \sqrt{3}\right]$

7.   (a) $[9, 25]$

(c) $(5 - \sqrt{22}, 5 - \sqrt{21}] \cup [5 + \sqrt{21}, 5 + \sqrt{22})$

12.   (b)   Let $t \in f(X) - f(Y)$. Then $t \in f(X)$, so there exists $x \in X$ such that $f(x) = t$. We note $x \notin Y$ since $t = f(x) \notin f(Y)$. Thus $x \in X - Y$ and, therefore, $t = f(x) \in f(X - Y)$.

13.   The converse is true. To prove f is one-to-one, let $x \neq y$ in A. Then $\{x\} \cap \{y\} = \varnothing$ and thus $f(\{x\} \cap \{y\}) = \varnothing$. By hypothesis, $f(\{x\} \cap \{y\}) = f(\{x\}) \cap f(\{y\}) = \{f(x)\} \cap \{f(y)\}$. Thus $f(x) \neq f(y)$.

16.   (a)   If f is one-to-one, then the induced function is one-to-one.

19.   (a)   E. The claim is not true. We cannot conclude $x \in X$ from $f(x) \in f(X)$.

Exercises 5.1

3.   *Hint:* Consider $f: A \to A \times \{x\}$ given by $f(a) = (a, x)$ for each $a \in A$.

4.   (a)   finite

(c)   finite

(e)   infinite

6.   (a)   Suppose A is finite. Since $A \cap B$ is a subset of A, $A \cap B$ is finite.

7.   (a)   *Hint:* Consider $f: \mathbf{N}_k \times \mathbf{N}_m \to \mathbf{N}_{km}$ given by $f(a, b) = (a - 1)m + b$. Use the Division Algorithm to show f is a one-to-one correspondence. This is a formal version of the Product Rule that tells us $\#(\mathbf{N}_k \times \mathbf{N}_m) = km$.

9.   (c)   not possible

10.   *Hint:* Write $A \cup B = (A - B) \cup B$ and $A = (A - B) \cup (A \cap B)$ and apply Theorem 5.6.

13.   *Hint:* Use the pigeonhole principle.

16.   (a)   *Hint:* Suppose f is not onto B and consider the range of f.

(b)   *Hint:* Suppose f is not one-to-one. Then A is not empty and since A and B are finite and $A \approx B$, there is some $n \in \mathbf{N}$ such that $\mathbf{N}_n \approx A$ and $B \approx$

$\mathbf{N}_n$. Use these facts to construct a function F from $\mathbf{N}_n$ onto $\mathbf{N}_n$ that is not one-to-one. Then for some x, y, $z \in \mathbf{N}_n$, $f(x) = f(y) = z$. Removing (y, z) from F produces a function from a proper subset of $\mathbf{N}_n$ onto $\mathbf{N}_n$. Now apply exercise 15.

17. Use induction on the number of elements in the domain.
18. (b) C. In case 2, it is not correct that $\mathbf{N}_k \cup \mathbf{N}_1 \approx \mathbf{N}_{k+1}$. In addition, if $x \in \mathbf{N}_k$, then $S \cup \{x\} \neq \mathbf{N}_k \cup \{x\} = \mathbf{N}_k$.

Exercises 5.2

1. (a) Let $f: \mathbf{N} \to D^+$ be given by $f(n) = 2n - 1$ for each $n \in \mathbf{N}$. We show that f is one-to-one and maps onto D^+. First, to show f is one-to-one, suppose $f(x) = f(y)$. Thus $2x - 1 = 2y - 1$, which implies $x = y$. Also, f maps onto D^+ since if d is an odd positive integer, then d has the form $d = 2r - 1$ for some $r \in \mathbf{N}$. But then $f(r) = d$.
   (e) *Hint:* Consider $f(x) = -(x + 12)$ with domain $\mathbf{N}$.
4. (a) *Hint:* Let $f(x) = 1/x$.
9. (a) $\mathbf{c}$
   (c) $\aleph_0$
   (e) $\mathbf{c}$
10. (b) Let m be the largest number in S.
    Then $\mathbf{N} - S = (\mathbf{N} - \mathbf{N}_m) \cup (\mathbf{N}_m - S)$. By part (a) $\mathbf{N} - \mathbf{N}_m$ is denumerable. Since $\mathbf{N}_m - S$ is a subset of $\mathbf{N}_m$, $\mathbf{N}_m - S$ is finite. Therefore, by Theorem 5.18, $\mathbf{N} - S$ is denumerable.
11. *Hint:* Mimic the proof that $(0, 1)$ is uncountable.
13. *Hint:* Use induction on n and Theorem 5.19.
15. (a) E. The main idea of the "proof" is that infinite subsets of $\mathbf{N}$ are denumerable. No justification for this is given.
    (c) E. The claim is false. Also "A and B are finite" is not a denial of "A and B are infinite."
    (d) E. Writing an infinite set A as $\{x_1, x_2, \dots\}$ is the same as assuming A is denumerable.

Exercises 5.3

1. We must show that $n \leq \aleph_0$ and $n \neq \aleph_0$. The inclusion map $i: \mathbf{N}_n \to \mathbf{N}$ is one-to-one; hence $\overline{\overline{\mathbf{N}_n}} \leq \overline{\overline{\mathbf{N}}}$. If $n = \aleph_0$, then $\overline{\overline{\mathbf{N}_n}} = \overline{\overline{\mathbf{N}}}$, and thus $\mathbf{N}_n \approx \mathbf{N}$. But $\mathbf{N}$ is not finite. Therefore, $n \neq \aleph_0$. Thus $n < \aleph_0$.
5. (a) $\overline{\overline{\varnothing}} < \overline{\overline{\{0\}}} < \overline{\overline{\{0, 1\}}} < \overline{\overline{\mathbf{Q}}} < \overline{\overline{(0, 1)}} = \overline{\overline{[0, 1]}} = \overline{\overline{\mathbf{R} - \mathbf{Q}}} = \overline{\overline{\mathbf{R}}} < \overline{\overline{\mathscr{P}(\mathbf{R})}} <$ $\overline{\overline{\mathscr{P}(\mathscr{P}(\mathbf{R}))}}$.
7. *Hint:* Give a proof by contradiction using Cantor's Theorem.
10. (b) not possible
14. (b) E. We have not defined or discussed properties of operations such as addition for cardinal numbers. Thus $\overline{\overline{C}} = \overline{\overline{B}} + \overline{\overline{(C - B)}}$ cannot be used unless C and B are finite sets. The claim is false.
    (d) A.

Exercises 5.4

5. Let $B \subseteq A$ with B infinite and A denumerable. Since $B \subseteq A$, $\overline{\overline{B}} \leqslant \overline{\overline{A}}$. Since A is denumerable, $\overline{\overline{A}} = \overline{\overline{\mathbf{N}}}$. Since B is infinite, B has a denumerable subset D by Theorem 5.23. Thus $\overline{\overline{A}} = \overline{\overline{\mathbf{N}}} = \overline{\overline{D}} \leqslant \overline{\overline{B}}$. By the Cantor-Schröder-Bernstein Theorem, $\overline{\overline{B}} = \overline{\overline{A}}$. Thus $B \approx A$.

8. *Hint:* Let $x \in A$. By Theorem 5.27, $A - \{x\}$ has a denumerable subset $\{a_n : n \in \mathbf{N}\}$. Construct a one-to-one correspondence between A and $A - \{x\}$.

9. (a) A.
    (c) A.

Exercises 5.5

1. (a) Let $A_n = \{n\}$ for each $n \in \mathbf{Z}$. Each A_n is finite and $\bigcup_{n \in \mathbf{Z}} A_n = \mathbf{Z}$, which is denumerable.

6. (a) *Hint:* Write $\mathbf{N} \times \mathbf{N} = \bigcup_{n \in \mathbf{N}} A_n$, where $A_n = \{(m, n) : m \in \mathbf{N}\}$ for each $n \in \mathbf{N}$. Show A_n is denumerable for all $n \in \mathbf{N}$ and then use Theorem 5.32.

7. *Hint:* Since $\mathbf{N} \subseteq \mathbf{Q}$, $\overline{\overline{\mathbf{N}}} \leqslant \overline{\overline{\mathbf{Q}}}$. On the other hand, a function $f : \mathbf{Q} \to \mathbf{Z} \times \mathbf{Z}$, given by $f(p/q) = (p, q)$, can be used to show $\overline{\overline{\mathbf{Q}}} \leqslant \overline{\overline{\mathbf{N}}}$. You must define f carefully to make it a function.

8. (b) *Hint:* Let $T = \{\{m, n\} : m, n \in \mathbf{N}$ with $m \neq n\}$. Use $g : \mathbf{N} \to T$, defined by $g(x) = \{x, x + 1\}$ for all $x \in \mathbf{N}$, to show $\overline{\overline{\mathbf{N}}} \leqslant \overline{\overline{T}}$. Then use $f : T \to \mathbf{N} \times \mathbf{N}$, given by $f(\{m, n\}) = (\min\{m, n\}, \max\{m, n\})$ for all $\{m, n\} \in T$, to show $\overline{\overline{T}} \leqslant \overline{\overline{\mathbf{N} \times \mathbf{N}}}$.

9. *Hint:* For each $n \in \mathbf{N}$, let T_n be the set of all sequences where all but n terms are 0. Show that T_n is denumerable [using exercise 8(d)] and that $S' = \bigcup_{n \in \mathbf{N}} T_n$.

11. (a) C. The proof is valid only when $f(1) = x$. In the case when $f(1) \neq x$, we must first redefine f: Let t be the unique element of $\mathbf{N}$ such that $f(t) = x$ and define $\hat{f} = (f - \{(1, f(1)), (t, x)\}) \cup \{(1, x), (t, f(1))\}$. Now let $g(n) = \hat{f}(n + 1)$ for all $n \in \mathbf{N}$.

Exercises 6.1

1. (a) yes
    (e) no
2. (a) not commutative
    (e) not an operation
3. (a) not associative
    (e) not an operation
8. *Hint:* Compute $e \circ f$.
9. (a) *Hint:* Compute $x \circ (a \circ y)$ and $(x \circ a) \circ y$.
15. $I(f + g) = \int_a^b (f + g)(x)\,dx = \int_a^b f(x)\,dx + \int_a^b g(x)\,dx = I(f) + I(g)$.

21. (a) Let $C, D \in \mathcal{P}(A)$. Then $f(C \cup D) = f(C) \cup f(D)$ by Theorem 4.16(b). Therefore, f is an OP mapping.

22. (b) E. The claim is false. One may pre-multiply (multiply on the left) or post-multiply both sides of an equation by equal quantities. Multiplying one side on the left and the other on the right is not allowed.

Exercises 6.2

1. (a)

$\cdot$	1	-1	i	$-i$
1	1	-1	i	$-i$
-1	-1	1	$-i$	i
i	i	$-i$	-1	1
$-i$	$-i$	i	1	-1

We see from the table that the set is closed under $\cdot$, 1 is the identity, and each element has an inverse. Also, $\cdot$ is associative.

   (d) *Hint:* $\varnothing$ is the identity.

2.

	e	u	v	w
e	e	u	v	w
u	u	v	w	e
v	v	w	e	u
w	w	e	u	v

4. (a) The group is abelian.
8. *Hint:* For $a, b \in G$, compute $a^2 b^2$ and $(ab)^2$ two ways.
9. *Hint:* In order to have both cancellation properties, every element must occur in every row and in every column of the table.
12. *Hint:* Let $a, b \in G$. For an element x such that $a * x = b$, try $a^{-1} * b$.
14. (b) E. A minor criticism is that no special case is needed for e. The fatal flaw is the use of the undefined division notation.

Exercises 6.3

3. (i) Suppose $a \equiv_m b$ and $c \equiv_m d$. Then m divides $a - b$ and $c - d$. Thus there exist integers k and ℓ such that $a - b = km$ and $c - d = \ell m$. Hence $(a + c) - (b + d) = (a - b) + (c - d) = km + \ell m = (k + \ell)m$. Then $k + \ell$ is an integer so m divides $(a + c) - (b + d)$. Therefore, $a + c \equiv_m b + d$.

6. (b) (143256)
7. (b) $\alpha^2 = (312)$ and $\alpha^3 = (123)$, the identity. Therefore, $\alpha^{-1} = \alpha^2$; $\alpha^4 = \alpha^3 \alpha = \alpha$; $\alpha^{50} = \alpha^{48}\alpha^2 = (\alpha^3)^{12}\alpha^2 = \alpha^2$; $\alpha^{51} = \alpha^{50}\alpha = \alpha^2\alpha = \alpha^3 = (123)$.
10. $10, 12, 2n$.
12. (a) The zero divisors are $0, 2, 3, 4, 6, 8, 9, 10$.
13. Since $(p - 1)(p - 1) = p^2 - 2p + 1 = p(p - 2) + 1$, $(p - 1)^2 \equiv_p 1$. Therefore $(p - 1)(p - 1) = 1$ in $\mathbf{Z}_p$, and hence $(p - 1)^{-1} = p - 1$.
14. (a) $x = 0, 4, 8, 12, 16$.
15. (b) C. There is no justification that $xy \neq 0$. One must begin differently by considering prime factorizations of a, b, and m.

Exercises 6.4

1. (a) $\{0\}, \mathbf{Z}_8, \{0, 4\}, \{0, 2, 4, 6\}$
   (d) There are six subgroups.
9. (a) Yes. Assume G is abelian and H is a subgroup of G. Suppose $x, y \in H$. Then $x, y \in G$. Therefore $xy = yx$.
11. (c) The order of 0 is 1.
    The elements 1, 3, 5, and 7 have order 8.
    The elements 2 and 6 have order 4.
    The order of 4 is 2.
13. N_a is not empty, because $ea = a = ae$, so $e \in N_a$. Let $x, y \in N_a$. Then $xa = ax$ and $ya = ay$. Multiplying both sides of the last equation by y^{-1}, we have $y^{-1}(ya)y^{-1} = y^{-1}(ay)y^{-1}$. Thus $(y^{-1}y)(ay^{-1}) = (y^{-1}a)(yy^{-1})$, or $ay^{-1} = y^{-1}a$. Therefore, $(xy^{-1})a = x(y^{-1}a) = x(ay^{-1}) = (xa)y^{-1} = (ax)y^{-1} = a(xy^{-1})$. This shows $xy^{-1} \in N_a$. Therefore, N_a is a subgroup of G, by Theorem 6.11.
16. The identity $e \in H$ because H is a group, and thus $e^{-1}ee \in K$. Thus K is not empty. Suppose $b, c \in K$. Then $b = a^{-1}h_1a$ and $c = a^{-1}h_2a$ for some $h_1, h_2 \in H$. Thus $bc^{-1} = (a^{-1}h_1a)(a^{-1}h_2a)^{-1} = (a^{-1}h_1a)(a^{-1}h_2^{-1}a) = a^{-1}h_1(aa^{-1})h_2^{-1}a = a^{-1}h_1h_2^{-1}a$. But H is a group, so $h_1h_2^{-1} \in H$. Thus $bc^{-1} \in K$. Therefore, K is a subgroup of G.
19. (a) 1, 3, 5, or 15
22. (c) a^{10}, a^{20}
25. *Hint:* Use exercise 21(a).

Exercises 6.5

1. $H = \{(213), (123)\}$. The left cosets of H are $(123)H = H$, $(132)H = \{(132), (312)\}$, and $(321)H = \{(321), (213)\}$.
4. Let G be a group of order 4. By Corollary 6.17, the order of an element of G must be 1, 2, or 4. The only element of order 1 is the identity. If G has an element of order 4, then G is cyclic. Otherwise, every element (except the identity) has order 2.
9. *Hint:* Let $G = (a)$, and suppose $n = mk$. Consider the element a^k.
10. (a) E. From $aH = bH$ we can conclude only that for some h_1 and $h_2 \in H$, $ah_1 = bh_2$.
    (c) A.

Exercises 6.6

1. (i) Suppose H is normal. Let $a \in G$. To show $a^{-1}Ha \subseteq H$, let $t \in a^{-1}Ha$. Then $t = a^{-1}ha$ for some $h \in H$. Thus $at = ha$. Since $ha \in Ha$, $at \in Ha$. By normality $Ha = aH$. Therefore, $at \in aH$, so there exists $k \in H$ such that $at = ak$. By cancellation, $t = k$. Thus $t \in H$.
   (ii) We let $x \in G$ and show that $xH = Hx$. Let $y \in xH$. Then $y = xh$ for some $h \in H$. But then $yx^{-1} = xhx^{-1}$. Since $xhx^{-1} \in xHx^{-1} \subseteq H$, $yx^{-1} \in H$.

Therefore, $yx^{-1} = k$ for some $k \in H$. Thus $y = kx$, which proves $y \in Hx$. Therefore, $xH \subseteq Hx$. A proof that $Hx \subseteq xH$ is similar.

2. *Hint:* Use exercise 1.
5. (b) *Hint:* If $G = (a)$, the group generated by $a \in G$, consider (aH), the group generated by the element aH in G/H. As a lemma prove that if $x \in G$, then $[xH]^n = x^nH$ for all $n \in \mathbf{Z}$.
8. *Hint:* See exercise 7.
12. (b) C. The proof is correct but a verification that $x(H \cap K) = xH \cap xK$ is not provided. Such a verification might not be required in an advanced class.

Exercises 6.7

5. *Hint:* See exercise 25 of section 6.4.
12. By Theorem 6.24 the homomorphic images of $\mathbf{Z}_8$ are

$$\mathbf{Z}_8/\{0\} \cong \mathbf{Z}_8$$
$$\mathbf{Z}_8/\mathbf{Z}_8 \cong \{0\}$$
$$\mathbf{Z}_8/\{0, 4\} \cong \mathbf{Z}_4$$
$$\mathbf{Z}_8/\{0, 2, 4, 6\} \cong \mathbf{Z}_2$$

Exercises 7.1

2. (a) supremum: 1; infimum: 0
   (c) supremum does not exist; infimum: 0
   (e) supremum: 1; infimum: $\frac{1}{3}$
   (g) supremum: 5; infimum: -1
4. (a) Let x and y be least upper bounds for A. Then x and y are upper bounds for A. Since y is an upper bound and x is a least upper bound $x \leq y$. Since x is an upper bound and y is a least upper bound $y \leq x$. Thus $x = y$.
9. (a) Let $s = \sup(A)$, $B = \{u: u$ is an upper bound for $A\}$ and $t = \inf(B)$. We must show $s = t$.
   (1) To show $t \leq s$ we note that since $s = \sup(A)$, s is an upper bound for A. Thus $s \in B$. Therefore, $t \leq s$.
   (2) To show $s \leq t$ we will show t is an upper bound for A. If t is not an upper bound for A, then there exists $a \in A$ with $a > t$. Let $\epsilon = (a - t)/2$. Since $t = \inf(B)$ and $t < t + \epsilon$, there exists $u \in B$ such that $u < t + \epsilon$. But $t + \epsilon < a$. Therefore, $u < a$, a contradiction, since $u \in B$ and $a \in A$.
12. (a) Let $m = \max\{\sup(A), \sup(B)\}$.
   (1) Since $A \subseteq A \cup B$, $\sup(A) \leq \sup(A \cup B)$. Also, $B \subseteq A \cup B$ implies $\sup(B) \leq \sup(A \cup B)$. Thus $m = \max\{\sup(A), \sup(B)\} \leq \sup(A \cup B)$.
   (2) It suffices to show m is an upper bound for $A \cup B$. Let $x \in A \cup B$. If $x \in A$, then $x \leq \sup(A) \leq m$. If $x \in B$, then $x \leq \sup(B) \leq m$. Thus m is an upper bound for $A \cup B$. Hence $\sup(A \cup B) \leq m$.
13. (a) E. The claim is true but $y = i + \epsilon/2$ might not be in A.

Exercises 7.2

1. (b) open
   (e) open
   (i) closed
2. *Hint:* Use De Morgan's Laws.
4. If $A = \{x_1, x_2, \ldots, x_n\}$ with $x_1 < x_2 < x_3 < \cdots < x_n$, then
   $$\widetilde{A} = (-\infty, x_1) \cup (x_1, x_2) \cup \cdots \cup (x_{n-1}, x_n) \cup (x_n, \infty).$$
   Thus $\widetilde{A}$ is open since it is a union of open sets. Therefore, A is closed.
10. (a) *Hint:* First show that an open cover of $A \cup B$ is an open cover of A and an open cover of B. A different proof uses the Heine-Borel Theorem.
13. (a) not compact (not bounded)
    (d) not compact (neither closed nor bounded)
    (g) not compact (not closed)
14. *Hints:* One proof uses directly the definition of compactness; the other uses the Heine-Borel Theorem.
18. (b) C. With the addition of O^* to the cover $\{O_\alpha : \alpha \in \Delta\}$ we are assured that there is a finite subcover of $\{O^*\} \cup \{O_\alpha : \alpha \in \Delta\}$, but not necessarily a subcover of $\{O_\alpha : \alpha \in \Delta\}$. Since $O^* = A - B$ is useless in a cover of B, it can be deleted from the subcover after it is used.

Exercises 7.3

1. (a) $\{\frac{1}{2}\}$
   (c) $\varnothing$
   (g) $\{0, 2\}$
3. *Hint:* First show for all $a \in A$, $z > a$. Then show z is an accumulation point of A by using Theorem 7.1.
6. *Hint:* Show $(A \cap B)' \subseteq A' \cap B'$ by using exercise 4(a).
8. (c) By parts (a) and (b), $c(A)$ is closed and contains A. Let $A \subseteq B$, and let B be closed. Then, by exercise 7, $A' \subseteq B$. Thus $c(A) = A \cup A' \subseteq B$.
9. (a) has no accumulation points
   (c) has accumulation points
12. $\{0, 1\}$
14. (a) E. *Hint:* See page 37 on the misuse of quantifiers.
    (c) E. $(\widetilde{B})'$ need not be a subset of $(\widetilde{B'})$.

Exercises 7.4

1. (a) $x_n \to 0$; for $\epsilon > 0$, use $N > 1/\epsilon$.
   (c) x diverges; use $\epsilon = 1$.
   (f) $x_n \to 0$; for $\epsilon > 0$, use $N > (2\epsilon)^{-2}$ and $\sqrt{n+1} - \sqrt{n} =$
   $$(\sqrt{n+1} - \sqrt{n})\left(\frac{\sqrt{n+1} + \sqrt{n}}{\sqrt{n+1} + \sqrt{n}}\right).$$
4. (a) Let $\epsilon > 0$. Then $\epsilon/2 > 0$. Since $x_n \to L$, there exists $N_1 \in \mathbf{N}$ such that if $n > N_1$, then $|x_n - L| < \epsilon/2$. Likewise, there exists $N_2 \in \mathbf{N}$ such that

$n > N_2$ implies $|y_n - M| < \epsilon/2$. Let $N_3 = \max\{N_1, N_2\}$, and assume $n > N_3$. Then $|(x_n + y_n) - (L + M)| = |(x_n - L) + (y_n - M)| \leq |x_n - L| + |y_n - M| < \epsilon/2 + \epsilon/2 = \epsilon$. Therefore, $x_n + y_n \rightarrow L + M$.

5. *Hint:* If y is bounded by a positive number B, use the definition of $x_n \rightarrow 0$ with ϵ/B.

6. (a) *Hint:* $||x_n| - |L|| \leq |x_n - L|$.

9. (a) *Hint:* Since $x_n \rightarrow L$, $|x_n| \rightarrow |L|$, by exercise 6(a). Now apply the definition of $|x_n| \rightarrow |L|$ with $\epsilon = |L|/2$.

11. (b) A. The proof uses exercise 4(b).

Index

Abelian group, 195
Accumulation point, 239
Adjacent, 114
Algebraic structure, 188
Algebraic system, 188
Antecedent, 7
Appel, Kenneth, 29
Associative operation, 189
Axiom of Choice, 180–181

Bernstein, Felix, 175
Biconditional sentence, 9
    translation of, 11
Bijection (see One-to-one correspondence)
Binary operation:
    definition, 189
    on sets, 49–50
    set closed under, 188
Binomial:
    coefficient, 81
    expansion, 84
Bolzano, Bernard, 239
Bolzano-Weierstrass Theorem, 240,
        249–251
Borel, Emile, 235
Boundary point, 237
Bounded-Monotone Sequence Theorem,
        249–251
Bounded sequence, 242
Bounded set, 227

Cantor, Georg, 168, 173, 174, 175
Cantor-Schroder-Bernstein Theorem,
        176
Cantor's Theorem, 175
Cardinal number, 159, 165, 168, 173
Cartesian product, 90
Cayley, Arthur, 219
Cayley table (see Operation table)
Cayley's Theorem, 219
Center of a group, 210
Closed:
    interval, 42
    path, 118
    ray, 42
    set, 231
Closure of a set, 241
Codomain, 126
Combination, 81
Commutative operation, 189
Compact, 235
Comparability Theorem, 179
Complement:
    of a set, 52
    of a digraph, 107
    of a graph, 116
Complete:
    digraph, 103
    field, 229, 249–251
    graph, 116
Component of a graph, 117

Composition:
    of a function, 134–136
    of a relation, 97
Conditional sentence, 6
    contrapositive of, 8
    converse of, 8
    translation of, 11–12
Congruence modulo m, 104, 201–203
Conjunction, 2
Connected graph, 120
Consequent, 7
Continuum, 168
Contradiction, 30
Contrapositive, 8
Converse, 8
Coset, 212–214
Countable set, 165, 183–185
Counterexample, 33
Cover of a set, 234
Cross product, 90
Cyclic group, 207

Denial, 5
Deductive reasoning, 1
Degree of a vertex, 114
Denumerable set, 165, 168–171, 181
Derived set, 239
De Morgan's Laws, 22, 53, 60
Digraph:
    definition, 94
    complement, 107
    complete, 103
Difference of sets, 50
Directed graph (see Digraph)
Disconnected graph, 120
Disjoint sets, 50
Disjunction, 2
Division Algorithm:
    for integers, 75
    for natural numbers, 74
Domain, 92

Edge, 94, 114
Edge set, 114
Empty set, 43
Equality:
    of functions, 131
    of sets, 46
Equivalence:
    class, 103
    relation, 103
Equivalent:
    open sentences, 15
    propositional forms, 4
    propositions, 4
    quantified sentences, 17
    sets, 158
Euclid, 28, 30
Euler, Leonard, 32
Exclusive or, 7
Existential quantifier, 15

Family of sets, 56
Field:
    complete, 229
    definition, 226
    ordered, 226
Finite set, 159
Function:
    canonical, 129, 207
    characteristic, 128
    constant, 128
    definition, 126
    extension of, 137
    greatest integer, 129
    identity, 129
    inclusion, 129
    one-to-one, 144–146
    onto, 142–144
    operation preserving, 180
    restriction of, 136, 147–148
    step, 128
    union of, 138, 147–148
Fundamental Theorem of Group
        Homomorphisms, 222

Galois, Evariste, 194
Generator, 207
Graph, 92
    complete, 116
    complement, 117
    component, 120
    connected, 120
    definition, 114
    disconnected, 120
    edge, 113–114
    isomorphism, 114
    null, 116
    order, 114
    path, 118
    simple, 114
    size, 114
    subgraph, 117
    trail, 118
    vertex, 113–114
    walk, 118
Greatest lower bound (see Infimum)
Group:
    abelian, 195
    cyclic, 207
    definition, 195
    isomorphic, 219–221
    octic, 201
    of permutations, 199
    quotient, 217
    symmetric, 199

Haken, Wolfgang, 29
Handshaking Lemma, 116
Heine, Edward, 249–251
Heine-Borel Theorem, 235–237
Homomorphic image, 197
Hypothesis of induction, 67

Identity:
    element, 189
    relation, 93
Image:
    of an element, 127
    of a set, 150
Incident, 114
Inclusive or (see Disjunction)
Indexed family, 58
Index:
    of a group, 214
    of a set, 58
Inductive:
    definition, 65
    reasoning, 1
    set, 65
Infimum, 228
Infinite set, 159
Injection (see One-to-one function)
Interior point, 231
Intersection:
    of two sets, 50
    over a family, 57, 59
Interval notation, 42
Inverse:
    element, 189
    of a function, 134, 145–146
    of a relation, 96
Inverse image of a set, 150
Isomorphism:
    of graphs, 114
    of groups, 219–221

Kernel, 206

Lagrange's Theorem, 214
Least upper bound (see Supremum)
Limit of a sequence, 243
Loop, 102, 114
Lower bound, 227

Mapping (see Function)
Metric, 133
Modus ponens, 22
Monotone sequence, 246
Multigraph, 114
Multiple of an element, 197

Negation, 2
Negative of an element, 197
Neighborhood, 231
Normalizer, 210
Normal subgroup, 216

Octic group, 201
One-to-one correspondence, 146, 158
One-to-one function, 144–146
Onto function, 142–144
Open:
    interval, 42
    ray, 42

Open: (continued)
  sentence, 14
  set, 231
Operation Preserving mapping, 190
Operation table, 188
Order:
  of an algebraic system, 188
  of an element, 208
  of a graph, 114
Ordered n-tuple, 90
Ordered pair, 89

Pairwise disjoint, 61
Partition, 109
Parity, 103
Pascal, Blaise, 84
Pascal's Triangle, 84–85
Path:
  closed, 118
  definition, 118
Permutation group, 200, 219–220
Permutation, 81, 199
Pigeonhole Principle, 162
Power set, 44
Pre-image under a function, 127
Principle of Complete Induction, 71
Principle of Inclusion and Exclusion, 79
Principle of Mathematical Induction, 64
Product Rule, 79
Proof:
  of biconditionals, 28–29
  by complete induction, 72
  by contradiction, 27–28, 32–33, 35
  by contrapositive, 26–27
  by exhaustion, 29
  by induction, 67–71
  by well ordering principle, 73–74
  constructive, 32
  definition, 21
  direct, 24–26, 34
Proposition:
  biconditional, 9
  compound, 2
  conditional, 7
  definition, 2
  denial, 5
  equivalent, 4
  simple, 2

Quantifier:
  existential, 15
  unique existential, 19
  universal, 15
Quotient group, 217

Range, 92
Reachable, 119

Reflexive relation, 101
Relation:
  antisymmetric, 107
  asymmetric, 107
  composite, 97
  definition, 91
  equivalence, 103
  identity, 93
  inverse, 96
  irreflexive, 107
  reflexive, 101
  symmetric, 101
  transitive, 101

Schröder, Ernest, 175
Sequence:
  bounded, 242
  convergent, 243
  decreasing, 246
  definition, 242
  divergent, 243
  increasing, 246
  limit of, 243
  monotone, 246
Set notation, 42
Subcover, 234
Subdigraph, 103
Subgroup:
  definition, 204
  identity, 205
  normal, 216
  proper, 205
Subset:
  definition, 43
  improper, 44
  proper, 44
Sum Rule, 77
Supremum, 228
Surjection (see Onto function)
Symmetric group, 199
Symmetric relation, 101
Symmetry, 200

Tautology, 21
Theorem, 21
Trail, 118
Transitive relation, 101
Truth set, 15

Uncountable set, 165, 167–168
Undecidable sentence, 30
Union:
  of two sets, 50
  over a family, 57, 59
Universal quantifier, 15
Universe of discourse, 15
Upper bound, 227

Vertex:
 distance between, 119
 initial, 117
 isolated, 116
 set, 114
 terminal, 117

Walk:
 definition, 117
 length, 117
Weierstrass, Karl 239
Well Ordering Principle, 73

List of Symbols

$P \wedge Q, P \vee Q, \sim P$ 2

$P \Rightarrow Q$ 7

$P \Leftrightarrow Q$ 9

$(\forall x)P(x), (\exists x)P(x)$ 15

$(\exists! x)P(x)$ 19

$x \in A$ 41

N, Z, Q, R   42

$[a, b], (a, b), (a, \infty), (-\infty, a)$ 42

$\varnothing$ 43

$A \subseteq B$ 43

$\mathscr{P}(A)$ 44

$A = B$ 46

$A \cup B, A \cap B, A - B$ 50

$\tilde{A}$ 52

$\bigcup_{A \in \mathscr{A}} A, \bigcap_{A \in \mathscr{A}} A$ 57

$\bigcup_{\alpha \in \Delta} A_\alpha, \bigcap_{\alpha \in \Delta} A_\alpha$ 59

$\bigcup_{i=j}^{k} A_i, \bigcap_{i=j}^{k} A_i$ 60

PMI   64

PCI   71

WOP   73

$\# A$ 77

$\binom{n}{r}$ 81

$A \times B$ 90

$a R b, a \mathrel{\rlap{R}{/}} b$ 92

$\text{Dom}(R), \text{Rng}(R)$ 92

I_A 93

R^{-1} 96

$S \circ R$ 97

$x/R, A/R$ 103

$\equiv_m$ 104

$\mathbf{Z}_m$ 104, 201

$\tilde{G}$ 117

$C(b)$ 120

$f: A \to B$ 126

χ_A 128

$[\![x]\!]$ 129

$f|_D$ 136

$f: A \xrightarrow{\text{onto}} B$ 142

$f: A \xrightarrow{1-1} B$ 144

$f(X), f^{-1}(Y)$ 150

$A \approx B, A \not\approx B$ 158

$\mathbf{N}_k$ 159

$\overline{\overline{A}}$ 159

$\aleph_0$ 165

$\mathbf{c}$ 168

$\mathbf{Q}^+$ 168

$\overline{\overline{A}} = \overline{\overline{B}}, \overline{\overline{A}} \leq \overline{\overline{B}}, \overline{\overline{A}} < \overline{\overline{B}}$ 174

$(A, *)$ 188

OP   190

S_n 199

$\mathbb{O}$ 201

$+_m, \cdot_m$ 202

$\ker(f)$ 206

(a) 207

xH, Hx 212

G/H 215

$\sup(A), \inf(A)$ 228

$\mathcal{N}(a, \delta)$ 231

A' 239

x_n 242

$\lim_{n \to \infty} x_n = L, x_n \to L$ 243